Methods in Human Cytogenetics

Edited by
H. G. Schwarzacher and U. Wolf

Coeditor of the English Version
E. Passarge

With Contributions by
W. Gey, W. Krone, S. Ohno, E. Passarge,
R. A. Pfeiffer, W. Schnedl,
H. G. Schwarzacher, M. Tolksdorf, U. Wolf

With 59 Figures

Springer-Verlag
Berlin Heidelberg GmbH 1974

Translated from a revision of the German edition "Methoden in der medizinischen Cytogenetik", edited by H. G. Schwarzacher and U. Wolf.

ISBN 978-3-642-65789-4   ISBN 978-3-642-65787-0 (eBook)
DOI 10.1007/978-3-642-65787-0

Preface

This volume was originally intended to be an English translation of the book *Methoden in der medizinischen Cytogenetik*, published in 1970. Just about then, however, a number of new techniques were introduced in human cytogenetics and soon acquired the utmost importance, particularly in clinical diagnosis, so that the English edition had to be considerably enlarged. As a result, there are now twelve chapters instead of eight, and two additional authors have been called upon, Dr. KRONE and Dr. SCHNEDL. In addition to the up-to-date presentation of conventional methods of cell culture and techniques for the preparation and identification of human chromosomes, this text covers the various techniques of producing banding patterns and applying them in chromosome identification. Further, it deals with the culture of amniotic fluid cells and gives instructions for handling tissue-culture cells for biochemical analysis; it thus meets the ever-increasing requirements of a modern cell-culture laboratory.

To paraphrase the aims of this book, we quote part of the preface to the German edition: "It was intended to collect the various methods so as to make them accessible for laboratory use. Furthermore, it is hoped that the reader faced with current research problems will be stimulated to modify and supplement the techniques described, instead of merely applying them automatically. In a rapidly developing field, some methods are still preliminary, and no final presentation seems possible."

In preparing the English edition, the German edition was first translated by Mr. YAIER COHEN, whose work is gratefully acknowledged. However, it would have been difficult to edit the book without the help of EBERHARD PASSARGE, who revised all chapters except the last two (XI and XII) and offered many valuable comments. We are much indebted to him.

Wien and Freiburg i. Br., March 1974

H. G. SCHWARZACHER
U. WOLF

Contents

Contents

List of Contributors

Editors:

Professor Dr. HANS GEORG SCHWARZACHER
Histologisch-Embryologisches Institut
der Universität
Schwarzspanierstraße 17
A-1090 Wien

Professor Dr. ULRICH WOLF
Institut für Humangenetik und Anthropologie
der Universität
7800 Freiburg i. Br.
Albertstraße 11

Coeditor:

Privatdozent Dr. EBERHARD PASSARGE
Institut für Humangenetik der Universität
2000 Hamburg 20
Martinistraße 52

Contributors:

Dr. WOLFGANG GEY
Facharzt für Kinderkrankheiten
3500 Kassel-Bettenhausen
Leipziger Str. 113

Professor Dr. WINFRID KRONE
Universität Ulm
Abteilung Humangenetik
7900 Ulm
Oberer Eselsberg

Professor Dr. SUSUMU OHNO
Dept. of Biology,
City of Hope National Medical Center
Duarte, CA 91010/USA

Professor Dr. RUDOLF ARTHUR PFEIFFER
Institut für Humangenetik der
Medizinischen Hochschule
2400 Lübeck
Ratzeburger Allee 160

Universitätsdozent Dr. WOLFGANG SCHNEDL
Histologisch-Embryologisches Institut
der Universität
Schwarzspanierstraße 17
A-1090 Wien

Professor Dr. MARLIS TOLKSDORF
Universitäts-Kinderklinik
2300 Kiel
Fröbelstraße 15 – 17

CHAPTER I

Cell Cultures from Blood and Bone Marrow

RUDOLF ARTHUR PFEIFFER

1. Introduction

The analysis of chromosomes from nucleated cells of the hemopoietic system, particularly lymphocytes, has become the main method in clinical cytogenetics, because of the ease with which material can be obtained by the physician.

The culture of lymphocytes from peripheral blood is inexpensive, simple to perform, reliable, and therefore suitable for routine work. Stimulated lymphocytes may enter metaphase within 48 hours in vitro thus permitting analysis of their first division.

The immune response of lymphocytes can be tested with unspecific mitogens. Mixed cultures of lymphocytes from different donors may be used as *in vitro* models of histocompatibility.

The direct effects of various agents (viruses, ionizing radiation, poisons, drugs) on chromosomes can thus be very easily assessed in vitro by using lymphocyte cultures. Since the mitotic rate in bone marrow is high, the cells in the metaphase can be prepared without culturing.

The effect of toxic substances on cells and chromosomes in vivo can relatively simply be tested on the bone marrow cell population (Schmid *et al.*, 1970; Arakaki *et al.*, 1970).

Leukemias result from neoplastic transformation of a particular cell type, and the karyotype can be determined during any stage of the disease.

2. Bone Marrow

No tissue or organ in the healthy adult organism contains as many dividing cells directly available for chromosome analysis as does bone marrow. The majority of metaphases observed in smears of bone marrow from healthy donors is presumably derived from the erythropoietic series: proerythroblasts, erythroblasts, and normoblasts. Less frequently encountered are dividing myeloblasts, promyelocytes, or the various forms of myelocytes. Reticulocyte and plasmacyte, and occasional monocyte and lymphocyte divisions are seen. Polyploid metaphases with 4n, 8 n or even 16 n mainly represent megakaryocytes.

According to Rohr (1960), 10–42% of erythroblasts and 6–11% of myeloblasts are in metaphase at any one time, but these figures vary with different diseases, the highest values being found for erythroblasts in hemolytic anemias. Killmann *et al.* (1962) calculated an average mitotic frequency of 8.86 per 1000 nucleated cells. Of these, 6.15 (69.4%) belonged to the erythrocyte, and 2.5 (28.2%) to the leukocyte series. Rhythmic diurnal fluctuations in mitotic frequency were not observed.

Chromosome preparation destroys the characteristic cytoplasmic structures and it is impossible to identify the dividing cells. Hence, the preparation of at least one marrow smear stained in the usual fashion (e.g. according to Pappenheim) of the preincubated or the incubated bone marrow, or both, is recommended for comparison. The cytogeneticist should be familiar with the morphology of the components of bone marrow.

2.1. Collection of Material

Aspiration of bone marrow is done from the tibia, the iliac crest (in children), the sternum, or from a vertebral process, using the standard methods reviewed by Berman (1953) or Rohr (1960). The presence of fat cells and marrow particles will indicate that bone marrow and not peripheral blood has been obtained.

2.2. Direct Preparation of Chromosomes from Cells in Metaphase or Prometaphase

The bone marrow is suspended in an isotonic salt solution (NaCl-glucose, normal saline, or balanced salt solutions prepared according to Tyrode, Hanks, Gey, Earle, etc.) containing heparin in order to prevent hemagglutination and to break up the marrow fragments and separate the fat cells from the bone marrow. The preparation may then be incubated, usually in the presence of colchicine. Separation of the different types of cells is not necessary.

2.2.1. Direct Preparation of Cells Immediately Following Biopsy

The aspirate is suspended as described above, and chromosome preparations are obtained directly. Sandberg *et al.* (1960) recommended washing the bone marrow in a cold solution of 0.6% glucose and 0.7% NaCl, leaving the cells in suspension for 10–15 min before preparation of the chromosomes, or suspending the bone marrow in cold isotonic Earle solution.

The mitotic index can be increased by injecting a mitotic inhibitor such as colchicine or its semisynthetic derivative Colcemid (Schär *et al.*, 1954) prior to the biopsy. Several authors (Bottura and Ferrari, 1960; Stewart, 1960; Kinlough *et al.*, 1961; Meighan and Stich, 1961) have applied this procedure to patients. The administration of mitotic inhibitors to patients should be limited to selected cases. However, for the study of animal bone marrow it can be very useful. In mice the mitotic index increased within one and a half hours after injection of colchicine (0.05–0.1 mg/kg body weight), the maximum being around 4 hours (Cardinali *et al.*, 1961). Vinblastine sulphate (Velban; Lilly) has also been shown to be a useful mitotic inhibitor in mice when injected at a single dose of 0.1 mg/kg of body weight before biopsy. The mitotic rate in the bone marrow of the rat is reported to increase after intraperitoneal injection of human blood group 0 plasma 16 hours before aspiration (De Vries and van Went, 1964). The induced hemolysis is followed by a compensatory hyperactivity.

2.2.2. Cell Preparation Following Short Term Incubation with Colchicine

Short-term cultures of suspensions of bone marrow grown in the presence of colchicine were developed by Lajtha (1952) and applied by Ford *et al.* (1958).

Details of the procedure:

Method a (Ford). The bone marrow is suspended in a prewarmed isotonic solution containing heparin (1:20,000) and mixed carefully by drawing it in and out of a syringe with a long needle without forming foam.

Saunders flasks (so-called "Universal Containers"), with a capacity of about 25 ml and screw-top lids with rubber seals, are suitable for all bone marrow and blood cultures. The rubber seals should be coated with collodion: a slightly viscous collodion solution derived from a commercially available stock solution is applied with a fine brush and dried at 100° C (unless the seals are siliconized or made from silicone). The rubber lid may be sterilised with 70 % ethyl alcohol, and then perforated with a sharp needle.

The cells in the closed flask are carefully centrifuged at 400–800 rpm., the supernatant is discarded and the sediment resuspended in Gey's salt solution and mixed well in order to ensure a reliable cell count. The suspension is then incubated in a 4:1 mixture of Gey's salt solution and human serum, devided into several subcultures of 3 ml each; 2000–4000 cells per mm³ are regarded as optimal. The time of incubation may vary from 5–15 hours, but 7 hours is usually optimal. Two hours before harvesting 0.1 ml of 0.04 % Colcemid in NaCl-glucose per ml of culture is added.

The following methods dispense with longer incubation periods:

Method b (Tjio and Whang, 1962). The bone marrow is washed in 0.85 % NaCl adjusted to pH 7.0 by 6.6×10^{-3} M phosphate, containing either colchicine or Colcemid at a concentration of 1 µg/ml, and allowed to stand for 1–2 hours at room temperature until processed.

Method c (Kiossoglou *et al.*, 1964; Forteza-Bover *et al.*, 1965). The aspirate is mixed with heparin and centrifuged at 300 rpm. for 5 min in a siliconised tube; the supernatant is discarded. The cells are resuspended in 3 ml of a buffered 0.85 % NaCl solution, pH 7.0–7.2, containing colchicine (solution 1:1000), and incubated at 37° C or at room temperature for $1–1^1/_2$ hours.

Method d. This procedure could be especially useful for bone marrow obtained post mortem. The culture medium consists of 40 % human serum, 30 % medium 199 (Difco), and 30 % Hanks' solution (Tolksdorf *et al.*, 1965). 5 ml of fetal calf extract is added to 100 ml of the mixture. Bone marrow cells are suspended in 3 ml of the medium with the addition of 0.5 ml phytohemagglutinin and 200 units of heparin, and kept in the refrigerator for 3 hours. After 24 hours the medium is pipetted off and replaced by fresh medium. The culture is harvested 48 hours after commencement.

The differences between these various procedures are minor and mostly concern the interval between initiation of suspension and preparation. With the exception of method d they all make use of colchicine. Sterile procedures are not necessary in any of them and leukocytes need not to be counted if no long-term culture is intended.

Investigations which compare these modifications systematically have not been published. The following factors appear to be important for obtaining an adequate rate of excellent metaphases: a) the pH and the composition of the medium used for washing, suspension, and incubation, b) the duration of incubation, c) the absolute number of dividing cells present in the specimen.

2.3. Storage of Bone Marrow Biopsies

It is still questionable whether the bone marrow must be prepared immediately after the biopsy or whether a delay and/or transportation is permissible. Kiossoglou *et al.* (1964) obtained good results even when the bone marrow had been stored for up to 24 hours after the biopsy, provided it was kept in a buffered solution at 4° C. Ford (1958) also reported successful storage in isotonic solution. It is essential, however, that the material is not allowed to dry out. The premitotic stages of leukemic cells are particularly sensitive and in general it is wise not to delay the preparation of bone marrow.

2.4. Comments on the Preparation of Chromosomes

In principle all the methods described in Chapter IV can be employed, and one can use bone marrow either directly suspended, or after short term incubations as described above.

Several authors prefer squash techniques for suspensions of bone marrow. Some of the essential steps are given below, as a supplement to the details given in Chapter IV. The preparation of chromosomes according to the method of Ford (1959, personal communication) is as follows:

a) hypotonic treatment with 0.95% Na citrate;

b) fixation with glacial acetic acid-alcohol (1:3);

c) hydrolysis with 1 N HCl at 60° C for 8 minutes;

d) Feulgen staining;

e) squashing of cells.

According to Sandberg *et al.* (1960), one volume of the suspension is mixed with four volumes of 0.44% Na citrate and left for 15 min. After careful centrifugation the cells are fixed with 50% ice cold acetic acid, which is later changed for glacial acetic acid-alcohol. The cells are stained in suspension using 1% orcein in 65% acetic acid and are then squashed.

For material which contains many fragments, Tjio and Whang (1962) recommend heating in a mixture of 1 N HCl and 2% orcein in 45% acetic acid, in the ratio 1:9, on a watch glass over a flame. The fragments are then squashed under the cover slip together with a drop of stain solution. Hsu and Patton (1969) suspend the bone marrow in any kind of growth medium, with 0.02 µg/ml Colcemid. After incubation for one hour the cells are centrifuged and suspended in a 1% citrate solution for

5 min, and then fixed immediately without stirring up (20 min). The supernatant fixative is drawn off and several drops of 2% acid orcein solution are added to the cell residue.

The cells are then squashed under a cover slip (see p. 75). Better results are occasionally achieved if the slide is heated over a gas flame prior to the second squashing.

In the case of myelofibrosis and myelosclerosis, if a cylinder of bone has been punched out or removed surgically, it must be cut up with scissors in an isotonic solution and the fragments rinsed thoroughly. It is advisable to draw off any blood which has collected in the bone cavity. Biopsy of the spleen and direct examination of dividing cells and the culture of suspension of cells may be attempted (Spiers and Baikie, 1965). Short-term incubation of peripheral blood cells without PHA is generally useful. Blast cells are expected to enter mitosis after incubation for 1–2 days. The number of cells/μl should be limited to 1–3000 and the medium should be buffered (McCoy 5a, RPMI series).

2.5. Long Term Cultures of Bone Marrow Cells

Under certain conditions suspensions of bone marrow cells can be induced to form a monolayer culture after approximately 1 week, when fibroblast-like clusters of cells are formed which can be treated as fibroblast cultures (for review see Lajtha, 1952; Berman et al., 1955; Woodliff, 1964).

Fibroblasts grown from bone marrow explants seem to be derived not only from stromal elements but – at least in leukemia – also from hemopoietic tissue. It is likely that the fibroblastic appearence of cells in culture is the morphological expression of their adaption to environmental conditions (Hentel and Hirschhorn, 1971).

Although it is unlikely that long term cultures of bone marrow cells will be important for routine studies of chromosomes, they may nevertheless be useful for certain biochemical, enzymatic, and cytological investigations. Lajtha (1952) stressed the problems of obtaining reproducible suspensions of bone marrow for cell kinetic studies owing to the difficulties of obtaining exactly comparable cultural conditions.

The following procedures are modifications of the method of Berman et al. (1955).

a) The bone marrow is incubated in bottles placed on their flat sides, in a mixture of 40% human umbilical cord serum, 2% chicken embryo extract, and 58% Hanks' solution; the pH of the mixture is adjusted to 7.6 with 1.4% $NaHCO_3$ solution. The culture fluid is changed when phenol red (0.002%) indicates a change of the pH. Cells which do not adhere to the glass can be used for subcultures. Cells adhering to the glass are treated as monolayer cultures (see Chapter II).

b) Chu and Giles (1959) treated bone marrow in Hanks' solution with heparin, 1:20000, at 4° C and then incubated $1-3 \times 10^5$ cells per ml in a medium consisting of 15% Eagle's solution, 20% human serum, and 5% ultracentrifuged embryo extract.

c) Fraccaro *et al.* (1960) incubated the suspension of bone marrow in Petri dishes, in a medium consisting of 40% human serum, 30% medium 199 (Difco), and 30% Hanks' solution, in 5% CO_2, with the Petri dishes in airtight plastic containers. After several passages (about 10–25 days) the optimal mitotic rate may be expected.

d) Farnes *et al.* (1963) suggested the incubation of fragments of bone marrow in a small moist chamber. The medium is composed of 80% NCTC-109 (Difco) and 20% human serum. Within 2–3 days the cells begin to proliferate and after 8 days a predominantly fibroblast-like monolayer may develop.

The following methods are particularly useful for the observation of living cells: covering the suspension of bone marrow with gauze in order to keep it moist (Reisner, 1959); culturing small amounts by the hanging-drop method (Berman, 1953; Boll and Fuchs, 1961); the use of agar (Pulvertaft and Jayne, 1953); the cultivation of bone marrow on a feeder layer (Bradley and Metcalf, 1966). With the feeder layer method, cell colonies develop rapidly and their growth can be stimulated by the addition of protein ultrafiltrate ($M = 60000$) from human urine, obtained by means of dialysis (Robinson *et al.*, 1969; Foster *et al.*, 1968).

If sufficient bone marrow cell suspension is available, it can be incubated like a leukocyte culture (see p. 10) with the addition of phyto-hemagglutinin. Lymphoblasts are generally present and capable of transformation, thus allowing the determination of the karyotypes of all blood-forming components.

3. Culture of Peripheral Blood Cells

The peripheral blood only contains mitotic cells under abnormal conditions, for instance, immature cells from the bone marrow in pernicious anemia, hemolytic anemias, leukemias, or in the dissemination of tumor cells.

Earlier attempts (Boll and Fuchs, 1961; Woodliff, 1964) to culture blood leukocytes in vitro by means of classical tissue culture techniques have shown that cells which are probably derived mostly from lympho-cytes and monocytes are capable of forming a monolayer similar to that of fibroblasts.

However, the large number of divisible cells and metaphases which are necessary for cytogenetic investigations can only be obtained by transformation after activation of immunocompetent cells. Resting lymphocytes, whose lifespan is estimated at 530 days (Sasaki and Norman, 1966; see also Fitzgerald, 1964; Nowell, 1965), or even years (Buckton et al., 1967), can be activated by specific antigens and certain other substances. P. C. Nowell (1960) observed, in a culture of blood cells which was prepared from human blood leukocytes according to the method of Osgood and Krippaehne (1955) ("gradient tissue culture") that the mitotic rate rose to an unusually high level. After testing various culture components Nowell concluded that the Bacto-phytohemagglutinin (Difco) used for separating leukocytes from erythrocytes must have been responsible for cell division. Phytohemagglutinin (PHA) is now considered to be the most reliable mitogen for human and animal lymphocytes.

For the properties and effects of PHA see p. 21.

The principle of lymphocyte culture is based on the incubation of blood, leukocytes, or lymphocytes alone in a suitable medium with the addition of PHA. Without PHA dividing cells are not observed within a reasonable time if blast cells (as in leukemia, mononucleosis) are absent. Most cells capable of division are lymphocytes (Bond et al., 1961; Astaldi, 1964); monocytes are of no numerical importance.

The problems associated with the various methods of lymphocyte culture are discussed below.

3.1. Whole Blood Cultures

The simplest method of lymphocyte culture is incubation of a small amount of whole blood without previous separation of the erythrocytes and granulocytes. This utilizes all available lymphocytes, which are usually 25–50% of the leukocyte count, i.e. 1,800–5,000 cells per mm^3 of blood. Whole blood culture methods, often called "micromethods" (due to the small amount of blood required), are described by Arakaki and Sparkes (1963); Tips et al. (1963); Hungerford (1965). Frøland (1962) separated erythrocytes also for microculture.

By techniques analogous to those of serology, lymphocytes from various donors can be grown simultaneously in "microplates". In this way different mitogens can be tested (Brody and Huntley, 1965).

Method. Blood, obtained by venipuncture or by puncturing fingertips, ear lobes or heels (in young children) after first carefully cleaning the skin with ethyl alcohol and drying is placed in a culture medium containing heparin. 2–3 and at most 10 drops of blood should suffice for 5–6 ml of medium. De Grouchy *et al.* (1964) incubated 0.5 ml blood in 15 ml medium (0.5 ml corresponds to 5–11 drops, depending on the caliber of the needle and the viscosity of the suspension).

The culture medium consists of an isotonic salt solution, animal or human serum or plasma, heparin, and antibiotics. Chicken embryo extract is not necessary. From the studies of Bishun (1967) the highest mitotic rate may be expected when the medium (medium 199) and human serum are mixed in the ratio 3:2. It falls sharply when the amount of serum is reduced. The majority of investigators favours a mixture of medium 199 and serum in a ratio of between 85:15 and 70:30.

A standard micromethod which gives reliable results, is set out in the appendix (p. 27). The essential difference between the whole blood method and the leukocyte culture method described in the next section (p. 10) is the presence of erythrocytes in the culture. This raises the question of whether erythrocytes influence the mitotic rate. It can be predicted that, due to the large number of erythrocytes and granulocytes,

1. the sources of energy are prematurely exhausted (glycolysis),
2. the pH will change because of increased metabolic activity,
3. the electrolyte composition of the medium will be altered due to cytolysis and alteration of cell permeability,
4. the inevitable hemolysis has a detrimental effect.

The medium should be restored after 3 days (Arakaki and Sparkes, 1963), if the culture is to be kept beyond this period. It should be remembered however, that the introduction of hemoglobin adds an additional buffer system to the phosphate, acetate, and protein buffers, since HbO_2 is more acid than Hb. The buffer capacity of Hb is very high in the physiological pH range.

Since the addition of heparin to PHA does not prevent the agglutination of erythrocytes in individual compact and solid clots, it must be expected that lymphocytes within the agglutinates will be under poor cultural conditions. In order to avoid agglutination the cultures should be shaken twice a day. Continuous agitation in roller drums may be effective but is not necessary for routine work.

The large number of erythrocytes may also have a detrimental effect, especially during fixation, because the erythrocytes very easily form clumps which must then be dispersed with the pipette. Opinions are divided as to whether the fixative should be placed as a layer on top of the suspension and permitted to act slowly, or whether the cells may be suspended during fixation. In the first case the clump could have to be dispersed later. A safer method consists of fixing the sediment drop by drop by drawing the entire suspension up into a clean pipette and letting it slowly down under the surface of the fixative. If the cells are fixed in 45% acetic acid (Williamson, 1965), the squash technique must be employed.

This method of whole blood culture has been made commercially available by Difco, Gibco and Behringwerke. *Difco*[1], "TC chromosome micro test kit" (No. 5060): 3 drops of blood from a lancet puncture are incubated in the freshly prepared complete medium. The individual components of the medium are reconstituted from powder and dissolved in the liquid medium supplied. It is advised that for transportion the blood should be suspended in the liquid solution only and the complete medium be made up before incubation. *Gibco*[2] also provides a chromosome kit (No. 6701; 6706) using a ready-prepared "Chromosome medium 1 A" which should be kept frozen ($-20°$ C) in screwcap culture tubes. It contains fetal calf serum, heparin, antibiotics, modified Earle's medium and phytohemagglutinin M or P. This company supplies a watery extract of *Phytolacca americana*. "Velban" (vincaleukoblastine) is supplied instead of colchicine.

Behringwerke AG[3] distribute a "chromosome kit", consisting of a lancet, a small flask of culture medium and PHA as well as "arresting fluid", which is added before completing the culture.

3.2. Leukocyte Culture after Separation of Erythrocytes

The "classic" method described by Moorhead *et al.* (1960) provides for a separation of leukocytes from venous blood. Its essential features are described in the following. 10 ml venous blood is mixed with a stock solution of commercial heparin with which the syringe is moistened prior to blood withdrawal and 0.2 ml PHA (Difco). The blood should be cooled (by contact of the containing vessel with icewater) for 30–60 min and is then immediately centrifuged at $5°$ C (25 g for 5–10 min; a refrigerated centrifuge is not necessary). The supernatant plasma, which contains the leukocytes, is removed with a syringe, placed in a sterile culture container, and carefully agitated, avoiding bubbles. A syringe with a long needle is advisable. The previously mentioned (p. 3) Saunders flasks (Universal Containers)[4] are excellent culture vessels. The leukocytes are now counted and the cultures set up at $1-2 \times 10^6$ cells per ml. As a rule the culture medium consists of 30–40% autologous

---

[1] Difco Laboratories. Detroit, Michigan 48201, USA (see brochure: Reagents, media and cell lines for tissue culture and virus propagation).

[2] Gibco (= Grand Island Biological Company) 3175 Staley Road, Grand Island, N.Y. 14072, USA.

[3] Behringwerke AG, 355 Marburg, Federal Republic of Germany.

[4] These inexpensive flasks can be substituted by plastic containers with similar dimensions screw-cap tubes or even prescription bottles. (Flow Laboratories, Ltd., Victoria Park, Heatherhouse Road, Irvine, Ayrshire; Greiner and Co., Nürtingen, W. Germany.)

plasma and 60–70% medium 199 or minimum essential medium, penicillin and streptomycin. The pH is set at 7.0–7.2 and deviations corrected by bubbling through CO_2 or adding NaOH. The culture is incubated for an average of 3 days.

The numerous modifications of this method result from the desire to achieve a maximum yield of lymphocytes and an optimal mitotic rate. The individual factors of importance are described in the following.

3.2.1. Separation of Erythrocytes

Spontaneous sedimentation of erythrocytes from heparinized whole blood yields about 14.5×10^3 leukocytes/mm³ in supernatant plasma; the sedimentation rate is optimal only with fresh blood. It is not merely dependent on the protein composition of the plasma, but also on the temperature; at 30–35° C it is relatively fast, but it is decelerated on cooling. The tube should be set up at an angle of 45°. The caliber of the tube is important. Mellman (1965) centrifuged heparinised venous blood without addition of PHA and reported leukocyte counts of $12–24 \times 10^3$ per mm³ in the supernatant plasma.

Sedimentation in the syringe itself simplifies extraction of the plasma supernatant; the needle is bent to 135°, which allows the plasma cell suspension to be dispensed under sterile conditions into a culture tube (Edwards and Young, 1961).

If only a small amount of blood is available, Edwards and Young (1961) recommend mixing the blood (0.5–1.0 ml) with a drop of PHA in a 2 ml syringe moistened with heparin, and filling it with serum. The syringe is centrifuged and the plasma supernatant ejected.

The sedimentation of erythrocytes can be improved by solutions of macromolecular substances, but it is necessary to replace these later with isotonic solutions.

For this procedure the following substances have been recommended:

Dextran. 10 ml heparinized blood are mixed with 2.5 ml of a 6% dextran solution (Macrodex, molecular weight 150,000). The leukocytes are centrifuged and resuspended in the nutrient medium (Anders *et al.*, 1966).

Ficoll. Blood cells and the leukocyte fraction can be separated in Ficoll, a high molecular weight sucrose which should be dialyzed to remove NaCl.

Noble *et al.* (1968) used 9 ml of a 35% solution overlayered with 5 ml blood previously mixed with 5 ml of Seligmann's solution and 5 mg EDTA as the anticoagulant. The suspension is centrifuged at 10,800 g for 30 min. Leukocytes suspended in the same balanced salt solution are centrifuged in a 28% Ficoll solution at 300 g for 15–20 min.

Polyvinylpyrrolidone. Ulrich and Moore (1966) prepared 10 ml blood with 40 ml buffered Earle's solution (pH 7.4), and 50 ml polyvinyl-pyrrolidone, molecular weight 40,000. The erythrocytes settle after about 1 hour, when the supernatant plasma is centrifuged and the sediment resuspended in Earle's solution. To remove the remaining erythrocytes, streptolysin 0 is added for 1 min, the mixture recentrifuged and the sediment washed and incubated in Earle's solution. Modification: 3.5% polyvinylpyrrolidone and venous blood in the same proportions.

Fibrinogen. 1.25 mg freshly prepared solution of 310 mg calf fibrinogen (fraction 1) in 10 ml water is mixed with 10 ml blood (Minor and Burnett, 1948).

If defibrinised blood is mixed with a 3% solution of gelatine ("plasma-gel") in a physiological salt solution in the ratio 3:1, the majority of erythrocytes and granulocytes sediment out after about 30 min at 37° C if siliconised vessels are used. The clear supernatant contains 90–99% lymphocytes. The gelatine solution (molecular weight 190,000) should not be heated and has to be filtered. Bubbles have to be avoided when mixing (Coulson and Chalmers, 1964).

The yields of leukocytes prepared by means of PHA, dextran and fibrinogen are reported by Skoog and Beck, 1956; Chen and Palmer, 1958). Gum acacia or bovine albumin seem to have gone out of use.

Another possibility of obtaining more leukocytes consists of rapid centrifugation of heparinised blood in a fine caliber tube (screw-cap centrifuge tube). The supernatant plasma column is used in setting up the culture and the top layer (buffy coat) carefully pipetted off. The process is the same as that used to increase the yield of leukocytes from heparinised or citrated blood. Maintenance of 1,500 rpm for 20 min has proved more satisfactory than the 4,000 rpm which has been suggested.

Separation of leukocytes is also possible by use of an unspecific (phyto)-hemagglutinin or a specific agglutinin with slow centrifugation. The erythrocyte agglutinates are heavier and sediment at a slower centri-fugation rate (of 300–400 rpm for 3–5 min) than the agglutinated leukocytes.

3.2.2. Separation of Mononuclear and Polymorphonuclear Leukocytes

Since granulocytes soon decompose in culture their metabolic activity is insignificant. Their persistence after several days is a measure of the efficacy of PHA since they generally occur in proportion inverse to the number of transformed lymphocytes. Whether or how, polymorphonuclear leukocytes inhibit the division of activated lymphocytes is not yet clear (Walker and Fowler, 1965). The separation of mononuclear and polymorphonuclear cells is therefore of interest for special purposes only.

A certain amount of stratification is to be expected on centrifuging heparinised blood in finely calibrated tubes, since the specific density of the lymphocytes, particularly however, of the thrombocytes, is lower than that of the other blood cell fractions, so an hour-glass shaped vessel was proposed (Kaijser, 1961).

A simple method (according to Gropp and Fischer, 1966) also consists of spreading out the leukocyte suspension in a flat layer at the bottom of a Petri dish or a culture vessel, e.g. a square flask (milk dilution bottle produced by Pyrex Company for use in bacteria count determination in milk); another method (Osgood, 1955) is to collect the leukocytes on a slide placed at an angle in the culture vessel. After about one hour the polymorphonuclear cells attach themselves fairly firmly to the glass, so the solution with the floating lymphocytes can be drawn off. Czerski et al. (1966) passed the leukocyte suspension through Petri dishes made of metacrylate which can be easily sterilised by ultra violet light.

Since granulocytes agglutinate in heparin solutions, lymphocytes in a leukocyte suspension can also be obtained by vigorous agitation and allowing the granulocyte aggregates to sink to the bottom (von Melen and Unger, 1967).

Rabinowitz (1964) separates leukocytes in glass columns; this method only requires amounts of suspension as small as 5 ml in 16 cm high columns packed with fiberglass (see also Lamvik, 1966).

A method of gaining a larger yield of lymphocytes is described by Agostini and Ideo (1965). A 5% EDTA solution is added to the blood in a ratio of 1:10. This mixture is diluted with a 0.83% NH_4Cl solution in the ratio 4:1 and centrifuged. The sediment is resuspended in the serum and incubated in a tube filled with cotton wool. After washing, lymphocytes alone remain.

Granulocytes and monocytes also adhere to glass beads (diameter 0.5 mm) filled into a column which is sealed with fiberglass. It is loaded with the plasma cell suspension. The degree of purity of the lymphocyte suspension is claimed to exceed that achieved by other methods (Ciresa and Huber, 1967).

Separation may be effected in a density gradient: Spriggs and Alexander (1960) report that in a mixture of bovine albumin and plasma as density gradient, a leukocyte suspension prepared with dextran can be separated. The upper layer contains monocytes and lymphocytes. Pertoft et al. (1968) prefer polyvinylpyrrolidone or silica gel.

Hulliger and Blazkovec (1967) describe a method for separating erythrocytes in which the blood is layered over a high density solution containing a mixture of methylcellulose and metrizoate; the erythrocytes clump at the interphase and sink to the bottom.

An elegant method for separating granulocytes is based upon their phagocytic properties. The leukocyte suspension is prepared with a double quantity of culture medium. About 0.5 g of extremely fine sterile iron powder filings is added. The mixture is agitated for a while in a shaker in a thermostat and left standing for a few minutes to permit surplus iron filings to sink. The container is placed on a strong magnet which accelerates the sedimentation of the iron filings. The supernatant suspension is drawn off and once more placed in a magnetic field for 5 min. The iron filings phagocytized by leukocytes drag these cells to the bottom of the container, so that the suspension finally consists of 95–99% mononuclear cells (Hastings et al., 1961).

Other methods requiring larger quantities of blood were introduced by Green and Solomon (1963) who mix about 80 cm³ blood with 6% dextran and EDTA and centrifuge in an "oil bottle". Closed systems for lymphocyte purification were proposed by Greenwalt et al. (1962), and by Woods (1970), who uses plastic blood bags. Heparinized blood is mixed 2:1 with Plasmagel. After the erythrocytes have sedimented, the plasma supernatant is agitated with nylon powder and a few glass beads.

3.2.3. Choice of Medium

Since Moorhead et al. (1960), medium 199 of Morton, Morgan and Parker seems to have been predominantly employed. However, other media are equally suitable for culturing if not even superior, e.g. Minimum Essential Medium (MEM Eagle), NCTC-805, and medium 1A from Gibco enriched with 15% serum and containing PHA, or McCoy 5a and the RPMI series of the Roswell Park Memorial Institute.

Serum or plasma, either from the same person, or mixed serum from different people[5], umbilical cord serum or heterologous serum, preferably from calf fetus, and less often from foals, will serve as protein.

Whenever possible, autologous plasma or serum is preferred (Johnson and Russell, 1965; Bloom and Iida, 1967). Incompatibility of blood and serum groups, which led to the suggestion that only AB Rh + serum should be used, is of no significance, since erythrocytes will be

---

[5] Leukocytes are destroyed by short-term heating at 56° C. Homologous serum should therefore be heated at this temperature shortly before use, so that no foreign leukocytes are admitted into the culture.

unspecifically agglutinated by PHA. The culture medium generally consists of serum and medium in the ratio 1:3, but ratios of 1:2–1:5 are acceptable. With a higher initial cell count, dilution is expedient although the remarks of Moorhead *et al.* (1960) with reference to optimal cell count of the inoculum need not be observed too strictly. In view of the sealed vessels an excessive number of cells would lead to too rapid a change of pH and exhaustion of energy sources. It is therefore desirable to maintain the cell density at between 1,000–2,000 per µl of culture. The quantity of medium to be added (in ml) can be calculated by a simple formula if the number of cells in the plasma suspension is known and an end concentration of 1,000/µl is desired: $\dfrac{\text{number of leukocytes}}{1,000} - 1 = x$ ml. With such a concentration the buffer systems present will compensate pH fluctuations, especially when erythrocytes are abundant. The buffer capacity of this type of solution is more effective in the acid range than in the alkali range.

If the mechanism of phytohemagglutinin action is connected with agglutination or with a substantial exchange between lymphocytes, then the cell density must be sufficiently high to allow such a process to occur. Whether a cell culture can only develop if there is a critical "cell density" is not known (Schindler, 1965).

According to Johnson and Russell (1965), in addition to the complex substances in nutrient medium (particularly in serum; antibiotics), others may cause the activation of lymphocytes without the presence of PHA ("background activity"). Leikin *et al.* (1968) also have occasionally observed spontaneous transformation in lymphocyte cultures from umbilical cord blood.

3.2.4. Supplements to Culture Medium

Inhibition of Coagulation. Since large numbers of leukocytes and lymphocytes are lost by clumping to erythrocytes, this should be inhibited. The amount of *heparin* added must be limited to the smallest effective quantity, although somewhat higher concentrations are fairly harmless, since heparin rapidly loses its activity. At very high concentrations, heparin induces hemolysis. Heparin can be placed either in the culture bottle or as a dry substance in the container into which the blood is drawn. Usually, however, a commercially available isotonic solution is used. It is calculated, that 200 units of heparin per 10 ml venous blood are necessary to inhibit coagulation for at least 24 hours, though this dose can be considerably exceeded without danger. Nevertheless, it should be noted that many of the commercial solutions are stabilized with phenol or another antiseptic, and therefore should not be employed. Depot-Vetren (Promonta) and Depot-Liquemin (Roche) widely used in the Federal Republic, are stabilized with phenol. P-hydroxybenzoic acid ester is the preservative in Vetren. If coagulation has already partly occurred, it is often possible to cultivate the free lymphocytes present in the serum after breaking up the clot.

Instead of heparin, several authors recommend EDTA (ethylene-diamine-tetra-acetic acid, Versene). The sodium salt is employed at 0.5–1.0 mg/ml blood. Kinlough *et al.* (1961) also attribute a mitostatic effect to EDTA.

Agglutination can also be partially inhibited by using siliconised syringes. This expense, however, is not justified considering that sili-conised articles cannot be cleaned together with nonsiliconised glass equipment.

Antibiotics. Antibiotics are usually added to the culture medium, generally penicillin and streptomycin at low concentration (Paul, 1970). A proprietary combination of penicillin and streptomycin may be used in quantities of 100–200 units of penicillin and 50–100 µg of streptomycin per ml.

Further supplements, e.g. additional glutamine, which is a component of various media, or vitamin B_{12}, are recommended by several authors, but do not appear to be necessary. Embryo extract has no application in leukocyte culture.

3.2.5. Storage of Culture Media

Several authors store the culture media at $-20°$ C until ready for use.

a) Tips *et al.* (1963). A mixture of 100 ml TC-199, 20,000 units of aqueous penicillin, 25 ml fresh human plasma or serum (blood group 0, Rh+) and 0.5 ml PHA-P (Difco) is made up, divided into aliquots of 5 ml and frozen (which destroys donor leukocytes).

b) Hungerford (1965). The medium consists of Eagle's solution (essential amino acids and vitamins) in double concentration in Earle's isotonic salt solution adjusted to pH 7 by adding 7.5% $NaHCO_3$. The solution is supplemented with 2 mM glutamine, 100 units of peni-cillin per ml, 100 µg streptomycin per ml, 7 µg Phenol Red per ml. Agammaglobulin serum from fetal calf or calf and phytohemagglutinin M are added to give 15% (serum) and 2% (PHA) in the final volume. 20,000 units of heparin per liter of completed medium are added. The medium is divided into aliquots of 5 ml and frozen at $-25°$ C to $-35°$ C when it should be stable for several weeks.

Razavi (1965) stores a medium made up from 80% MEM (with glutamine) and 20% human serum, heparin and PHA.

3.2.6. Duration of Culture

Mitoses are observable in the leukocyte culture after 2 days, although the mitotic rate only achieves optimal values after $2^1/_2$–3 days at constant temperature (37.5° C). Variations depend on the type of PHA used and on the temperature. Unexplainable individual differences are always being observed. It is recommended, therefore, that at least 2 simultaneous cultures be maintained for a period of 48–96 hours, or up to 120 hours in the cases of lymphocytic leukemia and Hodgkin's disease.

If unstable chromosome aberrations are to be examined, then the first wave of lymphocyte division must be interrupted after 48 hours (Bloom and Iida, 1967).

3.2.7. Increasing the Number of Mitoses

Colcemid is added 1–24 hours (usually 2–3 hours) before termination of a culture. According to an inquiry by Genest and Auger (1963) various authors do use colchicine (Merck), though Colcemid (Ciba) is usually used at 0.02 μg to 0.4 mg/ml of culture, corresponding to stock solutions of 0.00005% and 1:25,000. The stock solution can be made up by dissolving 4×1 mg tablets of Colcemid overnight in 100 ml distilled water or from sterile solutions for clinical application.

Vinblastine sulphate which is effective in a final concentration of 0.0075 μg/ml can be used instead of colcemid. Other mitotic inhibitors, or cortisone are not generally employed.

3.3. Storage and Transport of Blood Samples

Venous blood can be stored en route to a cytogenetics laboratory. It is unnecessary to prepare blood for culture directly after it has been drawn, and it is preferable to store the material 2 or more days at room temperature or in the refrigerator at 4° C. The reason for this is not clear.

When a further blood sample cannot be obtained, some blood should be reserved for a repeat culture in the event of failure of the first. Slight partial hemolysis sets in during storage, as well as coagulation due to inactivation of the heparin. K^+ is released from the destroyed erythrocytes. However, Petrakis and Politis (1967) demonstrated that lymphocytes can be cultivated from blood which was 14 days old and kept in an ACD stabilizer, so it is not necessary to store the leukocyte fraction in autologous serum.

Venous blood is best transported in sterile universal containers previously moistened with heparin. The blood should be drawn into a heparinized syringe, and injected into a bottle through the rubber seal without opening the screw top (avoid excessive pressure!). The bottle should be dispatched inside a wooden block and a padded bag. Blood may also be injected into a culture medium to which heparin has been added, particularly where there are only small amounts of blood, as with newborn babies or small children when excessive heparin could have a toxic effect. Phytohemagglutinin is added after receipt at the laboratory and prior to incubation. It is not necessary for the sender to separate leukocytes (Moore et al., 1966 and Uchida and Ray, 1966). After complete removal of the erythrocytes lymphocytes can be kept deep-frozen ($-80°$ C) in a medium containing dimethyl sulfoxide (Pegg, 1965). For methods of cell conservation see p. 50.

3.4. Procedure with Leukemias

To examine leukemic cells, bone marrow should be examined directly whenever possible (see above). Short term cultures (8–48 hours) or whole blood yield equally useful results, provided leukemic cells (blast cells) are present in the blood stream. Usually a small amount of bone marrow is sufficient. If there is enough bone marrow available it is expedient to incubate several parallel cultures in one of the media recommended by Moore et al. (1966), i.e. NCTC-109 (Difco), McCoy-5a, RPMI-906, RPMI-1630 and RPMI-1640 (Gibco) (not medium 199), in order to have sufficient material for supplementary investigations. The use of PHA is not recommended as there is the risk that several mitoses are from activated normal lymphocytes. If PHA is used, however, the cultures must be terminated at the point at which no or very few lymphocytes are present, i.e. after 24–36 hours at the latest.

3.5. Long Term Cultures from Peripheral Blood Cells

Leukocytes have been successfully cultivated from healthy and diseased donors in several laboratories and kept in culture over a long period. By means of suspension cultures in bottles or shake cultures, large quantities of cells can be produced.

Leukocyte cell lines have been established by Dunham et al. (1963) and Martin et al. (1966). In these cases the investigators commenced with leukocyte mixtures of various healthy donors. Medium 199 with 20% human serum and later with 10% fetal calf serum was used.

Monolayer cultures can be established on cover slips in Petri dishes. Prempree and Merz (1966) allow leukocytes (cell density $1-2 \times 10^6$ per 10 ml Eagle's medium, 15% human plasma) to settle at the bottom of milk dilution bottles and change the medium twice a week.

Berman and Stulberg (1962) place leukocyte suspensions in Leighton tubes and change the medium (40% calf serum!) every fourth day.

Lucas *et al.* (1966) suspended the top cell layers of previously settled leukocytes from a case of chronic myeloid leukemia in separate aliquots of MEM (Eagle's) with 20% fetal calf serum (with penicillin, streptomycin, aureomycin, kanamycin, glutamine). Depending on the pH, up to 50% of the medium was changed every 3–5 days. Zur Hausen (1967) achieved good results with medium No. 1629 (Baltimore Biological Laboratories) with addition of 10% fetal calf serum, glutamine, penicillin and streptomycin. Subcultures with fresh medium were set up weekly (cell density 1×10^5 cells/ml).

Pope (1967) and Glade *et al.* (1968) reported on their experiences with lymphocyte culturing from patients with infectious mononucleosis. The lymphoid cell type which developed between 21 and 40 days persisted for a period of months; the cells resembled activated lymphocytes. Doubling times ranged from 48–80 hours. In continuous suspension cultures derived from patients with the Chediak-Higashi syndrome the typical cytoplasmic inclusions were retained (Blume *et al.*, 1969).

Systematic investigations of optimal culture media for tumour cell lines from tumor cells circulating in the blood were carried out by Moore *et al.* (1966). No single medium seems to be equally suitable for all cell types. The best results were obtained with RPMI-1630 and RPMI-1640 ($2-15 \times 10^6$ cells/ml). When a change in pH occurred, the volume of the medium was increased by about 15% and up to 100–250 mg/100 ml glucose added. Large quantities of material in containers of up to 100 liters can be cultivated in this way.

Since the initiation of long-term suspension cultures is rather uncertain, Broder *et al.* (1970) use a highly purified non-hemagglutinating PHA (Wellcome, No. E119) at a dose of 0.025 µg/ml added at the beginning of the primary culture or repeatedly throughout the first 6 weeks.

Colony cultures on agar-gel medium are a new approach to the study of the growth potential and functions of leukocytes. This technique was introduced by Bradley and Metcalf (1966), Senn *et al.* (1967) and others, starting with bone marrow cells. The cells are immobilized in a soft agar-gel containing a nutrient medium complemented with human serum (Senn *et al.*) or in methylcellulose (0.8%) with McCoy-5a medium supplemented with 10% fetal calf serum and 2% deionized bovine serum albumin (Iscove et al., 1971). This suspension is placed over a "feeder" layer which consists of a nutrient medium, serum, agar and cells from fetal kidney tubules (Senn et al., 1967) or of a "conditioned medium" prepared from human peripheral leukocytes cultured for 7 days in 0.5% agar containing McCoy 5a medium and 12.5% fetal calf serum (Iscove et al.). Colonies of peroxidase-positive granulocytes are observed after 7–10 days' incubation. In leukemia and preleukemic states, colony formation is absent or reduced (Senn and Pinkerton, 1972).

Peripheral leukocytes plated into plastic Petri dishes over a layer of white blood cells in McCoy's 5a medium, immobilized with 0.3% agar, are capable of division and granulocytic maturation in vitro (Kurnick and Robinson, 1971).

3.6. Lymphocytes from Lymph Nodes, Thymus, and Spleen

Lymphocytes from post-mortem or biopsied spleen, or from lymph nodules or thymus can be cultured like blood lymphocytes. The tissue sample, excised with as little contamination as possible is washed in medium 199 or in a Ca^{++}- and Mg^{++}-free isotonic solution containing glucose, streptomycin and penicillin. Capsule residues are removed, the tissue is broken up into small fragments with scissors or a fine scalpel and mixed with a pipette. The larger pieces are allowed to sediment. Conen and Erkman (1964) then immediately add PHA, first centrifuge the larger agglutinates and incubate the sediment from the supernatant in medium 199 with 20% fetal calf serum.

Baker and Atkin (1963) count the cells (even if cell clumps have formed) and adjust the cell density of the culture at 1,000–2,000/ml Medium 199 with 25–30% A or 0 serum, with 0.2 ml PHA added to each 10 ml portion of culture.

These adaptable methods are especially suitable for routine use in pathology. The danger of contamination is small, and the karyotype can be determined in 3–5 days. Heart blood taken post-mortem can also be cultured (Mold, 1966) but the chances of success decrease considerably if the aspirate is not taken within a few hours after clinical death owing to bacterial contamination. Fibroblast cultures may also be set up from bone marrow sampled at post-mortem. PHA also activates lymphocytes from fetal thymus or spleen (Bain and Gauld, 1964).

In the first $3^{1}/_{2}$ hours post-mortem, heart blood can be cultivated according to Harrod and Cohen (1969). The mitotic rate is influenced by temperature, drugs, attempts to rescuscitate etc.

4. Phytohemagglutinin: Properties and Mode of Action

In leukocyte suspensions prepared according to Osgood and Krippaehne (1955), i.e. by agglutinating and centrifuging erythrocytes with phyto-hemagglutinin, mitoses occur in large numbers after 2–3 days incubation (Nowell, 1960). This is due to PHA, a mucoprotein in an aqueous solution extracted from *Phaseolus vulgaris* or *Phaseolus coccineus*, which has been known since the beginning of the century (Krüpe, 1956; Robbins, 1964; Dechary, 1968).

Both erythrocytes of all blood groups and leukocytes are agglutinated by PHA. These activities can be separated through adsorption of PHA by erythrocytes or even through a single washing in protein-free physio-logical saline. In serum several washings are required. Leukocyte ag-glutination and mitogenic activity cannot be separated (Barkhan and Ballas, 1963; Kolodny and Hirschhorn, 1964; Nordman *et al.*, 1964; Rivera and Mueller, 1966).

The most important property of PHA for cytogenetic investigations is its mitogenic action. It activates immunocompetent lymphocytes and elicits cell division. The lymphocyte may be considered as a highly re-pressed cell with limited cytoplasm, a slow rate of RNA and protein synthesis and restricted ability to undergo cell division (Imrie and Mueller, 1968). Studies with antiserum to PHA indicate that lympho-cytes do not need continuous exposure after initially having been stimulated by PHA. DNA synthesis is PHA-independent after several cell divisions but can be restimulated by fresh PHA (Younkin, 1972). Different authors have claimed that a specific factor released by activated lymphocytes can stimulate DNA synthesis in other lymphocytes from the same or genetically similar donors (Powles *et al.*, 1971). After only 5 hours incubation, fairly large single monocytoid cells with a less dense nucleus, nucleolus and basophilic cytoplasm can be observed. After 48 hours these cells predominate largely. Degenerative changes appear first in the neutrophils and later in the eosinophils and basophilic granulocytes (loss of motility, cytoplasmic vacuolation, nuclear pycnosis). As the nuclei disintegrate, the total number of cells falls to about 45% of its initial value within 24 hours. The mononuclear cells can be divided into three classes; small pycnotic lymphocytes; large cells with basophilic cytoplasm and light-colored loosely structured nuclear chromatin; and those which occupy an intermediate position (the majority of blast-like cells appear to originate from these). The Golgi apparatus appears to be well developed in these cells, ribosomes and mitochondria are increased, cytoplasm is pyroninophilic. The cells demonstrate an amoeboid motility and often take the form of a hand mirror; they show a tendency to attach themselves to a substrate. The first mitoses occur after about 48 hours. After about 72 hours, the activated lymphocytes often represent 90% of all cells. 6–8 days after initiation of the culture, degenerative changes also occur in these cells (McKinney et al., 1962; Cooper et al., 1963; Coulson and Chalmers, 1964; McFarland and Heilman, 1965; Marshall and Roberts, 1965).

Regression of granulocytes is considerably slower without addition of PHA although even then after 3–4 days the culture consists almost entirely of small mononuclear cells. After 8–10 days polynuclear macrophages predominate; these may congregate in fibroblast-like formations.

The rate of DNA and RNA synthesis or production of proteins measured by incorporation of radiolabeled precursors can serve as a measure of the "activation" of the cells. While only 0.1% of the cells are found to be marked without the use of PHA, with PHA the proportion of these cells rises to 45%. The mitotic rate can amount to 1% per hour at 72 hours.

The transformation rate can be determined microscopically by counting the activated lymphocytes, the number of mitoses or the autoradiographically (^{3}H) labeled cell nuclei. Newly formed DNA can be identified following incorporation of ^{3}H-thymidine, newly formed RNA following incorporation of ^{3}H-uridine and protein by means of labeled amino acids (Schram, 1963; Tormey and Mueller, 1965; Kay, 1968).

The following arguments support the assumption that the cells transformed by PHA are activated lymphocytes:

a) The granulocytes visibly disintegrate.

b) The rate of DNA synthesis depends on a high initial number of cells. If the population of activated cells originated from monocytes, a cell cycle of 5 hours instead of about 18.5 hours would have to be assumed to explain the final cell number (Cave, 1965).

Sasaki and Norman (1966) calculated a generation time of 22 hours. If the transformation of the small lymphocytes is completed after 24 to 48 hours, most of the lymphocytes may be found in the second mitosis after 72 hours. Opossum lymphocytes stimulated with PHA have a maximal mitotic rate at 46 hours (Schneider and Rieke, 1968).

c) Blast-like cells do not occur in congenital alymphocytosis.

There seems to be no threshold value for the efficacy of PHA. The number of activated cells increases with the PHA concentration until all immunocompetent lymphocytes are activated. The effect of PHA is independent of blood group, sex, presence of heparin and corticosteroids. The blastogenic response was found to be reduced or even inhibited by 1-asparaginase, prednisone, methotrexate, 5-fluorouracil, puromycin, actinomycin D, amantidin hydrochloride and chloroquine (see Ohno and Hersh, 1970). A maximal activation has been determined at PHA concentrations of 2.0–85 µg protein/10 ml of culture, depending on the

Phaseolus species from which it was derived, and its purity (Tanaka *et al.*, 1963; McKinney, 1964; Tormey and Mueller, 1965).

In the blast-like cells, a high level of succinic and lactic acid dehydrogenase activity can be histochemically demonstrated indicating anaerobic glycolysis, while esterases and phosphorylases are only present as traces. The cells are peroxidase negative. Schiff and Sudan Black reactions are negative. Pronounced pyroninophilia indicates increased RNA content in the cytoplasm. The cell nuclei become larger and looser so that they appear less heterochromatic (Quaglino *et al.*, 1962; Fischer and Gropp, 1966). Hirschhorn *et al.* (1969) and Cooper *et al.* (1969) have studied the appearance of hydrolase-rich granules and the increase of lysosomal enzyme activity.

After staining with acridine orange, an increase in the nucleolar and the cytoplasmic RNA can be demonstrated by fluorescence microscopy (Yamamoto, 1966). Newly formed proteins can be demonstrated by means of fluorescein-labeled antibodies (Hirschhorn, 1965).

PHA acts in the same way on lymphocytes of lymph nodules, spleen and thymus. A small percentage of lymphocytes is activated in patients with lymphocytic leukemia (Bernard *et al.*, 1964), indicating that the malignant cells are not immunocompetent and that individual mitoses represent normal lymphocytes. In Hodgkin's disease the response to PHA is significantly reduced but there is no correlation with the clinical course (Corder *et al.*, 1972).

There is also depressed activation in acquired and congenital hypogammaglobulinemia.

Stimulation of lymphocytes by PHA is said to diminish with the age of the individual. Since even fetal cells can be activated, previous exposure does not appear to be necessary, though the mitotic activity during the first two years of life seems to be low (Matsaniotis *et al.*, 1967).

Fluorescent microscopy of labeled PHA demonstrates that PHA is present exclusively in cell nuclei (Michalowski *et al.*, 1964).

Leucocytes can be activated by various other substances:

a) by rabbit antibodies to human leukocytes (Gräsbeck *et al.*, 1964);

b) by RNA from lymphocytes activated with PHA;

c) by toxins [tuberculin (Pearmain *et al.*, 1963), diphtheria toxoid, typhus-paratyphus toxoids, streptolysin, staphylococcus toxins (Ling *et al.*, 1965) etc.];

d) by mixing leucocytes from individuals of dissimilar genotype. Activation is poor when the donors are closely related and absent in the case of monozygotic twins who have an identical antigen pool (Bain *et al.*, 1964).

This experiment can be considered as a simple *in vitro* model of a histocompatibility test if mitogenic factors of the medium (proteins, antibiotics) are ignored. Mitotic rate and DNA synthesis are a measure of the incompatibility of the donor. To recognize reciprocal stimulation effects it is necessary to repress the DNA synthesis of leukocytes from one or the other of the donors by means of actinomycin (Bach and Voynow, 1966). Retarded DNA synthesis rate in relation to the PHA stimulation is a characteristic of these cultures. The maximum mitotic rate is attained only after 4–6 days (Bach and Hirschhorn, 1964; Hirschhorn *et al.*, 1963).

PHA is regarded by most authors as an antibody whose primary effect is the aggregation of the cells.

Preparation of PHA

PHA was extracted in the following manner from *Phaseolus vulgaris* and *Phaseolus coccineus* by Li and Osgood (1949): 200 g seed are soaked in 1,000 ml 0.85% sodium chloride solution for 24 hours, macerated, and then left for 3 hours at room temperature. After the particles have been centrifuged off the supernatant is filtered and the solution is adjusted to a pH of 1.3–1.6 with NaOH.

The solution is filtered once more and sterilized through a Seitz filter.

The mitogenic effectiveness of the *Phaseolus* species must be known. It is expedient to commence with small quantities, hull the seeds and grind them in a coffee grinder or a mixer. The Seitz filtrate is kept in deep-freeze as it is only fully active for about 4 weeks.

The hemagglutination titer in leukocyte cultures provides no measure of the mitogenic effectivenes. The commercial bean varieties exhibit considerable differences, even total discrepancy with regard to mitogenic and agglutinogenic components. The majority of *Phaseolus* varieties are unsuitable. Those with bright red and white seeds may contain the mitogenic factor in adequate quantities, especially the red blooming ones.

The crude extract of *Phaseolus* was marketed as Bactophyto-hemagglutinin by the Difco Company up to 1961. An unpurified extract is currently supplied by Burroughs Wellcome while Difco supplies a mucoprotein (form M) which contains a polysaccharide fraction and a hemagglutinating euglobulin; Difco also supplies a pure protein (euglobulin, form P). The hemagglutinating and mitogenic action of the P form is stronger than that of the M form. The procedures for purifying the *Phaseolus* proteins were described by Rigas and Osgood (1955), Punnett *et al.* (1962), Genest (1963), Börjeson *et al.* (1964), Takahashi *et al.* (1967) amongst others. They are based on a repeated precipitation in ammonium sulfate and separation in columns if required. By means of electrophoretic fractionation of proteins from *Phaseolus* several fractions with antigenic or mitogenic effect can be obtained (Allen *et al.*, 1969).

Other plant mitogens are still being tested. An aqueous extract of *Phytolacca americana* (pokeweed) is distributed by Gibco (Catalogue No. 536). The biological and physicochemical properties of the pokeweed mitogen and the response of lymphocytes were studied by Börjeson *et al.* (1966), and Chessin *et al.* (1966). The constant and higher level of efficacy of extracts from seeds and especially from roots is emphasized. Aqueous solutions of *Wistaria floribunda* also have a mitogenic property (Barker and Farnes, 1967). Hirschhorn (1965) also found that extracts of *Robinia pseudoacacia* and *Vicia Faba* are mitogenic. Variations are constantly being observed in the activity of preparations from Difco and Wellcome, indicating that they are not standardized. Batches should be tested on several standard microcultures. Transformation of lymphocytes is effected by concanavalin[6] A (Con A), a lectin isolated from the jack bean (Douglas *et al.*, 1969). It binds to a saccharide-containing site in the cell membrane. Activation is proportional to the concentration of Con A and can be suppressed by methyl-alpha-D-mannoside (Powell and Leon, 1970). The mitogenic action of $HgCl_2$ on lymphocytes was demonstrated by Schöpf and Schulz (1967), and by Schöpf and Nagy (1970). The cells were set up in the usual way and incubated for 5 days with 10 µg $HgCl_2$/ml culture medium. Fiskesjö (1970) avoids use of cysteine-containing media since S-H groups react with mercury preparations. Zn^{++} in a concentration of 2.5×10^{-4} M is also capable of stimulating lymphocytes (Kirchner and Rühl, 1970).

At the same time there is also a question of a model test for hypersensitivity to drugs, heavy metals etc., although in the case of a mercuric chloride, sensitizing should not be necessary.

Action of PHA in Other Species

The hemagglutinating[7] and mitogenic action of PHA is not confined specifically to human lymphocytes, but it does exhibit quantitative and qualitative differences in other species.

Since only small quantities of blood are usually available from small animals and the leukocytes are not separated, it is of no importance whether erythrocytes agglutinate. In cattle, they do not although the mitogenic factor is fully active. PHA stimulation of lymphocytes for chromosomal analysis has been successful in, for example:

Primates (Egozcue and Egozcue, 1966; Sanders and Humason, 1964);
Dog and cat (Ford, 1965);
Hamster (Galton and Holt, 1963);

---

[6] Concanavalin A is distributed e.g. by Calbiochem (catalogue No. 234567) and Miles-Yeda, Ltd.
[7] Erythrocyte agglutination was demonstrated in the dog, cat, rabbit, guinea pig, mouse, rat, sheep, horse, pig, hen, duck and frog.

Cattle (Crossley and Clarke, 1962);

Marsupials (Shaw and Krooth, 1964);

Goats, pigs (Evans, 1965);

Domestic fowls (Newcomer and Donelly, 1964; see Ling, 1968);

Crocodiles, lizards, snakes, toads (Beçak et al., 1962; Ullerich, 1966; see esp. Ling, 1966).

Although Nichols and Levan (1962) as well as Rieke and Schwarz (1964) obtained satisfactory results with rodents, other investigators consider the yield of mitoses insufficient. Systematic investigations with phytagglutinins or toxoids have not been reported. Willard et al. (1965) tested various modifications of culture methods for mouse leukocytes without consistent results. Buckton and Nettesheim (1968) recommend Leighton tubes placed horizontally at a 5° angle and pokeweed as mitogen. DNA synthesis was maximal at 52 hours, the first mitoses having appeared at 45 hours after incubation. Mercuric chloride was found to stimulate the lymphocytes of guinea pigs, rats and rabbits (Pauly et al., 1969) at a concentration of 3.5×10^{-5} M. Prasad and Bushong (1970) directly prepared dividing peripheral mouse leukocytes without prior culturing after having injected colchicine (2 µg/gm) 24 hours prior to sacrifice. Williams and Ray (1965) used the following method: Rats and rabbits received 2×1 ml intraperitoneal doses of AB serum or albumin at an interval of 14 days, following which suspensions of leukocytes from the animals were incubated with and without AB serum or albumin. The highest mitotic rate was observed when both AB serum and albumin were present. Metcalf (1965) demonstrated that rat plasma has a toxic effect on rat, but not on mice or golden hamster, lymphocytes in culture, so specific antigens are necessary. The proliferation of lymphocytes is presumed to be linked with antigen antibody reaction. Since one of the difficulties in activating lymphocytes arises from rather fast plasma coagulation, only small quantities of venous (or heart) blood should be added to the medium (Watson et al., 1966).

Appendix

Laboratory Procedures

1. Chromosome Preparation from Bone Marrow

A.1. 5–10 drops of bone marrow aspirate are suspended in prewarmed 5–10 ml medium 199 or any other balanced salt solution (McCoy 5a, Hanks' solution). The medium contains 50–100 units of heparin and 0.5 ml of an 0.4% solution of Colcemid. If little material is

available, the aspirating syringe should be rinsed out with the isotonic salt solution being used.
2. Incubation at 37° C for 90 min (optional).
3. Chromosome preparations (see also Chapter IV, p. 71). The following procedure has proved effective:
   a) Centrifuge the suspension in conical pointed centrifuge tubes for 5 min at 800 rpm.
   b) Suspend sediment in 4 ml hypotonic solution (Hanks' 1 part + distilled water 3 parts, or 1% Na citrate, or KCl 0.075 M) and allow to stand 20 min at room temperature.
   c) Centrifuge. Draw off cell-free supernatant.
   d) Pipette sediment into freshly prepared fixative (methanol to acetic acid 3:1). Allow to stand for 30 min in a refrigerator.
   e) Change fixative at least 5 times.
   f) Chromosome preparation by air drying (see Chapter IV, p. 73).
B.1. Carefully suspend bone marrow in Hanks' solution. Centrifuge at 500 rpm in closed universal containers.
2. Draw off supernatant. Incubate sediment in McCoy 5a medium. (Cell concentration approximately 2,000–3,000 mm^3).
3. Prepare subcultures 8, (16), 24 and 48 hours. Colchicine has been added 3 hours prior to the use of each sub-culture inoculum.
C.1. Bone marrow is suspended in a prewarmed balanced salt solution (pref. McCoy 5a) containing heparin.
2. The suspension is prepared immediately (A 3a–f).

2. Whole Blood Culture (Microculture)

1. 2–5 drops of venous blood from the finger tips or heels (desinfect the skin with alcohol; allow to dry) in previously prepared medium consisting of : 6 ml McCoy 5a medium/2 ml fetal calf serum, autologous or homologous serum or human plasma, addition of heparin (50–100 units) and PHA 0.1 ml.
2. Incubation in tightly closed universal containers (also Erlenmeyer flasks or similar) at 37° C for 60–72 hours.
3. 2 hours before harvesting, addition of Colcemid (0.3 ml of a 0.04% solution, giving a final concentration of 12 µg/ml).
4. Chromosome preparations according to the instructions in Chapter IV, p. 78.

3. Leukocyte Culture

1. 10–20 ml venous blood is taken with a syringe containing 100–200 units of heparin (e.g. Vetren 1 ml).
2. Place syringe at an angle with the needle upwards until about 2–4 ml plasma supernatant has formed (about 120 min).

3. Insert a sterile needle bent to 135 degrees and carefully inject plasma containing the cells into a sterile Universal Container.
4. Mix the suspension carefully by sucking alternating in and out.
5. Cell count. Dilution to 1,000–2,000 leukocytes/mm³ with medium 199. Exact regulation of the cell count is not absolutely necessary. If no exceptional fluctuations in the leucocyte count are suspected the supernatant plasma can simply be diluted 1:3 with medium 199. With a high cell count, in order to prevent the ratio of plasma to medium 199 from falling below 1:3, the remaining blood must be rapidly centrifuged, the cell-free plasma collected, and added to the culture in suitable quantities.
6. Division into individual portions each of 5–8 ml. Addition of 0.1 ml PHA to each portion.
7. Incubation in tightly sealed universal containers (or similar vessels) at 37° C, 60–72 hours.
8. 2 hours before harvesting, addition of Colcemid (0.3 ml of a 0.04% solution, giving a final concentration of 12 µg/ml).
9. Chromosome preparations according to the instructions in Chapter IV, p. 78.

Modification

a) Allow heparinized venous blood to stand in universal containers in the refrigerator (not below 4° C); add PHA 0.2/10 ml blood 45 min before setting up culture.

b) Centrifuge, 400 rpm, 5 min.

c) Draw off plasma and upper cell layer.

d) Cell count.

e) Dilute and set up with medium 199 or similar.

References

Agostini, A., Ideo, G.: Separation of large numbers of lymphocytes from human blood. Experientia (Basel) **21**, 82 (1965).

Allen, L. W., Svenson, R. H., Yachnin, St.: Purification of mitogenic proteins derived from Phaseolus vulgaris: isolation of potent and weak phytohemagglutinins possessing mitogenic activity. Proc. nat. Acad. Sci. (Wash.) **63**, 334 (1969).

Anders, J. M., Moores, E. C., Emanuel, R.: Chromosome preparation from leucocyte culture. A simplified method for collecting samples by post. J. med. Genet. **3**, 74 (1966).

Arakaki, D. T., Schmid, W.: Chemical mutagenesis. The Chinese hamster bone marrow as an in vivo test system. II. Correlation with in vitro results on Chinese hamster fibroblasts, human fibroblasts and human lymphocytes. Humangenetik **11**, 119 (1971).

Arakaki, D. T., Sparkes, R. S.: Microtechnique for culturing leucocytes from whole blood. Cytogenetics 2, 57 (1963).

Astaldi, G.: Phytohämagglutinin und das Problem des Lymphocyten. Med. Klin. 59, 368 (1964).

Bach, F., Hirschhorn, K.: Lymphocyte interaction: a potential histocompatibility test in vitro. Science 143, 813 (1964).

Bach, F., Voynow, N. K.: One-way stimulation in mixed leukocyte cultures. Science 153, 545 (1966).

Bain, A. D., Gauld, I. K.: The use of thymus and spleen in the demonstration of chromosomes postmortem in foetuses and infants. Brit. J. exp. Path. 45, 530 (1964).

Bain, B., Vas, M. R., Lowenstein, L.: The development of large immature mononuclear cells in mixed leukocyte cultures. Blood 23, 108 (1964).

Baker, M. C., Atkin, N. B.: Short-term culture of lymphoid tissue for chromosome cultures. Lancet 1963 I, 1164.

Barker, B. F., Farnes, P.: Mitogenic property of Wistaria floribunda seeds. Nature (Lond.) 215, 569 (1967).

Barker, B. F., Farnes, P., Fanger, H.: Mitogenic activity in Phytolacca americana (pokeweed). Lancet 1965 I, 170.

Barkhan, P., Ballas, A.: Phytohaemagglutinin: Separation of haemagglutinating and mitogenic principles. Nature (Lond.) 200, 141 (1963).

Beçak, W., Beçak, M. L., Nazareth, H. R. S.: Karyotypic studies of two species of South American snakes (Boa constrictor amarali and Bothrops jararaca). Cytogenetics 1, 305 (1962).

Beckman, L., Fichtelius, K. E., Finley, S. C., Finley, W. H., Lindahl-Kiessling, K.: On the effect of mitogenic plant extracts (phyto-haemagglutinin) on human white blood cells cultivated in vitro. Hereditas (Lund) 48, 619 (1962).

Belpomme, D., Minowada, J., Moore, G. E.: Are some human lymphoblastoid cell lines established from leukemic tissues actually derived from normal leukocytes? Cancer (Philad.) 30, 282 (1972).

Berman, L.: A review of methods for aspiration and biopsy of bone marrow. Amer. J. clin. Path. 23, 384 (1953).

Berman, L., Stulberg, C. S.: Primary cultures of macrophages from normal peripheral blood. Lab. Invest. 11, 1322 (1962).

Berman, L., Stulberg, C. S., Ruddle, F. H.: Long-term tissue cultures of human bone marrow. A report of isolation of a strain of cells resembling epithelial cells from bone marrow of a patient with carcinoma of the lung. Blood 10, 896 (1955).

Bernard, C., Geraldes, A., Boiron, M.: Action de la phytohemagglutinine "in vitro" sur les lymphocytes de leucemies lymphoides chroniques. Nouv. Rev. franç. Hémat. 4, 69 (1964).

Bishun, N. P.: Comparison of mitotic growth rates of human capillary whole blood cultured in a variety of media. J. med. Genet. 4, 41 (1967).

Bloom, A. D., Iida, Sh.: Two-day leukocyte cultures for human chromosome studies. Jap. J. hum. Genet. 12, 38 (1967).

Blume, R. S., Glade, P. R., Gralnick, H. R., Chessin, L. N., Haase, A. T., Wolff, S. M.: The Chediak-Higashi syndrome: Continuous suspension cultures derived from peripheral blood. Blood 33, 821 (1969).

Börjeson, J., Bouveng, R., Gardell, S., Norden, A., Thunell, St.: Purification of the mitosis-stimulating factor from Phaseolus vulgaris. Biochim. biophys. Acta (Amst.) 82, 158 (1964).

Börjeson, J., Reisfeld, R., Chessin, L. N., Welsh, P. D., Douglas, St. D.: Studies on human peripheral blood lymphocytes in vitro. I. Biological and physicochemical properties of the pokeweed mitogen. J. exp. Med. **124**, 859 (1966).

Boll, I., Fuchs, G.: Vereinfachtes Verfahren zur kurzfristigen Kultivierung von menschlichem Knochenmark in vitro. Blut **7**, 257 (1961).

Bond, V. P., Fliedner, T. M., Cronkite, E. P., Rubini, J. R., Brecher, Schork, P. K.: Proliferative potentials of bone marrow and blood cells studied by in vitro uptake of H_3-thymidine. Acta haemat. (Basel) **21**, 1 (1961).

Bottura, C., Farrari, I.: A simplified method for the study of chromosomes in man. Nature (Lond.) **186**, 904 (1960).

Bradley, T. R., Metcalf, D.: The growth of mouse bone marrow cells in vitro. Aust. J. exp. Biol. med. Sci. **44**, 287 (1966).

Broder, S. W., Glade, P. R., Hirschhorn, K.: Establishment of longterm lines from small aliquots of normal lymphocytes. Blood **35**, 539 (1970).

Brody, J. A., Huntley, B.: Human lymphocytes cultured in microplates. Nature (Lond.) **208**, 1232 (1965).

Buckton, K. E., Nettesheim, P.: In vitro and in vivo culture of mouse peripheral blood for chromosome preparations. Proc. Soc. exp. Biol. (N.Y.) **128**, 1106 (1968).

Buckton, K. E., Smith, P. G., Court Brown, W. M.: The estimation of lymphocyte lifespan from studies on males treated with X-rays for ankylosing spondylitis. In: Evans, H. J., Court Brown, W. M., McLean, A. S., eds., Human radiation cytogenetics, p. 106–114. Amsterdam: North Holland Publ. Co. 1967.

Cardinali, G., Cardinali, G., Agrifoglio, M. F.: The colchicine method in the study of bone marrow cell proliferation. Blood **18**, 328 (1961).

Cave, M. D.: The reverse patterns of thymidine-^{3}H incorporation in human chromosomes. Hereditas (Lund) **54**, 338 (1965).

Chen, H. P., Palmer, G. K.: A method for isolating leukocytes. Amer. J. clin. Path. **30**, 567 (1958).

Chessin, L. N., Börjeson, J., Welsh, P. D., Douglas, St. D., Cooper, H. L.: Studies on human peripheral blood lymphocytes in vitro. II. Morphological and biochemical studies on the transformation of lymphocytes by pokeweed mitogen. J. exp. Med. **124**, 873 (1966).

Chu, E. H. Y., Giles, N. H.: Human chromosome complements in normal somatic cells in culture. Amer. J. hum. Genet. **11**, 63 (1959).

Ciresa, M., Huber, H.: Die Isolierung funktionsfähiger Blutlymphozyten. Acta haemat. (Basel) **38**, 300 (1967).

Conen, P. E., Erkman, B.: Necropsy spleen samples for chromosome cultures. Lancet **1964 I**, 665.

Cooper, E. H., Barkhan, P., Hale, A. J.: Observations on the proliferation of human leucocytes cultured with phytohaemagglutinin. Brit. J. Haemat. **9**, 101 (1963).

Cooper, H. L., Hirschhorn, R., Hirschhorn, K., Rubin, A.: Biochemical and synthetic activities of cultured lymphocytes. Proc. 3rd Ann. Leucocyte Culture Conf. Ed. W. O. Rieke, p. 647. New York: Appleton-Century-Crofts 1969.

Corder, M. P., Young, R. C., Brown, R. S., DeVita, V. T.: Phytohemagglutinin-induced lymphocyte transformation: The relationship to prognosis of Hodgkin's disease. Blood **39**, 595 (1972).

Coulson, A. S., Chalmers, D. G.: Separation of viable lymphocytes from human blood. Lancet 1964 I, 468.

Coulson, A. S., Chalmers, D. G.: Effects of phytohaemagglutinin on leucocytes. Lancet 1964 II, 819.

Crossley, R., Clarke, G.: The application of tissue-culture technique to the chromosomal analysis of Bos taurus. Genet. Res. 3, 167 (1962).

Czerski, P., Szmigielski, S., Litwin, J.: Simple methods for obtaining purified suspensions of lymphocytes. Vox Sang. (Basel) 11, 734 (1966).

Dechary, J. M.: Phytohemagglutinins. A survey of recent progress. Vox Sang. (Basel) 15, 401 (1968).

Douglas, St. D., Kamin, R. M., Davis, W. C., Fudenberg, H. H.: Biochemical and morphologic aspects of phytomitogens: Jack Bean, Wax Bean, Pokeweed, and Phytohemagglutinin. Proc. 3rd Ann. Leucocyte Culture Conf. Ed. W. O. Rieke, p. 607. New York: Appleton-Century-Crofts 1969.

Dunham, W. B., Ewing, F. M., Parker, M. V.: Culture of leukocytes after storage in serum. Proc. Soc. exp. Biol. (N.Y.) 114, 234 (1963).

Edwards, J. H.: Chromosome analysis from capillary blood. Cytogenetics 1, 90 (1962).

Edwards, J. H., Young, R. B.: Chromosome analysis from small volumes of blood. Lancet 1961 II, 49 (1961).

Egozcue, J., Egozcue, M. V. de: Simplified cultures and chromosome preparations of primate leukocytes. Stain Technol. 41, 173 (1966).

Evans, H. J.: A simple microtechnique for obtaining human chromosome preparations with some comments on DNA replication in sex chromosomes of the goat, cow and pig. Exp. Cell Res. 38, 511 (1965).

Farnes, P., Baker, B. E., Fanger, H.: A technique for chromosome study of human bone marrow fibroblast-like cells. Exp. Cell Res. 29, 86 (1963).

Fischer, R., Gropp, A.: Cytochemie des Lymphocyten in vitro. Klin. Wschr. 44, 733 (1966).

Fiskesjö, G.: The effect of two organic mercury compounds on human leucocytes in vitro. Hereditas (Lund) 64, 142 (1970).

Fitzgerald, P. H.: The immunological role and long life-span of small lymphocytes. J. theor. Biol. 6, 12 (1964).

Ford, C. E., Jacobs, P. A., Lajtha, L. G.: Human somatic chromosomes. Nature (Lond.) 181, 1565 (1958).

Ford, L.: Leukocyte culture and chromosome preparations from adult dog blood. Stain Technol. 40, 317 (1965).

Forteza Bover, G., Baguena Candela, R.: Technica para el estudio cromosomico por un metodo directo de las celulas obtenidas mediante puncion de los ganglios infaticos. Med. esp. 54, 26 (1965).

Foster, R., Metcalf, D., Robinson, W. A., Bradley, T. R.: Bone marrow colony stimulating activity in human sera. Brit. J. Haemat. 15, 147 (1968).

Fraccaro, M., Kaijser, K., Lindsten, J.: Somatic chromosome complement in continuously cultured cells of two individuals with gonadal dysgenesis. Ann. hum. Genet. 24, 45 (1960).

Frøland, A.: A micromethod for chromosome analysis on peripheral blood cultures. Lancet 1962 II, 1281.

Galton, M., Holt, S. F.: Culture of peripheral blood leucocytes of the golden hamster. Proc. Soc. exp. Biol. (N.Y.) 144, 218 (1963).

Genest, P.: Production of a semi-purified phytohaemagglutinin (mucoprotein) of high potency for the study of chromosomes of leucocytes. Lancet 1963 I, 838.

Genest, P., Auger, C.: Observations on the technique for the study of human chromosomes by the culture of leukocytes from peripheral blood. Canad. med. Ass. J. **88**, 302 (1963).

Glade, Ph. R., Kasel, J. A., Moses, H. L., Whang-Peng, J., Hoffman, P. F., Kammermeyer, J. K., Chessin, L. N.: Infectious mononucleosis: Continuous suspension culture of peripheral blood leukocytes. Nature (Lond.) **217**, 564 (1968).

Gräsbeck, R., Nordman, C. T., Chapelle, A. de la: The leucocyte-mitogenic effect of serum from rabbits immunized with human leucocytes. Acta med. scand., Suppl. **412**, 39 (1964).

Green, I., Solomon, W.: Separation of human lymphocytes and monocytes using an "oil bottle". J. clin. Path. **16**, 180 (1963).

Greenwalt, T. J., Gajewski, M., McKenna, J. L.: A new method for preparing buffy coat-poor blood. Transfusion (Philad.) **2**, 221 (1962).

Gropp, A., Fischer, R.: Ergebnisse der Züchtung von Lymphocyten in vitro. Klin. Wschr. **44**, 665 (1966).

Grouchy, J. de, Roubin, M., Passage, E.: Microtechnique pour l'étude des chromosomes humains à partir d'une culture de leucocytes sanguins. Ann. Génét. (Paris) **7**, 45 (1964).

Harrod, E. K., Cohen, M. M.: Cytogenetic analysis of postmortem cultured leukocytes. Pediatrics **44**, 128 (1969).

Hastings, J., Freedman, St., Rendon, O., Cooper, H. L., Hirschhorn, K.: Culture of human white cells using differential leucocyte separation. Nature (Lond.) **192**, 1214 (1961).

Hausen, H. zur: Chromosomal changes of similar nature in seven established cell lines derived from the peripheral blood of patients with leukemia. J. nat. Cancer Inst. **38**, 683 (1967).

Heilmeyer, L., Begemann, H.: Atlas der klinischen Hämatologie und Zytologie. Berlin-Göttingen-Heidelberg: Springer 1955.

Hentel, J., Hirschhorn, K.: The origin of some bone marrow fibroblasts. Blood **38**, 81 (1971).

Hirschhorn, K.: The mitogenic effect of different substances on lymphocytes in tissue culture. Congr. Internat. Soc. Haemat. **10**, 6 (1965).

Hirschhorn, K., Bach, F., Kolodny, R. L.: Immune response and mitosis of human peripheral blood lymphocytes in vitro. Science **142**, 1185 (1963).

Hirschhorn, R., Brittinger, G., Hirschhorn, K., Troll, W., Weissman, G.: Effect of phytohemagglutinin stimulation upon hydrolytic enzymes and nuclear template activity. Proc. 3rd Ann. Leucocyte Culture Conf. Ed. W. O. Rieke, p. 639. New York: Appleton-Century-Crofts 1969.

Hsu, T. C., Patton, J. L.: Bone marrow preparations for chromosome studies. In: Comparative mammalian cytogenetics (ed. K. Benirschke). Berlin-Heidelberg-New York: Springer 1969.

Hulliger, L., Blazkovec, A. A.: A simple and efficient method of separating peripheral-blood leucocytes for in vitro studies. Lancet **1967 I**, 1304.

Hungerford, D. A.: Leukocytes cultured from small inocula of whole blood and the preparation of metaphase chromosomes by treatment with hypotonic KCl. Stain Technol. **40**, 333 (1965).

Imrie, R. C., Mueller, G. C.: Release of a lymphocyte growth promoter in leucocyte cultures. Nature (Lond.) **219**, 1277 (1968).

Iscove, N. N., Senn, J. S., Till, J. E., McCulloch, E. A.: Colony formation by normal and leukemic human marrow cells in culture: Effect of conditioned medium from human leukocytes. Blood **37**, 1 (1971).

Johnson, G. J., Russell, P. S.: Reaction of human lymphocytes in culture to components of the medium. Nature (Lond.) **208**, 343 (1965).

Kaijser, K.: Container for cultivating blood for chromosome studies. Lancet **1961 II**, 1362.

Kay, J. E.: Early effects of phytohaemagglutinin on lymphocyte RNA synthesis. Europ. J. Biochem. **4**, 225 (1968).

Killmann, S. A., Cronkite, E. P., Fliedner, T. M., Bond, V. P.: Mitotic indices of human bone marrow. I. Number and cytologic distribution of mitosis. Blood **19**, 743 (1962).

Kinlough, M. A., Robson, H. N., Hayman, D. L.: A simplified method for the study of chromosomes in man. Nature (Lond.) **189**, 420 (1961).

Kinlough, M. A., Robson, H. N., Hayman, D. L.: Study of chromosomes in human leukaemia by a direct method. Brit. med. J. **1961 II**, 1052.

Kiossoglou, K. A., Mitus, W. J., Dameshek, W.: A direct method for chromosome studies of human bone marrow. Amer. J. clin. Path. **41**, 183 (1964).

Kirchner, H., Rühl, H.: Stimulation of human peripheral lymphocytes by Zn^{++} in vitro. Exp. Cell Res. **61**, 229 (1970).

Kolodny, R. L., Hirschhorn, K.: Properties of phytohaemagglutinin. Nature (Lond.) **201**, 715 (1964).

Krüpe, M.: Blutgruppenspezifische pflanzliche Eiweißkörper. Stuttgart: Enke 1956.

Kurnick, J. E., Robinson, W. A.: Colony growth of human peripheral white blood cells in vitro. Blood **37**, 136 (1971).

Lajtha, L. G.: Culture of human bone marrow in vitro. The reversibility between normoblastic and megaloblastic series of cells. J. clin. Path. **5**, 67 (1952).

Lajtha, L. G.: Bone marrow in culture. In: Cells and tissues in culture, ed. E. N. Willmer, vol. 2. London-New York: Academic Press 1965.

Lamvik, J. O.: Separation of lymphocytes from human blood. Acta haemat. (Basel) **35**, 294 (1966).

Leikin, S., Mochir-Fatemi, F., Park, K.: Blast transformation of lymphocytes from newborn human infants. J. Pediat. **72**, 510 (1968).

Li, J. G., Osgood, E. E.: A method for the rapid separation of leucocytes and nucleated erythrocytes from blood or marrow with a phytohemagglutinin from red beans. Blood **4**, 670 (1949).

Ling, N. R.: Lymphocyte stimulation. Amsterdam: North-Holland 1968.

Ling, N. R., Spicer, E., James, K., Williamson, N.: The activation of human peripheral lymphocytes by products of staphylococci. Brit. J. Haemat. **11**, 421 (1965).

Lucas, L. S., Whang, J. J. K., Tjio, J. H., Manaker, R. A.: Continuous cell culture from a patient with chronic myelogenous leukemia. I. Propagation and presence of Philadelphia chromosome. J. nat. Cancer Inst. **37**, 753 (1966).

Mac Kinney, A. A.: Dose-response curve of phytohaemagglutinin in tissue culture of normal human leucocytes. Nature (Lond.) **204**, 1002 (1964).

Mac Kinney, A. A., Stohlman, F., Brecher, G.: The kinetics of cell proliferation in cultures of human peripheral blood. Blood **19**, 349 (1962).

Marshall, W. H., Roberts, K. B.: Continuous cinematography of human lymphcytes cultured with a phytohaemagglutinin including observations on cell division and interphase. Quart. J. exp. Physiol. **50**, 361 (1965).

Martin, G. M., Sprague, C., Dunham, W. B.: Chromosomal analysis of "leukocyte" cell lines. Lab. Invest. **15**, 692 (1966).

Matsaniotis, N., Economou-Mavrou, C., Tsenghi, C.: Low mitotic activity of peripheral lymphocytes during the first two years of life. Arch. Dis. Childh. **42**, 549 (1967).

McFarland, W., Heilman, D. H.: Lymphocyte foot appendage: Its role in lymphocyte function and in immunological reactions. Nature (Lond.) **205**, 887 (1965).

Meighan, S. Sp., Stich, H. F.: Simplified technique for examination of chromosomes in the bone marrow of man. Canad. med. Ass. J. **84**, 1004 (1961).

Melen, J. van, Unger, P.: Simple lymphocyte separation. Lancet **1967 II**, 313.

Mellman, W. J.: Human peripheral blood leucocyte cultures. In: Human chromosome methodology, ed. J. J. Yunis. New York-London: Academic Press 1965.

Metcalf, D., Stanley, E. R.: Haematological effects in mice of partially purified colony stimulating factor (CSF) prepared from human urine. Brit. J. Haemat. **21**, 481 (1971).

Metcalf, W. K.: Some experiments on the phytohaeamgglutinin of leucocytes from rats and other mammals. Exp. Cell Res. **40**, 490 (1965).

Michalowski, A., Jasinska, A., Brozosko, W. J., Nowoslawski, A.: Cellular localization of the mitogenic principle of phytohaemagglutinin in leukocyte cultures. Exp. Cell Res. **34**, 117 (1964).

Minor, A. H., Burnett, L.: A method for obtaining living leukocytes from human peripheral blood by acceleration of erythrocyte sedimentation. Blood **3**, 799 (1948).

Mold, J. W.: Chromosomes after death. Lancet **1966 II**, 107.

Moore, G. E., Ito, E., Ulrich, K., Sandberg, A. A.: Culture of human leukemia cells. Cancer (Philad.) **19**, 713 (1966).

Moorhead, P. S., Nowell, P. C., Mellman, W. J., Battips, D. M., Hungerford, D. A.: Chromosome preparations of leucocytes cultured from human peripheral blood. Exp. Cell Res. **20**, 613 (1960).

Nichols, W. W., Levan, A.: Chromosome preparations by the blood tissue culture technique in various laboratory animals. Blood **20**, 106 (1962).

Noble, P. B., Cutts, J. H., Carroll, K. K.: Ficoll flotation for the separation of blood leukocyte types. Blood **31**, 66 (1968).

Nordman, C. T., Chapelle, A. de la, Gräsbeck, R.: The interrelations of erythroagglutinating, leucoagglutinating and leucocyte mitogenic activities in Phaseolus vulgaris phytohaemagglutinin. Acta med. scand., Suppl. **412**, 49 (1964).

Nowell, P. C.: Phytohaemagglutinin: an initiator of mitosis in cultures of normal human leukocytes. Cancer Res. **20**, 462 (1960).

Nowell, P. C.: Unstable chromosome changes in tuberculin-stimulated leukocyte cultures from irradiated patients. Evidence for immunologically committed, long-lived lymphocytes in human blood. Blood **26**, 798 (1965).

Ohno, R., Hersh, E. M.: The inhibition of lymphocyte blastogenesis. Blood **35**, 250 (1970).

Ohnuki, Y.: Demonstration of the spiral structure of human chromosomes. Nature (Lond.) **208**, 916 (1965).

Osgood, E. E., Brownlee, I. E.: Culture of human marrow: details of a simple method. J. Amer. med. Ass. **108**, 1793 (1937).

Osgood, E. E., Krippaehne, M. L.: The gradient tissue culture method. Exp. Cell Res. **9**, 116 (1955).

Paul, J.: Cell and tissue culture, 4. ed. Edinburgh: Livingstone 1970.

Pauly, J. L., Caron, G. A., Suskind, R. R.: Blast transformation of lymphocytes from Guinea pigs, rats, and rabbits induced by mercuric chloride in
vitro. J. Cell Biol. **40**, 847 (1969).

Pearmain, G., Lycette, R. R., Fitzgerald, P. H.: Tuberculin-induced mitosis in
peripheral blood leucocytes. Lancet **1963** I, 637.

Pegg, P. J.: The preservation of leucocytes for cytogenetic and cytochemical
studies. Brit. J. Haemat. **11**, 586 (1965).

Pertoft, H., Bäck, O., Lindahl-Kiessling, K.: Separation of various blood cells
in colloidal silica-polyvinylpyrrolidone gradients. Exp. Cell Res. **50**, 355
(1968).

Petrakis, N. L., Politis, G.: Prolonged viability of mitotically competent mononuclear leukocytes in stored whole blood. New Engl. J. Med. **267**, 286
(1962).

Pope, J. H.: Establishment of cell lines from peripheral leucocytes in infectious
mononucleosis. Nature (Lond.) **216**, 810 (1967).

Powell, A. E., Leon, M. A.: Reversible interaction of human lymphocytes
with the mitogen concanavalin A. Exp. Cell Res. **62**, 315 (1970).

Powles, R., Balchin, L., Currie, G. A., Alexander, P.: Specific autostimulating
factor released by lymphocytes. Nature (Lond.) **231**, 161 (1971).

Prasad, N., Bushong, St. C.: Direct preparation of chromosomes from mouse
peripheral blood. Acta cytol. (Philad.) **14**, 523 (1970).

Prempree, Th., Merz, T.: Continuous culture of normal human leucocytes
from peripheral blood. Nature (Lond.) **212**, 1576 (1966).

Pulvertaft, R. J. V., Jayne, W. H. W.: Agar cultures of exsudates. J. clin. Path.
10, 390 (1953).

Punnett, Th., Punnett, H. H., Kaufmann, B. N.: Preparation of a crude leucocyte growth factor from phaseolus vulgaris. Lancet **1962** I, 1359.

Quaglino, D., Hayhoe, F. G. J., Flemans, R. J.: Cytochemical observations of
the effect of phytohaemagglutinin in short term tissue cultures. Nature
(Lond.) **196**, 338 (1962).

Rabinowitz, Y.: Separation of lymphocytes, polymorphonuclear leukocytes
and monocytes on glass columns, including tissue culture observations.
Blood **23**, 811 (1964).

Razavi, L.: An inexpensive and simple method for preparing chromosome
spreads. Proc. Soc. exp. Biol. (N.Y.) **118**, 717 (1965).

Reisner, E. H., Jr.: Tissue culture of bone marrow. Ann. N.Y. Acad. Sci. **77**,
487 (1959).

Rigas, D. A., Johnson, E. A.: Studies on the phytohaemagglutinin of
Phaseolus vulgaris and its mitogenicity. Ann. N.Y. Acad. Sci. **113**, 2800
(1964).

Rigas, D. A., Osgood, E. E.: Purification and properties of the phytohaemagglutinin of Phaseolus vulgaris. J. biol. Chem. **212**, 607 (1955).

Rivera, A., Mueller, G. C.: Differentiation of the biological activities of
phytohaemagglutinin affecting leucocytes. Nature (Lond.) **212**, 1207 (1966).

Robbins, J. H.: Tissue culture studies of the human lymphocyte. Science **146**,
1648 (1964).

Robinson, W. A., Stanley, E. R., Metcalf, D.: Stimulation of bone marrow
colony growth in vitro by human urine. Blood, **33**, 396 (1969).

Rohr, K.: Das menschliche Knochenmark, 3. Aufl. Stuttgart: Thieme 1960.

Sandberg, A. A., Crosswhite, L. H., Gordy, E.: Trisomy of a large chromosome. Association with mental retardation. J. Amer. med. Ass. **174** 221
(1960).

Sandberg, A. A., Ishihara, T., Crosswhite, L. H., Hauschka, T. S.: Chromosomal dichotomy in blood and marrow of acute leukemia. Cancer Res. **22**, 748 (1962).

Sandberg, A. A., Ishihara, T., Crosswhite, L. H., Hauschka, T. S.: Comparison of chromosome constitution in chronic myelocytic leukemia and other myeloproliferative disorders. Blood **20**, 393 (1962).

Sandberg, A. A., Kikuchi, Y., Crosswhite, L. H.: Mitotic ability of leukemic leukocytes in chronic myelocytic leukemia. Cancer Res. **24**, 1468 (1964).

Sandberg, A. A., Koepf, G. F., Crosswhite, Hauschka, T. S.: Chromosome constitution of human marrow in various development and blood disorders. Amer. J. hum. Genet. **12**, 231 (1960).

Sanders, Ph. C., Humason, G. L.: Culture and slide preparations of leukocytes from blood of Macaca. Stain Technol. **39**, 209 (1964).

Sarkany, I.: Lymphocyte transformation in drug hypersensitivity. Lancet **1967 I**, 743.

Sasaki, M. S., Norman, A.: Proliferation of human lymphocytes in culture. Nature (Lond.) **210**, 913 (1966).

Schär, B., Loustalot, P., Gross, F.: Demecolcin (Substanz F), ein neues aus Colchicum autumnale isoliertes Alkaloid mit starker antimitotischer Wirkung. Klin. Wschr. **32**, 49 (1954).

Schindler, R.: Die tierische Zelle in Zellkultur. Berlin-Heidelberg-New York: Springer 1965.

Schmid, W., Arakaki, D. T., Breslau, N. A., Culbertson, J. C.: Chemical mutagenesis. The Chinese hamster bone marrow as an in vivo test system. I. Cytogenetic results on basic aspects of the methodology, obtained with alkylating agents. Humangenetik **11**, 103 (1971).

Schneider, L. K., Rieke, W. O.: Opossum lymphocytes in vitro: a valuable tool for cytogenetic investigations. Cytogenetics **7**, 1 (1968).

Schöpf, E., Nagy, G.: Fine structural features of lymphocyte transformation induced by mercuric salts. Acta haemat. (Basel) **43**, 73 (1970).

Schoepf, E., Schulz, K. H.: Mitogenesis by mercuric chloride. Lancet **1967 II**, 840.

Schram, E.: Organic scintillation detectors. Amsterdam: Elsevier 1963.

Senn, J. S., McCulloch, E. A., Till, J. E.: Comparison of colony-forming ability of normal and leukaemic human marrow in cell culture. Lancet **1967 II**, 597.

Senn, J. S., Pinkerton, P. H.: Defective in vitro colony formation by human bone marrow preceding overt leukaemia. Brit. J. Haemat. **23**, 277 (1972).

Shaw, M. W., Krooth, R. S.: The chromosomes of the Tasmanian Rat-Kangoroo (Tridactylis apicalis). Cytogenetics **3**, 19 (1964).

Skoog, W., Beck, W. S.: Studies on the fibrinogen, dextran and phytohemagglutinin methods of isolating leukocytes. Blood **11**, 436 (1956).

Spiers, A. S. D., Baikie, A. G.: Chronic granulocytic leukaemia: Demonstration of the Philadelphia chromosome in cultures of spleen cells. Nature (Lond.) **208**, 497 (1965).

Spriggs, A. I., Alexander, R. E.: An albumin gradient method for separating the different white cells of blood, applied to the concentration of circulation tumour cells. Nature (Lond.) **188**, 863 (1960).

Steinberger, A., Smith, K. D., Steinberger, E., Perloff, W. H.: Chromosomal analysis from small volumes of peripheral blood. J. A. Einstein med. Cent. **12**, 5 (1964).

Stewart, J. S. S.: Chromosome analysis. Lancet **1960 II**, 651.

Takahashi, T., Ramachandramurthy, P., Liener, I. E.: Some physical and chemical properties of a phytohemagglutinin isolated from Phaseolus vulgaris. Biochim. biophys. Acta (Amst.) 133, 123 (1967).

Tanaka, Y., Epstein, L. B., Brecher, G., Stohlman, F.: Transformation of lymphocytes in cultures of human peripheral blood. Blood 22, 614 (1963).

Thierfelder, G.: A method for the isolation of human lymphocytes. Vox Sang. (Basel) 9, 447 (1964).

Tips, R. L., Smith, G. S., Meyer, D. L., Ushijima, R. N.: Karyotype analysis of leucocytes as a practical laboratory procedure. Tex. Rep. Biol. Med. 21, 581 (1963).

Tjio, J. H., Whang, J.: Chromosome preparations of bone marrow cells without prior in vitro culture or in vivo colchicine administration. Stain Technol. 37, 17 (1962).

Tjio, J. H., Whang, J.: Direct chromosome preparations of bone marrow cells. In: Human chromosome methodology, ed. J. J. Yunis. New York-London: Academic Press 1965.

Tolksdorf, N., Wiedemann, H. R., Hansen, H. G., Lehmann, W.: Pätau-Syndrom mit Trisomie D_1 und D/D-Translokation. Med. Welt 1965, 2304.

Tormey, D. C., Mueller, G. C.: An assay for the mitogenic activity of phytohemagglutinin preparations. Blood 26, 569 (1965).

Uchida, I. A., Ray, M.: Mail-order chromosome analysis. Canad. med. Ass. J. 94, 649 (1966).

Ullerich, F. H.: Karyotyp und DNS-Gehalt von Bufo bufo, B. viridis, B. bufo × B. viridis und B. calamita (Amphibia, Anura). Chromosoma (Berl.) 18, 316 (1966).

Ulrich, K., Moore, G. E.: Separation of viable leukocytes from normal human blood. Acta heamat. (Basel) 35, 338 (1966).

Vries, G. F. de, Went, J. J. van: The production of abundant metaphases in bone marrow of adult rats by hemolytic stimulation. Stain Technol. 39, 52 (1964).

Walker, R. I., Fowler, I.: Granulocyte inhibition of human peripheral blood lymphocyte growth in vitro. Exp. Cell Res. 38, 379 (1965).

Watson, E. D., Blumenthal, H. T., Hutton, W. E.: A method for the culture of leucocytes of the Guinea pig (Cavia cobaya) with karyotype analysis. Cytogenetics 5, 179 (1966).

Willard, H. G., Hoppe, I. B. H., Nettesheim, P.: Mouse leukocytes in culture. Proc. Soc. exp. Biol. (N.Y.) 118, 993 (1965).

Williams, T. W., Ray, M.: A method for culturing leucocytes of rats and rabbits. Cytogenetics 4, 365 (1965).

Williamson, S. M.: The leukocyte method for the study of human chromosomes. Canad. J. med. Technol. 24, 13 (1962).

Woodliff, H. J.: Blood and bone marrow cell culture. London: Eyre & Spottiswoode 1964.

Woods, A. H.: A closed system for large-scale lymphocyte purification. Blood 35, 39 (1970).

Yamamoto, H.: Reversible transformation of lymphocytes in human leucocyte cultures. Nature (Lond.) 212, 997 (1966).

Younkin, L. H.: In vitro response of lymphocytes to phytohemagglutinin (PHA) as studied with antiserum to PHA. I. Initiation period, daughter-cell proliferation, and restimulation. Exp. Cell Res. 75, 1 (1972).

CHAPTER II

Cell Cultures from Tissue Explants

ULRICH WOLF

With 1 Figure

1. Introduction

Cell cultures can be prepared from the most diverse tissues and organs. The basic procedure is as follows: Tissue specimens (biopsy or autopsy material) are taken from the body, reduced to small pieces, and set up in vitro in a nutrient medium. There are a number of methods available to prevent explant fragments from floating in the nutrient medium. In these socalled primary cultures cells divide which are usually morphologically heterogeneous. Depending on the source of material, epithelial or fibroblast-like cells at first predominate. After some time, at the latest after the first subcultures, only fibroblast-like cells are found. When a dense cell layer has formed, the tissue fragments are isolated from the proliferated cells. A subculture is then set up, from which further subcultures are derived after cell multiplication. To obtain mitoses for chromosome analysis, a cell culture in the most vigorous cell growth phase (the logarithmic phase) is processed according to the directions given in Chapter IV.

2. Areas of Application

For routine diagnosis of clinical cases, chromosome analysis of lymphocyte mitoses (see Chapter I) or the determination of X- and Y-chromatin from epithelial cells of buccal mucosa (see Chapter IX), or both, may often be sufficient. However, other tissues or cell systems might well yield results that deviate from those obtained in lymphocytes, so whether the blood culture method is adequate for diagnosis in individual cases can only be decided with reference to the clinical picture, the medical history of the family, and the cytogenetic findings from lymphocytes and epithelial cells of buccal mucosa. If chromosomal analysis of peripheral blood does not yield an unequivocal result, or if there are doubts about the completeness of diagnosis, the examination of other tissues or cell systems is necessary. In general, an unusual finding in one cell or tissue type should be confirmed by the examination of at least one other cell type.

Specific tissue differences in chromosome diagnoses are to be particularly expected in mosaic cases (see Chapter VIII) and diseases of the hemopoietic system (see Chapter I). Mosaics frequently exhibit varying proportions of the individual stem lines in different tissues (Penrose and Smith, 1966). In particular, the proportion of anomalous fibroblasts can often be substantially higher than the proportion of anomalous lymphocytes, so that fibroblast culture offers better prospects for detecting mosaics. Occasionally an abnormal cell line may be found in one tissue (e.g. fibroblasts) while another (e.g. blood cells) has a normal karyotype.

In diseases of the hemopoietic system, chromosome aberrations are usually confined to cells originating from this system. In these cases, it is necessary to exclude a generalised chromosomal aberration by analysis of fibroblasts.

Fibroblast culture is also useful for the examination of fetal tissue and autopsy material, when only solid tissue is usually available.

Fibroblast culture sometimes provides an alternative clinical diagnostic method in cases in which lymphocyte cultures repeatedly fail owing to adverse blood conditions (e.g. agammaglobulinemia, leukopenia).

Refined methods of chromosome identification, particularly autoradiography (see Chapter VII) and the various banding techniques (see Chapters V and VI) often require a large number of mitoses for study. Since fibroblast culture permits considerable proliferation of the same initial material (30–50 cell generations), this method is therefore preferable to blood culture in pertaining cases.

Cells in interphase may also be investigated in cell cultures of solid tissue, e.g. results of sex chromatin determinations are often more unequivocal than those of buccal smears (see Chapter IX). Autosomal heterochromatin, interphase fluorescence and nucleoli can also be studied in cell cultures.

Certain experimental investigations, for example, studies of selection and resistance, radiation or virus effects are often not feasible in blood cells, and fibroblast cultures are preferable. As outlined in Chapter XII, somatic cell genetics mainly employs fibroblast cultures, particularly in certain biochemical genetic analyses. The methods described here can be employed on a wide range of mammals.

3. Material

Cell cultures can be obtained from any body tissue. However, in view of the need for technically simple preparation and most readily growing material, certain tissues have proved particularly suitable. From a living donor a skin biopsy is usually taken. A portion of fascia lata used to be biopsied because this tissue grows particularly rapidly in vitro; nowadays this method is scarcely used at all in routine work. Autopsied skin material may be used until about 48 hours after death, thereafter it is not recommended because of the danger of infection. Muscle, fascia, peritoneum and pericardium samples can be simply obtained under sterile conditions and have proved themselves reliable in culture.

In mammals, the lung is a particularly suitable source of culture material, because cell growth from this tissue is relatively rapid. However, if material is not fresh it is usually infected.

4. Introduction to Methods

A tissue culture laboratory is required. Work must be performed under strictly aseptic conditions. The requirements for equipment and procedures in human tissue cell culture for chromosome analysis do not vary substantially from those of ordinary tissue culture. For the principles of tissue culture, the appropriate texts should be consulted (e.g. Hsu, 1972; Kruse and Patterson, 1973; Merchant *et al.*, 1965; Parker, 1961; Paul, 1972; Penso and Balducci, 1963; Rothblat and Cristofalo, 1972; White, 1963; Willmer, 1965). Details of equipment and nutrient solutions will be mentioned in the special sections.

A large number of methods are available for obtaining cell cultures. The tissue is dissociated while setting up, thereby inducing growth of individual cells as a single layer on a firm surface (monolayer), or explants are first set up from which cells emigrate, divide, and finally also form a single layer of individual cells. The explants must be caused adhere to the surface, because explants floating in the culture fluid do not yield sufficient cells.

Methods in which the original tissue material is dissociated, usually result in a monolayer culture more rapidly than do explant methods. However, most routine samples of tissue for culture are so rich in collagen fibres that dissociation does not provide sufficient free cells for culturing. Epithelial tissues are better suited for this method (e.g. from kidney). Puck *et al.* (1960) describe a method producing a cell culture directly from a skin biopsy (see below).

Dissociation methods for human chromosome preparation play a subordinate role, so this procedure will only be briefly described. Only those explant methods which are most frequently used in obtaining chromosome preparations from fibroblasts are mentioned.

The classic procedure is the plasma method. Small explant fragments are made to adhere to a surface covered with plasma, and supplied with medium. Good growth is generally achieved. Under certain circumstances, the plasma clot can interfere with subsequent processing and evaluation. Since plasma became commercially available, the somewhat laborious method of obtaining plasma from chicks is no longer necessary.

The standard methods which we use to achieve primary growth are culture of explants under perforated cellophane and between two cover slips in the "sandwich" method. Both methods complement each other satisfactorily in their ability to produce sufficient cell growth and can be employed in parallel or as alternatives.

All these methods aim at obtaining the highest possible number of mitoses. The cellophane method requires a longer culture time before producing subcultures from which mitoses can be obtained, though the method has the advantage of a higher yield. The plasma and sandwich methods produce mitoses from the primary culture in a shorter time but often in smaller quantities. Each method allows to proliferate cell material from the primary explant or from subcultures over a longer period and to obtain mitoses repeatedly. To increase the yield of mitoses it is recommended that divisions be partially synchronized by handling of the cultures at regular time intervals. Before harvesting, the addition of a cytostaticum (colchicine, Colcemid, Velban) is recommended to block mitoses in the metaphase (see Chapter I).

We shall discuss these methods in detail, in particular:

Techniques for taking tissue samples;

Setting up and culturing of the material obtained;

Harvesting the culture for chromosome preparation.

The areas where each technique is applied will be mentioned. The further processing of cultures for chromosome analysis is described in Chapter IV.

5. Techniques for Collecting Tissue

Skin biopsy can be carried out on various parts of the body. In small children we prefer the outside of the upper thigh or between the shoulder blades, in adults between the shoulder blades or the ventral side of the upper or lower arm. The area of skin must be thoroughly cleaned before biopsy. It can also be shaved beforehand, but this is not necessary. Vigorous washing of the area twice with soap is required, followed by a rinsing with water or disinfectant. The disinfectant must not contain any iodine or mercury. The skin is then rubbed down several times with 70% alcohol, using fresh alcohol each time. Various methods are used to obtain biopsy material, some of which are presented here (strictly aseptic conditions must be maintained throughout the procedure):

a) Lift the skin with a needle or a pair of pointed forceps and cut off with a scalpel close under the needle (forceps) (Hsu and Kellog, 1960; Harnden, 1960).

b) Pinch a section of skin with small forceps so that a short strip projects; wait about 1 min until the squeezed part of skin becomes insensitive; remove the projecting strip with a scalpel.

c) Puncture out a cylindrical portion of skin with a skin punch of caliber 2–3 mm by revolving the punch under light pressure; grasp the cylindrical portion with pointed forceps and cut off with scissors.

d) A section of skin about 1 × 0.2 cm is abraded with a skin fraise and covered with a sterile plaster. After 72 hours the resulting scab is used as starting material for cell culture. The skin section is washed with 70% alcohol and removed with forceps. The regions at the nape of the neck or behind the ear are particularly suitable. The area should be locally anesthetised. The same section of skin can be used once again after a further 72 hours. The advantage of this method is that by reducing the explant to small fragments a cell suspension can be immediately placed in the culture, and that the scab grows especially fast (Puck *et al.*, 1960).

Many authors use local anesthesia, others consider it detrimental to cell growth. In any event, the techniques mentioned are fairly painless even without local anesthesia.

The skin sample taken, should not be less than 2–3 mm in diameter; a somewhat larger section is preferable. Care should be taken to assure that the biopsy contains the germinal layer of the skin, but no subcutaneous tissue; the wound should bleed lightly from capillaries after taking the sample. In order to be sure that sufficient tissue material is on hand, it is recommended to take another sample. A parallel culture can then be set up, thus reducing the danger of loss through contamination.

Directly after taking samples, the biopsy material is placed in a sterile tube with medium or in a Petri dish with muslin soaked in medium. Before setting up the explant in the tissue culture, the material can be placed in the refrigerator for 2–3 days or even left standing at room temperature. Even after storing for 5 days growth can be achieved, there is no difficulty in transporting skin biopsies over long distances. As a rule however, fresh tissue starts growing more rapidly.

Material from small children grows better in culture than that from older people, although there is no age limit. At autopsy, fascia, muscle and other tissues may be taken. They manifest growth in vitro even 5 days *post mortem*. Nevertheless, a tissue sample should be taken as soon as possible *post mortem*.

6. Setting up Tissue Explants and Culturing Cells

Tissue taken from the body can be dissociated enzymatically before the culture is set up so that the primary culture starts from a cell suspension, or tissue can be cut into small cubes from which cells proliferate and can be brought into suspension when the first subculture is set up.

6.1. Enzymatic Dissociation of Tissue Explants

This method can be applied only occasionally, because not all tissues can be sufficiently dissociated into single cells. The preferred starting material is kidney; fetal tissue (abortus material) can also be used.

A detailed description of dissociation methods can be found e.g. in Parker (1961). In our laboratory the following simplified procedure for obtaining cell cultures from kidney is used: The tissue sample is cut up with fine scissors and repeatedly washed in physiological saline to eliminate erythrocytes; the tissue should then be of a pulpy consistency. It is placed in an Erlenmeyer flask with a magnet of appropriate size and mixed with a 4–5 fold volume of a 0.25–0.5% prewarmed trypsin suspension. This mixture is placed on the electric stirrer for 15–20 min and agitated with sufficient vigor for thorough stirring; care should be taken to avoid formation of foam. The suspension is allowed to stand after agitating for 1 min so that larger tissue fragments can settle; the supernatant is then pipetted into centrifuge tubes. After centrifuging, the residue is mixed with culture medium and placed in a larger or smaller culture bottle depending on the quantity; for example, 0.2 ml of packed cell centrifuge material is set up in a milk dilution bottle.

The culture is first incubated for 2 days; during this time numerous cells grow on the glass surface. The medium is then changed and the cell fragments which remained in suspension are discarded with the old medium. On the third day, the cells may cover up to three quarters of the glass surface. The primary culture is heterogeneous in its cell composition. If it is from kidney, epithelial cells predominate in the culture for the first few days; they generally survive one more passage but are subsequently overgrown by fibroblast-like cells.

6.2. Setting up Solid Explants

The tissue sample is transferred to a Petri dish and moistened with medium. Fatty tissue adhering to the sample is removed with a scalpel. If necessary, the material is rinsed with Hank's solution and the liquid then pipetted off. If there is a possibility that the tissue sample is not perfectly sterile, antibiotics should be added to the Hanks's solution at ten times the concentration normally used in culture medium. The tissue is then cut up with scissors and forceps (or with two scalpels) into cubes of 0.5–1 mm. Care must be taken to ensure that the edges of the cubes are cleanly cut. The tissue material thus cut up is now suspended in a small volume of complete culture medium (about 2 ml) and transferred into culture bottles with a long pipette. It is just important that the explant is adhered to the bottom of the flask, otherwise it will float in the culture medium. There are various methods available to ensure this (see below). The initial use of too much fluid is the most common mistake in setting up a cell culture.

Every freshly set up culture should be gassed depending on the medium, with 5–10% CO_2 in air, which generally improves outgrowth of cells. A sterile pipette stoppered with cotton wool at its neck and connected with the gas by tubing may be used for this purpose.

6.2.1. Direct Deposition on the Glass Surface

The simplest way to attach tissue fragments is to place them on the bottom of an appropriate culture vessel together with a very small amount of medium so that the fragments cannot float around. The explants are well distributed with a Pasteur pipette and the culture is then incubated. The medium has to be replaced daily. After outgrowth of cells around the explants, the latter may be shifted with the pipette to another position so that a new colony can be formed. If a large part of the bottom is covered with cells, they may be redistributed by trypsinization and further treated as described under 6.3.

6.2.2. Plasma Method (Lejeune et al., 1959; Harnden, 1960; Harnden and Brunton, 1965)

One drop of plasma and one drop of embryo extract are mixed and drawn together with an isolated tissue fragment into a bent Pasteur pipette; the contents of the pipette are deposited at the bottom of a culture flask

(T-flask or milk dilution bottle) or on a cover slip which has been placed on the bottom of a culture bottle (Leighton tube or the above-mentioned bottles). The same procedure applies to the other explants. The surface (i.e. cover glass or bottom of flask) can also first be thinly spread with plasma with the aid of a bent pipette, the explant placed on this, and a suitable amount of embryo extract added. Alternatively, the tissue pieces can be transferred into embryo extract and then deposited on the bottom prepared with plasma. The culture bottles are first incubated while sealed (at room temperature or 37° C) until the plasma has coagulated (from a few minutes to several hours), when culture medium is added. After a few days the first cells begin to proliferate. Further treatment follows the procedures described under 6.2.3. (for cover glass cultures) or 6.3. (for flask bottom cultures).

6.2.3. "Sandwich" Method

The explants are made to adhere by placing them between two cover slips. This so-called "sandwich" can then be placed together with medium in tubes, flasks or Petri dishes and incubated.

In our laboratory we proceed as follows:

A tube with one side flattened on the lower half (tissue culture tube of the type "Institut Pasteur" or Leighton tube) and containing 2 cover slips (12×35 mm) is prepared and closed with an aluminum foil cap. Since the cover slips must be subsequently taken out again, they can also be separately sterilized in larger numbers, e.g. in small Petri dishes. Non-toxic silicon stoppers of suitable length are autoclaved separately. For setting up the explants, the cover slips are removed from the tube with forceps and placed in a Petri dish. 2–3 tissue pieces are deposited dry on one cover slip, and the other cover slip is placed on top and lightly pressed. This sandwich is then placed in the tube, about 2 ml of medium is added, the tube sealed with a silicon stopper, and incubated at 37° C. It is recommended that several such tubes from each tissue sample be set up in parallel. After a few days the culture is examined under an inverted microscope. After about 6–8 days a halo of cells has grown around each explant. When this halo has reached a diameter several times that of the explant, the first passage may take place. The sandwich is placed in a dry Petri dish, the upper cover slip removed with forceps and incubated with medium in a fresh tube. The explants are removed from the lower cover slip with pointed forceps (the cells which have proliferated remain on the glass surface) and placed on a fresh cover slip which is once more processed as a sandwich. The cover slip with the proliferated cells is also incubated in a fresh tube.

The explants can be repeatedly transplanted if it is desirable to continue the culture; they will continue to proliferate in the course of numerous passages. If necessary, the edges of the explants should be trimmed.

The cover slips which are left after transplanting explants are incubated until they are fully covered with cells. The medium is changed every 3–4 days. At subculture, both cultures are treated with pre-warmed trypsin (15 min), the cell suspensions are pooled and centrifuged. The cell residue is set up in a larger flask (Carrel flask). Subcultures can now be carried out in larger flasks if necessary.

Variation: Cover slip culture *in situ*

Cultures set up according to the plasma or the sandwich methods can also be processed without subculturing (Lejeune *et al.*, 1959). This procedure has the advantage that preparations can be ready for chromosome analysis after only 8–10 days (see p. 50). On the other hand the small number of mitoses per cover slip usually available after this procedure can be a disadvantage, so that under certain circumstances it may be necessary to start a new stem culture and harvest continuously over a longer period of time.

6.2.4. Cellophane Method (Evans and Earle, 1947; Hsu and Kellogg, 1960; Manojlovic and Hienz, 1967)

Cleaned perforated cellophane (see Appendix) is cut into pieces to fit the bottom of the flask. Milk dilution bottles (150 ml) with a screw top have proved suitable in our laboratory. The flasks, each containing 1–2 ml distilled water, are autoclaved and sealed with a cotton wool stopper; the lids are separately autoclaved. Before proceeding, the water is poured out of the bottle, and the bottle rinsed with a little medium. The pieces of tissue are put under the cellophane with a long pipette and distributed on the floor of the flask (about 1 cm apart).

Variation. Pieces of tissue can also be placed on the side of the bottle opposite the cellophane and the cellophane then placed on top with the pipette: in doing so the edge of the cellophane which faces the opening is pressed down on the explant side; the rest of the cellophane strip can then be easily removed from the glass.

The cellophane has to be pressed down with the pipette to press the pieces on to the glass. If the flasks mentioned above are used, add 8 ml medium, seal, and incubate. The first growth is usually observed after 6–10 days; it sometimes takes several weeks before the explants begin to proliferate, although this period can vary according to the quantity of explants, origin, and age of the tissue. The medium is not changed until the end of the first week; then about every 4–7 days, depending on the pH of the medium. If a flask rapidly becomes alkaline, the pH can be adjusted by addition of CO_2 from a 5 ml syringe.

6.3. Subculturing

In cultures with outgrown cells on the bottom of the flask, the first sub-culture should be carried out when about half the bottom of the flask is covered with cells. Using milk dilution bottles of 150 ml, this normally occurs after 3–4 weeks. The culture is then treated with prewarmed 0.25% trypsin until the cells are detached (10–15 min). If the cells do not detach from the glass, then the time of trypsin action should not be extended, nor should the concentration of trypsin be increased. Instead, the rest of the medium should be removed by rinsing with an isotonic salt solution before trypsinizing. The use of Ca- and Mg-free trypsin solution is recommended. The cell suspension is lightly shaken or mixed by pipetting so that as few cells or cell clumps as possible are left behind; the suspension is then poured into a centrifuge tube, centrifuged, and the residue incubated in a fresh flask. The explants begin to grow once more in the primary flask.

If the quantity of cells thus obtained is low, the first subculture should be set up in a small flask (e.g. Carrel flask, T 15 or T 30 flask). The yield from two or more primary cultures can be combined into one subculture. Owing to the low plating efficiency of diploid cells, an in-sufficient initial density (less than about 2×10^4 cells/ml) leads to poor growth if not to absolute failure.

Variation (Hsu and Kellogg, 1960).

After trypsinizing the cellophane is grasped with forceps, lightly shaken to rinse off the adhering cells, removed from the flask, and discarded. The suspension is mixed by pipetting in order to loosen all cells and reduce cell clumps. The primary explants are then included in the subculture.

The trypsinized cells settle and attach again to the glass surface within about 1 hour. The subculture should be allowed to grow until the bottom of the flask is covered with a monolayer. It can then be further subcultured. After trypsinizing, the cell material is distributed in one or more fresh flasks according to the quantity of material on hand.

In order to be able to apply different procedures to cells which have proliferated under the same conditions, cover slips are placed at the bottom of the culture flask or Petri dish (in a CO_2 incubator, see 6.4.) before the suspension is added. To prevent the cover slips from floating, they can be stuck to the bottom of the flask with a drop of plasma. The individual cover slip cultures can then be harvested at different times (e.g. for autoradiography) or prepared in different ways (e.g. comparison between sex chromatin and karyotype).

6.4. Culturing in a CO_2 Incubator

The culture bottles should not be air tight. Work is normally carried out with Petri dishes (glass or plastic) which are easier to handle. The incubator is supplied with a mixture of air and 5% CO_2. Owing to the humidity of the atmosphere in the CO_2 incubator, a high level of sterility must be maintained. If a culture becomes contaminated, the entire incubator must be disinfected. Primary cultures in Petri dishes can be set up and subcultured according to one of the methods described above. The subcultures, however, are often first incubated in the CO_2 incubator. The used medium is very easily drawn off by using a pipette attached to a water suction pump.

6.5. Abortus Material

Tissue samples from abortion material can be cultivated according to the various methods outlined here. The particular technical problems presented by this material as regards tissue culture and chromosome analysis were discussed in detail (Geneva Conference, 1966).

Some remarks may be made here as to the choice of tissue material. Various parts of an abortus can demonstrate considerable variation in growth. If a well-preserved fetus is available, it may, depending on size, either be set up as a whole or samples can be taken from all three germ layers. Umbilical cord is also a suitable material. However, often the fetus is not available, when portions of the placenta should be taken. Amnion and chorion have proved particularly suitable for tissue culture. Due to the mixture of fetal and maternal constituents in the placenta, it is necessary to confirm histologically the fetal origin of the tissue to be set up. Due to the higher risk of contamination in abortus material as compared to other tissues, a thorough washing in antibiotic solution of ten times the usual concentration is advisable. Only about half of all cultures from fetal tissue grow, for reasons which are not yet clear.

7. Harvest of Cultures for Chromosome Preparation

To obtain the highest possible number of metaphases, the culture must be in the logarithmic growth phase. This is usually achieved by subculturing a dense homogeneously grown monolayer culture and harvesting the subculture at some time within the following 48 hours. It is useful to determine the time of harvest by checking the culture under the microscope. Mitoses can be recognized by the round shape of cells. Declining cell growth and mitotic rate is indicated, if the spindle-shaped interphase cells appear sharply contoured and if cell clumps are visible. For the purposes of chromosomal diagnosis, it is not necessary to synchronize the cells by any special means. A regular sequence of medium change and subculture induces a partial synchronization which yields a sufficient number of mitoses at an appropriate time.

Harvest of cultures from the bottom of the flask: When the bottom of the flask is almost covered with a solid monolayer, it is trypsinized and the cell material is distributed into two fresh flasks. About 24 to 48 hours later enough mitoses are usually present (monitored microscopically) to allow harvest of the culture. 0.5–2.0 µg colchicine per ml medium is added for about 2–6 hours; we routinely use 4 hours. Cultures are then trypsinized and prepared further, either together or separately, depending on the cell yield. If numerous mitoses are visible at the time of harvest, a monolayer culture of 30 cm² is sufficient for preparation. If only 2 subcultures are available from a sample, only one flask should be harvested while the other is held in reserve and further subcultured as required.

Harvest of cover slip cultures in situ: If a monolayer culture has grown on cover slips, the procedure is the same as in harvesting from the bottom of a flask, except that the former is not trypsinized; instead, the cover slips are taken directly out of the culture medium and placed in a hypotonic solution (see Chapter IV). The fastest method for obtaining chromosome preparations from a tissue sample uses a primary culture on cover slips. Once a dense growth surrounds the explants, they are transplanted and the cover slips with the outgrown cells are incubated for a further 1–2 days with fresh medium. During this period the cells proliferate into the center of the cell halo where the explant was previously situated. Most mitoses are found here and on the periphery of the halo. The culture is colchicinized after 24–48 hours and processed without trypsinization. Lejeune *et al.* (1959) use this method without colchicine. They stimulate with nutrient (a few drops of embryo extract) 24–48 hours after removing the explant and harvest the culture after another 16 hours.

8. Long Term Preservation of Tissue Culture Cells

The long term conservation of cultured cells is achieved by deep freezing, for which several methods are available. Cells are usually preserved at

−196° C in liquid nitrogen or at −70° C in dry ice or a deep freezer. For successful conservation a number of aspects must be considered. Prior to freezing a cryoprotective agent must be added to the culture medium, usually glycerol or dimethylsulfoxide (DMSO) (both at 5–10% final concentration). The survival rate after freezing and thawing is progressively higher with younger cultures. Cells in logarithmic growth survive better than others. The cell density is also important. As a rule, cells in monolayer from one milk dilution or Kolle flask are suspended in 1 ml freezing medium. The cell density should be around 1 to 5×10^6 cells per ml. Uncontaminated (including contamination with mycoplasma [PPLO]) cells only should be frozen.

Gentle trypsinization is recommended prior to freezing, followed by careful washing. Several (at least 5–6) parallel cultures should be frozen. To test whether living cells have been frozen, a test sample should be thawed after a few days to test survival and viability. It is advisable to keep a culture of the original material until this test has been done. Freezing should be gradual between +4° C and about −70° C. A step-wise cooling of 1.5–2° C per min has proved to be optimal for diploid fibroblasts. Once cells have been frozen successfully the length of storage seems to be unlimited at temperatures below −96° C (Paul, 1972).

Thawing should be as rapid as possible. Ampules that are not completely closed will explode upon thawing as a result of trapped liquid nitrogen. It is therefore mandatory to wear protective goggles during thawing. The freezing medium should be replaced immediately, especially when DMSO has been used. Prior to use, the fresh medium should be gassed with CO_2 in air to guarantee optimal pH. Cells normally settle and attach to the bottom of the flask within 24–48 hours if the initial cell density was high (5×10^6 cells per ml or more). First subculture will not normally be possible prior to one week. General growth characteristics of a culture are not influenced by the freezing procedure if freezing was done under optimal conditions.

Appendix I

Summary of Some of the Methods

A. Explants under Perforated Cellophane

Preparation of Culture Flasks

Boil perforated cellophane overnight in acetone (reflux condenser); rinse twice, 10 min each time in ether, absolute alcohol and distilled water.

Cut up cellophane into pieces which will fit into the bottles to be used (e.g. 150 ml milk dilution bottles). Place cellophane pieces into culture bottles.

1 ml distilled water per bottle, cotton wool stopper. Autoclave.

Setting up the Culture

Place explant in Petri dish with Hanks' solution or Eagle's MEM and rinse; remove fat tissue; change medium as often as necessary.

Pipette medium off and cut tissue into cubes of about 1 mm³ with iris scissors.

Pour off water from culture bottle.

Moisten cellophane with a little medium; spread medium over the surface and pour off.

Take tissue pieces with a Pasteur pipette and distribute well under cellophane, – or place pieces on a side of bottle free of cellophane and cover over with cellophane.

Spread tissue pieces further by pressing cellophane onto wall of bottle with a pipette.

Pour off the rest of the medium.

Add 8 ml of culture medium.

Composition of culture medium (100 ml):

80 or 90 ml TC MEM Eagle,

20 or 10 ml fetal calf serum

1 ml 5% glutamine

1 ml penicillin-streptomycin (see Appendix II)

1 ml phenol red solution

3–4 weeks incubation at 37° C.

No medium change during the first week, but every 4 days to one week thereafter.

Subcultures

Pipette or pour off medium.

Wash with normal saline or Ca- and Mg-free balanced salt solution (e.g. Hanks').

Trypsinize with 0.25% trypsin (10–15 min)

Mix cell suspension by pipetting

Place cell suspension in a conical centrifuge tube; add 8 ml medium to primary culture with explants; reincubate.

Centrifuge the cell suspension at 700–1,000 r.p.m., 5–10 min; pour off.

Reincubate residue with fresh medium in a Carrel flask or a milk dilution bottle, according to the number of cells.

Harvest of Cultures

Allow the subculture to grow to an almost dense monolayer. Further subculture 1:1 or 1:2 etc.

Allow cells to grow 1–2 days after final subculture. Check cell density and mitotic activity microscopically. When enough cells in mitosis occur at a loose cell density, add

0.1 ml of a 0.002% solution of colchicine to 1 ml medium (about 2 μg/ml final concentration).

Incubate 2–4 hours.

Pour off medium and replace with trypsin (as above). Centrifuge.

Process cell pellet (see Chapter IV).

B. Cover Slip Culture, "Sandwich" Method

Preparation of Culture Tubes

2 rectangular cover slips are placed in a Pasteur or a Leighton tube; the cover slips fit the size of the flattened side of the tube; seal with aluminum foil; sterilise dry. Autoclave silicon stoppers.

Setting up the Culture

Prepare tissue sample as in A.

Tip the cover slips in the tube against the aluminum cap. Remove cap.

Take out one cover slip and place in a dry Petri dish. Place 2–3 pieces of tissue on the cover slip. Place second cover slip on this and press lightly. Replace the "sandwich" in the tube on the flat side.

Add 2 ml medium (as in A), seal.

Incubate at 37°C, 6–8 days, then change medium every 3 days. Check through the microscope every 2 days.

Passaging

Pour off medium after a dense halo of fibroblasts has grown. Place "sandwich" in dry Petri dish.

Remove upper cover slip. Transfer explants with pointed forceps to a fresh cover slip and continue this as a further "sandwich" culture.

Incubate both cover slips if cells grow on them, each in a fresh tube with culture medium.

Subculture

Allow a dense monolayer to grow on cover slips.

Trypsinize (as with A).

Combine contents of both tubes and centrifuge.

Pour off trypsin, resuspend pellet with medium and set up in Carrel flask or milk dilution bottle.

Harvesting Cultures: as with A

Variation: After removing explants, incubate cover slips separately with fresh medium.

Observe for mitotic activity after 24–48 hours.

Add colchicine (as with A).

Harvest culture without trypsinizing, and complete process (see Chapter IV).

C. Cell Preservation

Terminate culture at exponential growth phase if possible, i.e. 24–48 hours after the last subculture.

Detach cells with trypsin as usual.

Centrifuge, rinse cell residue in medium, and centrifuge again.

Add 1 ml of 5% glycerol.

Glycerol is autoclaved before use; a 5% solution is set up in complete medium. DMSO does not have to be autoclaved.

Transfer cell suspension (syringe, pipette) in a 1.2 ml ampule (Wheaton Glass Co., Millville, New Jersey, Cat. No. 12523). Avoid wetting the neck of the ampule.

Pik up ampule with (cooled) forceps and close it by smelting with an oxygen burner. Place ampule into refrigerator at $+4$ to $6°$ C for 3 to 6 hours to permit equilibrium between freezing medium and cells, before they are placed into a deep freezer of $-70°$ C or on dry ice. It is recommended to avoid direct contact with dry ice, but rather to place the ampules into a carrier vessel, e.g. a Packard scintillation counting vial (von Böhmer et al., 1973).

For storage in liquid nitrogen the following procedure is recommended:

In our routine, the freezing stopper of Union Carbide BF-5 is used which takes 9 ampules. Depending on the number of ampules to be preserved, the stopper remains $1-1^{1}/_{2}$ hours in the start position upon the container with liquid nitrogen.

For stepwise freezing screw down the stopper according to the scheme given in Fig. 1.

Frozen ampules are transferred to the aluminum holder, and placed in the storage container. Each holder should contain only ampules with cell material of common origin. The holder should be previously marked and its location recorded.

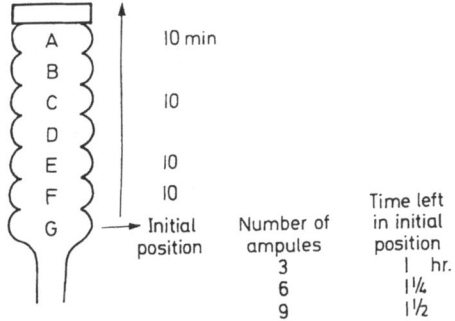

Fig. 1. Freezing stopper BF-5 of Union Carbide Co. Depending on the number of ampules, the stopper is left in the initial position as indicated. It is subsequently screwed down according to the schedule

Thawing

Remove ampules cautiously from the holder.

Place ampules immediately in water bath of 37° C.

(Beware of explosions!)

Shake, until the freezing medium has thawed.

Open ampule, transfer contents into a sterile centrifuge tube.

Add 2 ml of fetal calf serum and 2 ml of medium.

Centrifuge, resuspend pellet of cells in complete nutrient medium.

The fresh medium should be gassed before use with CO_2 in air (1:9) for 10 min.

Incubate in a Carrell or T flask.

Appendix II

Catalog of Some Reagents, Media and Culture Vessels

Cellophane, perforated. Microbiological Associates, 4813 Bethesda Ave, Bethesda, Md., U.S.A., Cat. No. 18–900.

Colcemid, Difco[1]. 1 ampule of 1 mg/ml. Final concentration: 1 µg/ml medium. In the literature the end concentration is given as low as 0.01 µg/ml medium.

Colchicine. E. Merck, Darmstadt, West Germany; Ampules of 0.1 g. Final concentration: 2 µg/ml medium. In the literature the final concentration is given as low as 0.5 µg/ml medium, or lower.

---

[1].Products of Difco Laboratories, Detroit, Michigan, USA are supplied by numerous dealers.

Eagle's MEM. "TC Minimal Medium Eagle, dried", Difco, Code 5675. 9.06 g to 972 ml double distilled water. Stir until completely dissolved. Add 2 ml of a 10% $CaCl_2$ solution. Heat medium to 37° C, adjust pH to 7.2–7.4 with a 10% $NaHCO_3$ solution (about 10 ml). Bring final volume up to 1,000 ml with double distilled water. Filter sterilization.

Fetal Calf Serum. Is produced by several Companies (Difco; Microbiological Associates; Grand Island Biological Co., 3175 Staley Road, Grand Island, New York, 14072; Flow Labs. Manchester, England; Colorado Biologicals, Denver, Colo., U.S.A., amongst others).

Glutamine. "TC Glutamine 5%" Difco, Code 5789.
Add 10 ml Hanks' solution to ampules.

Hanks' Solution. "TC-Hanks' Solution dried", Difco. Code 5775. Add 9.9 g to 1 liter double distilled water and stir till completely dissolved. Heat to 37° C; add 3.5 ml of a 10% $NaHCO_3$ solution, pH 7.2 to 7.4 (Phenol Red indicator becomes orange-red). Filter sterilization.

Mycostatin (against fungus infection). Squibb & Sons, New York, N.Y. 10022, U.S.A., 500,000 units.

Suspend contents of ampule in 5 ml Hanks' solution and place in 45 ml Hanks' solution. Store in deep freeze. Use 0.1 ml per 10 ml medium. End concentration: 100 units/ml medium. The preparation is applied for 5–6 hours or longer in the case of fungus infection.

Pasteur Tubes. "Tubes a cultures de tissu du type Institut Pasteur". Ets. Verrefer, 75, Rue St. Jacques, Paris 5ᵉ, France.

Penicillin-Streptomycin. Penicillin G, Hoechst, 10^6 units; Streptomycin, Bayer, 1 g (in the U.S. from e.g. Wyeth Laboratories, Philadelphia, Pa. 19101).

Both substances are placed in 100 ml Hanks' solution. Final concentration: 100 units penicillin/ml medium; 0.1 mg streptomycin/ml medium.

Petri Dishes. Dishes with an etching on the top of the lid from: Bellco Glass, Inc. Type: "Sealable type culture dish", 65 × 15 mm, Cat. No. 15–065. Plastic: Falcon Plastics (Los Angeles). Plastic tissue culture dishes 60 × 15 mm, Cat. No. 3002.

The Companies of Bellco and Falcon Plastics are represented in Europe by Tecnomara AG, Zürich, Rieterstrasse 48, Switzerland.

Phenol Red. "TC Phenol Red Solution 1%", Difco, Code 5358. Contents of ampule added to 9 ml Hanks' solution.

Trypsin. "Bacto-Trypsin" Difco, Code 0153.

Suspend contents of ampule (0.5 g) in 10 ml Hanks' solution; of this, 5 ml, and 2 ml of a 2.8% $NaHCO_3$ solution are added to 95 ml Hanks' solution. Final concentration: 0.25%.

Velban. Vinblastine-sulphate; E. Lilly Comp., Indianapolis, Ind. U.S.A.

Final concentration: 0.0075 µg/ml medium.

Milk Dilution Bottle. Bottles with screw caps are particularly easy to handle.

Pyrex "Milk dilution bottle, plain, with screw cap", Cat. No. 1367. Obtainable from: Fisher Scientific Co., New York, U.S.A. or Zürich, Zeitweg 67, Switzerland. The company of Schott and Gen., Mainz, West Germany, supplies milk dilution bottles made of Pyrex glass, but without a screw top; a useful stopper is the silicon stopper.

References

Böhmer, H. von, Wöhler, W., Wendel, U., Passarge, E., Rüdiger, H. W.: Studies on the optimal cooling rate for freezing human diploid fibroblasts. Exp. Cell Res. **79**, 496–498 (1973).

Evans, V. J., Earle, W. R.: The use of perforated cellophane for the growth of cells in tissue culture. J. nat. Cancer Inst. **8**, 103–119 (1947).

Geneva Conference: Standardization of procedures for chromosome studies in abortion. Bull. Wld Hlth Org. **34**, 765–782 (1966).

Harnden, D. G.: A human skin culture technique used for cytological examinations. Brit. J. exp. Path. **41**, 31 (1960).

Harnden, D. G., Brunton, S.: The skin culture technique. In: Human chromosome methodology (J. J. Yunis, ed.), p. 57–73. New York-London: Academic Press 1965.

Hsu, T. C.: Procedures for mammalian chromosome preparations. In: D. M. Prescott (ed.), Methods in cell physiology, Vol. V, p. 1–36. New York-London: Academic Press 1972.

Hsu, T. C., Kellogg, D. S.: Primary cultivation and continuous propagation *in vitro* of tissues from small biopsy specimens. J. nat. Cancer Inst. **25**, 221–235 (1960).

Kruse, P. F., Jr., Patterson, K. M., Jr. (eds.): Tissue Culture. Methods and Applications. New York and London: Academic Press 1973.

Lejeune, J., Gautier, M., Turpin, R.: Les chromosomes humains en culture de tissus. C. R. Acad. Sci. (Paris) **248**, 602–603 (1959).

Manojlovic, N., Hienz, H. A.: Zur Methode der primären Explantation und Kultivierung menschlicher Gewebe mit Hilfe von perforiertem Cellophan. Mikroskopie **22**, 70–80 (1967).

Merchant, D. J., Kahn, R. H., Murphy, W. H.: Handbook of cell and organ culture, 2nd ed. Minneapolis: Burgess Publ. Co. 1965.

Parker, R. C.: Methods of tissue culture, 3. ed. New York: P. B. Hoeber Inc., Med. Div. Harper Broth. 1961.

Paul, J.: Cell and tissue culture, 4. ed. Edinburgh-London: Livingstone Ltd. 1972.

Penrose, L. S., Smith, G. F.: Down's anomaly, p. 133ff. London: Churchill Ltd. 1966.

Penso, G., Balducci, D.: Tissue cultures in biological research. Amsterdam-London-New York: Elsevier Publ. Co. 1963.

Puck, T. T., Robinson, A., Tjio, J. H.: Familial primary amenorrhoea due to
    testicular feminization: a human gene affecting sex differentiation. Proc.
    Soc. exp. Biol. (N.Y.) **103**, 192–196 (1960).
Rothblat, G. H., Cristofalo, V. J. (eds.): Growth, nutrition and metabolism
    of cells in culture. Vol. I/II. New York and London: Academic Press 1972.
White, P. R.: The cultivation of animal and plant cells, 2nd. New York:
    Ronald Press 1963.
Willmer, E. N. ed.: Cells and tissues in culture, methods, biology and physio-
    logy, vol. I/II/III. London-New York: Academic Press 1965.

CHAPTER III

Culture and Preparation of Cells from Amniotic Fluid

Ulrich Wolf

With 3 Figures

1. Introduction

The analysis of fetal chromosomes is the most common occasion of prenatal diagnosis (reviews by Bergsma, 1971; Dorfman, 1972; Emery, 1973; Harris, 1972; Milunsky, 1973; Nadler, 1972), owing to the frequency of chromosomal aberrations, nearly all of which can be recognized. Prenatal cytogenetic diagnosis is particularly indicated in the following cases:

1. Above average maternal age (more than 35 or 40 years),
2. Familial chromosomal translocations (Robertsonian or otherwise),
3. Pregnancy following the birth of a child with a spontaneous chromosomal aberration,
4. Certain X-linked diseases.

Furthermore, the culture of amniotic fluid cells is a special prerequisite for the biochemical diagnosis of a number of inborn errors of metabolism.

The diagnosis of sex chromatin in amniotic cells dates back to the middle fifties (Serr *et al.*, 1955; Fuchs and Riis, 1956). The first successful chromosomal analyses were published in 1966 (Steele and Breg, 1966; Thiede *et al.*, 1966). Tissue culture techniques have been substantially improved since then, although the approaches used by different workers in a way vary more than other methods described in this book.

A standard method therefore can not yet be given. The objective is a diagnosis in the shortest possible time with the smallest possible number of cells, as well as to achieve a high percentage of successful cultures. Since some of these goals are mutually exclusive, a compromise is necessary. Many investigators favor the rate of success which is mainly dependent on experience. Critical investigations in prenatal diagnosis can be pursued only after substantial experience with disposable material (Edwards, 1970).

The success rate may depend upon:

1. Gestational age,
2. Volume of amniotic fluid,
3. Amount of viable cells.

These parameters may vary independently within certain limits (Hahnemann, 1972).

The *amount of amniotic fluid* increases during gestation. By the end of the first trimester, it is approximately 50 ml; after fifteen weeks it reaches 125 ml, and it increases thereafter by about 50 ml every week (Fuchs, 1966). The withdrawal of 10–20 ml (or even 40 ml used by some workers) after the fifteenth week does not appear to have any harmful effect upon the fetus.

The *amount of cells* in any sample is uniformly low before the 12th week; sometimes hardly any are found. The cell count increases progressively up to the 15th week, and thereafter it increases markedly. There is a simultaneous increase of viable cells, of which the proportion between the 12th and the 20th weeks is about 50%, though it subsequently decreases (Hahnemann, 1972).

Most investigators agree that the 16th week is the most favorable time for amniocentesis. At this stage, a considerable amount of amniotic fluid may be obtained (10–40 ml), and the proportion of viable cells is appropriate for a diagnosis in less than fourteen days. The rate of success may approach 100% at this time (Therkelsen *et al.*, 1972).

The *morphology of cells* found in each extraction is heterogeneous. The fetal origin of the cells in the amniotic fluid is, however, certain (Van Leeuwen *et al.*, 1965; Votta *et al.*, 1968). It is hypothesized that the cells may be derived from: amnion, skin, and the genitourinary, alimentary, and respiratory tracts (Valenti and Kehaty, 1969). Cells show different sizes and morphology, and develop as fibroblastic or epithelial-like cells in culture (Figs. 2 and 3). Most cells show karyopyknosis and are of the "squamous" type (Fig. 1).

There are also *tetraploid cells* in the amniotic fluid which are of amniotic origin (Klinger and Schwarzacher, 1960); these may give rise to tetraploid clones. The proportion of tetraploid cells lies within 0 to 100%, depending on the culture (Milunsky *et al.*, 1971). Tetraploidy is diagnostically insignificant for it is wholly compatible with the birth of normal children. Polyploid fetuses are usually spontaneously aborted.

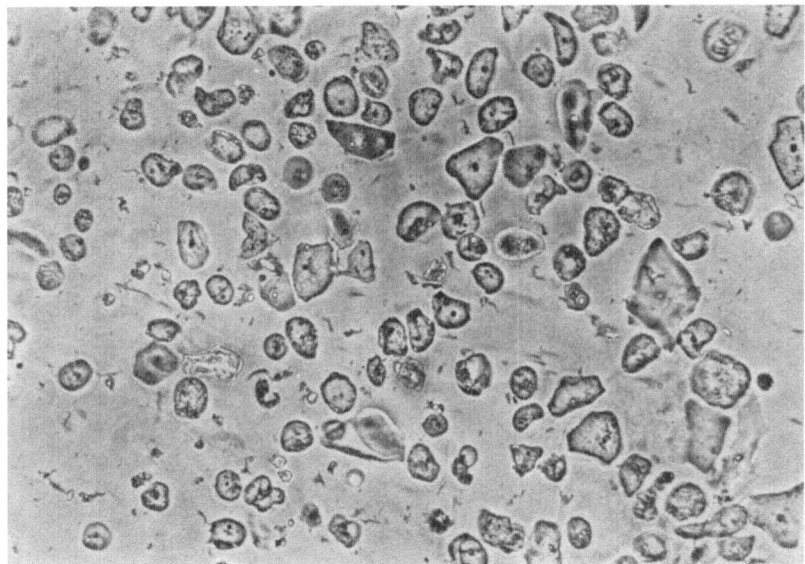

Fig. 1. Squamous cells in diluted amniotic fluid immediately after primary incubation (courtesy of Dr. H. W. Hoehn, Seattle)

Twin pregnancies represent a source of error which can only be eliminated with difficulty. Since most twin fetuses have separate amniotic sacs, one should attempt to sample both when twin pregnancy is recognized.

Contamination with maternal cells. The extracted fluid almost always contains erythrocytes which are often only microscopically demonstrable, but rarely in large amounts. The blood contaminating the culture is probably always of maternal origin and it normally exerts no effect upon the cell culture. Although gross contamination (bloody sample) may affect the success of cultures (Lisgar *et al.*, 1970; Hahnemann, 1972). Some authors separate the erythrocytes by adding ammonium chloride (Lee *et al.*, 1970).

The effect of contamination with maternal macrophages is controversial. Nadler and Gerbie (1970) maintain that they die after one week in culture. According to the experience of others, however, they survive for at least two weeks, but divide only during the first week of culture. Depending on clonal morphology, a chromosome preparation should not, therefore, be undertaken earlier than 2 weeks after sampling. Trypsinization will eliminate mature macrophages which remain attached to the glass surface.

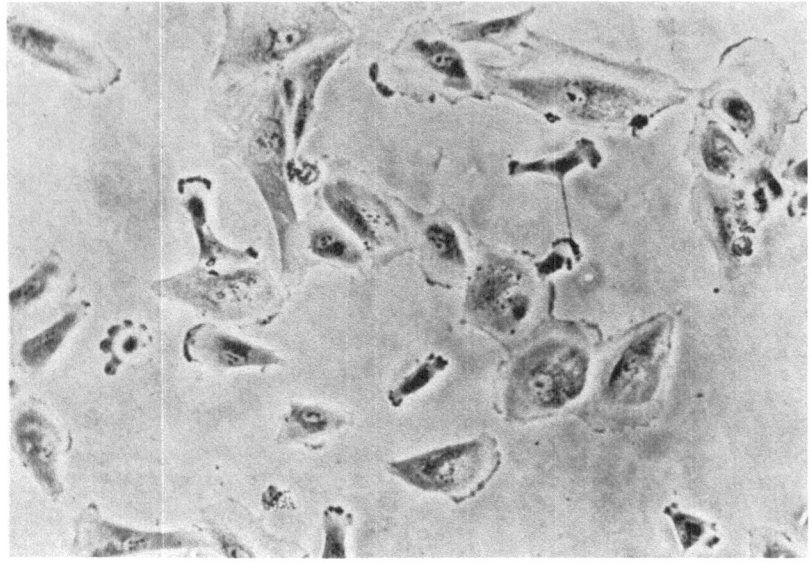

a

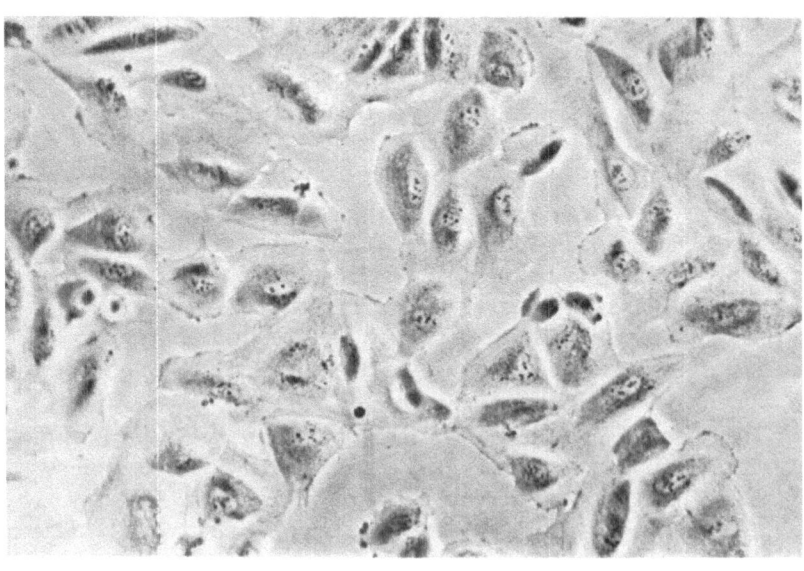

b

Fig. 2. a Epitheloid cell type, primary clone, day 8. b Epitheloid cell type, confluent culture, 2nd passage, day 28 (courtesy of Dr. H. W. Hoehn, Seattle)

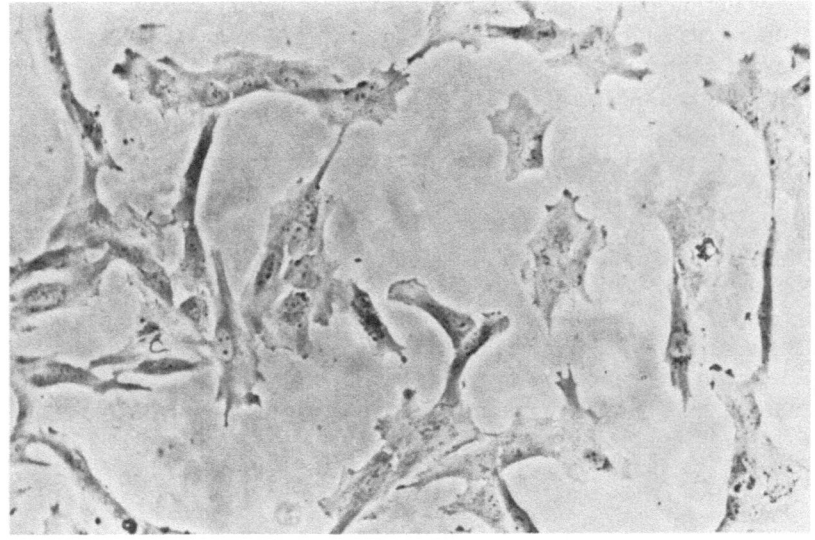

a

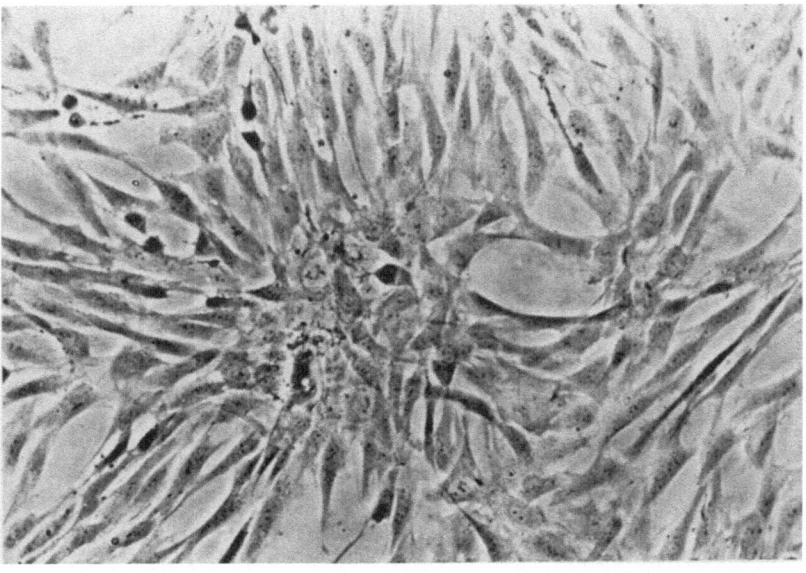

b

Fig. 3. a Fibroblastic cell type, primary clone, day 8. b Fibroblastic cell type, confluent culture, 2nd passage, day 20 (courtesy of Dr. H. W. Hoehn, Seattle)

Maternal tissue could also be introduced during sampling, though such pieces of tissue may be recognized from their morphology. To prevent contamination with maternal cell material, the following steps may be taken (Macintyre, 1971):

1. The first one or two ml of an amniocentesis should be aspirated with a separate syringe. This may then be discarded or cultivated separately.

2. The risk of false diagnosis is considerably diminished by setting up several parallel cultures, and analysing at least two cultures in case the karyotype turns out to be identical with that of the mother.

There are methods available for distinguishing cells of maternal origin. Philip *et al.* (1972) and Jonasson *et al.* (1972) used polymorphic fluorescent chromosome markers, and Jonasson *et al.* demonstrated that the HL-A (histocompatibility) system can be useful for the identification of fetal cells.

2. Collecting and Transportation of Amniotic Fluid

According to the literature, the smallest amount of amniotic fluid which allowed a successful culture was 1 ml (Milunsky *et al.*, 1972). Nevertheless, one should aim at 10 ml of amniotic fluid to set up cultures. To prevent loss of cells, the amniotic fluid may be delivered into siliconized glass or plastic tubes. Conical tubes should be employed if the cells are to be centrifuged. The sample may be immediately distributed into at least two tubes to avoid problems in case a tube breaks during transportation or centrifugation. The amniotic fluid does not have to be set up immediately. Shipment may succeed at room temperature. Even after several days storage growth may be obtained. Wahlstrøm *et al.* (1970), found that half the cells are still alive after 5 days.

3. X- and Y-Chromatin

The large amount of pycnotic and desquamous cells may impede the diagnosis of X-chromatin in cells from the amniotic fluid. Maternal cells may also be analysed together. The Y-chromatin is usually more reliable, but a preparation occasionally does not enable any diagnosis. The demonstration of Y- or X-chromatin is unequivocal evidence of the presence of a Y or a heteropycnotic X chromosome, though its absence is inconclusive (Rook *et al.*, 1971). In general, therefore, the diagnosis of sex should be confirmed by karyotype analysis.

To prepare slides for sex chromatin determination, 5 ml of amniotic fluid is centrifuged at 800 to 1,000 rpm to collect cells. Alternatively those cells which remain in the supernatant after a culture has been set up may be used. For preparation and staining, see Chapter IX. The frequency of X- or Y-chromatin may differ from other cells. For example, Rook *et al.* (1971) found Y-chromatin in only 3–9% of cells of fetuses confirmed to be male. Abbo and Zellweger (1970) set an upper limit of 5% for X-chromatin in female cells.

4. Cell Culture

The different culture methods fall into two groups: those in which the amniotic fluid is included in the culture medium, and those in which the cells have been previously separated. All methods depend upon several (at least two) parallel cultures being set up, since the chances of success are thereby greatly increased. The following indications of methods and media are only general, owing to constant advances in methodology, such as the use of growth factors (Rüdiger *et al.* (1974).

Setting up Cultures in Amniotic Fluid

Gray *et al.* (1971) have recommended a method in which the amniotic fluid itself serves as culture medium. Freshly obtained amniotic fluid is gently shaken. Two milliliters are delivered into a culture bottle with a pipette. After incubation for several days, the medium is changed. Autologous or heterologous cell-free amniotic fluid may be used for this purpose. Pure amniotic fluid may be stored without damage for over 4 weeks at room temperature and stored frozen for a longer time.

Setting up Cultures with Cell Sediment

Five to ten milliliters of amniotic fluid are centrifuged for each culture. In contrast to peripheral blood cultures, the cells are centrifuged at low speeds. Accounts in the literature vary from $42 \times g$ (Valenti and Kehaty, 1969) to $250 \times g$ (Lisgar *et al.*, 1970) for 5–10 min. The highest value given is $700–800 \times g$ for 5 min (Abbo and Zellweger, 1970). The damaging effect of higher speeds was demonstrated by Gray *et al.* (1971). The cell-free supernatant may be pipetted off and stored. It may be immediately frozen for biochemical analysis, or it may be maintained at room temperature if it is shortly to be used as culture medium.

The sediment is resuspended with two milliliters of fetal bovine serum or with medium. The culture is set up in a Petri dish (35×10 mm) in a CO_2 atmosphere or gassed with 5–10% CO_2 in air and sealed in a Leighton tube.

Setting up Cultures with Medium Including Amniotic Fluid

Since centrifugation may be harmful to cells and manipulation of vessels may also produce loss of cells, many authors prefer to add culture medium to the amniotic fluid (e.g. Knörr-Gärtner and Härle, 1972; Hoehn, see Appendix). To this effect, medium is introduced into culture bottles of appropriate size and a CO_2-air mixture is added. The proportion of medium to amniotic fluid should lie between 1:1 and 2:1, although the proportion of medium may be raised considerably. The amniotic fluid is delivered into culture bottles previously set up and containing a corresponding amount of medium.

Culture Maintenance

To arrive at a diagnosis within the shortest possible time, most authors insert a cover slip into the culture bottle and later withdraw it to prepare the attached cells *in situ*. Standard culture media as well as amniotic fluid may be used as culture fluid. No two laboratories seem to follow the same procedure. Good results may be obtained with a certain medium, and new supplies of the same may be found faulty. Different batches of the same type of medium may have effects as different as those with different media. The serum is of utmost importance. Since its quality is also variable, any attempt to correct a growth failure of cultures should start with checking the serum.

Examples of media are:

1. Gibco BME-diploid medium with 20% fetal bovine serum and 10% human AB serum (Therkelsen *et al.*, 1972).

2. Eagle's MEM with 30% fetal bovine serum (Abbo and Zellweger, 1970).

3. Ham's F-10 with 20–30% fetal bovine serum (W. Schmid, personal communication).

An inoculum in a culture bottle frequently produces less than 10 cell colonies. Uniform morphology of a colony can indicate clonal origin (Figs. 2 and 3).

After two to four days incubation, the medium may be changed and the supernatant utilized for setting up another culture. One can keep the same medium also for as long as 6 or 7 days.

The cells are trypsinized *in situ* after the clones have reached a diameter of approximately 1 mm (7–14 days), thus allowing a regular distribution of the cells without changing the culture vessel. The medium is exchanged for a trypsin solution which is poured off immediately after rinsing the culture, then a few drops of fresh trypsin are added which are left to act for a few minutes upon the cells. Afterwards, the trypsin is diluted by adding medium. The medium may be exchanged entirely when the cells again attach to the glass (within several hours).

Termination of Culture and Preparation

Colchicine treatment should be undertaken 1–3 days after trypsinization (microscopic monitoring of mitoses should be carried out). In the case of cover slip cultures, the cover slip should first be transferred into a new vessel so that the old vessel serves as reserve culture. One may follow the standard procedure if the preparation is done by the suspension technique. In cover slip cultures, mitoses can easily be lost because they are loosely attached to the cover slip. This possibility may be avoided by cautious handling. One should always attempt to obtain mitoses from primary cultures on cover slips. Should one be unsuccessful, cells growing in the flask itself can be utilized.

With the cover slips, hypotonic treatment and fixation should also be carried out carefully. The fluid should always be introduced at the side of the cover slip, preferably behind a protective cotton plain or wire gauze through which the fluid may flow freely (W. Schmid, personal communication). To assure even fixation, the medium should be slowly diluted and replaced with fixative. All fixative is subsequently exchanged with fresh fixative and pipetted off after an appropriate period of time. Finally, air-dried slides may be prepared and stained according to standard methods.

Appendix

(provided by Dr. Holger W. Hoehn, Dept. of Pathology, University of Washington, Seattle).

Protocol for Amniotic Fluid Cultures

1. Culture Vessels. Commercial tissue culture type plastic ware offers sterility and good optical properties. Glass surfaces (e.g. coverslips in

Petri-dishes, disposable slide-chambers) appear to promote the growth of different proportions of cell types.

2. Culture Media and Conditions. Lots of commercially prepared media and sera should be pre-tested for growth promoting capacity by determination of dilute plating efficiencies of diploid human fibroblasts in their early growth phase. For example, media and culture-conditions could be standardized to support macroscopic clone formation of at least 5% of cell input after 12-day incubation periods with change of medium at day 7 or 8.

In this laboratory, the amniotic fluid cultures are maintained in a non-commercial Dulbecco-Vogt modification of Eagle's medium. Selected lots of heat-inactivated ($56°$ C for 30 min) fetal calf sera (to prevent mycoplasma infection) are added to give a final concentration of 16–20%. 27 mM concentrations of bicarbonate are employed in a 5% CO_2 atmosphere; 100 units/ml of penicillin are added routinely.

3. Initiation. Within the shortest possible time after amniocentesis uncentrifuged aliquots of amniotic fluid are distributed into culture containers. The following proportions of amniotic fluid to culture media and container surface areas are used in this laboratory:

Amniotic fluid (ml)	Media (ml)	Container (cm^2)	Replicas
1	4	25[a]	2
2	3	25[a]	2
5	10	≈ 56[b]	2

[a] Falcon Plastics 25 cm^2 Culture Flask 3012.
[b] Lab-Tek 10 cm Square Petri dishes, # 4021, containing three 2.5×7.5 cm sterile glass slides (acid-washed).
There is no special treatment for bloody specimens.

4. Culture Maintenance and Harvest. The cultures are left undisturbed for 4–5 days. There is a complete change of media at days 5–6. The following procedures are carried out between days 8 and 12 with culture *flasks:*

Flasks showing more than 3 mitotically active colonies of the size of the microscopic field at a $30 \times$ or less magnification are briefly trypsinized to disperse colonies. Cultures yielding sparse clonal growth may be improved by pooling of cells from parallel containers at this point.

2–4 days later the cultures are exposed to 0.04 µg/ml Colcemid for 5–6 hours. The duration of the subsequent hypotonic treatment is adjusted to the prevalent cell type; epitheloid types remain 16–17 min (exclusive of centrifugation) in a 3:1 solution of 0.9% sodium citrate: 0.07 N KCl. Fibroblast-like cells are treated for as long as 20–22 min. Fixation thereafter is in 3:1 methanol:acetic acid, and slide preparations are made according to standard suspension techniques.

The square Petri dishes with *glass slides* receive a feeding every second day from day 4 after initiation. When clones are of appropriate size and mitotic activity (between days 8 and 10), and depending on the type of diagnosis desired (e.g.Sex-determination, complete banding study, etc.), Colcemid is added for 5 hours prior to hypotonic treatment with 15 ml of a prewarmed solution of 10:1 H_2O:fetal calf serum. After 30 minutes (time may be adjusted to clonal type) equal amounts of fixative are pipetted to the hypotonic under precautions to avoid severe turbulence; the mixture is left for 5 min, and thereafter is replaced by complete fresh fixative for 60 min. Slides are left to air dry after two fixative washings.

Acknowledgement. Drs. Margareta Mikkelsen and Holger W. Hoehn kindly commented on the manuscript.

References

Abbo, G., Zellweger, H.: Prenatal determination of fetal sex and chromosomal complement. Lancet **1970** I, 216–217.

Bergsma, D. (ed.): Intrauterine diagnosis. Birth defects: Original article series VII/5. The National Foundation — March of Dimes (U.S.A.), 1971.

Dorfman, A. (ed.): Antenatal diagnosis. Chicago: Chicago Univ. Press 1972.

Edwards, J. H.: Uses of amniocentesis. Lancet **1970** I, 608–609.

Emery, A. E. H. (ed.): Antenatal diagnosis of genetic disease. Edinburgh London: Churchill Livingstone 1973.

Fuchs, F.: Volume of amniotic fluid at various stages of pregnancy. Clin. Obstet. Gynec. **9**, 449–460 (1966).

Fuchs, F., Riis, P.: Antenatal sex determination. Nature (Lond.) **177**, 330 (1956).

Gray, C., Davidson, R. G., Cohen, M. M.: A simplified technique for the culture of amniotic fluid cells. J. Pediat. **79**, 119–122 (1971).

Hahnemann, N.: Possibility of culturing foetal cells at early stages of pregnancy. Clin. Genet. **3**, 286–293 (1972).

Harris, M. (ed.): Early diagnosis of human genetic defects. Fogarty Internat. Center Proc. No 6, 1972.

Jonasson, J., Lindsten, J., Lundborg, R., Kissmeyer-Nielsen, F., Lamm, L. U., Petersen, G. B., Therkelsen, A. J.: HL-A antigens and heteromorphic characters of chromosomes in prenatal paternity investigation. Nature (Lond.) **236**, 312–313 (1972).

Klinger, H. P., Schwarzacher, H. G.: The sex chromatin and heterochromatin bodies in human diploid and polyploid nuclei. J. biophys. biochem. Cytol. **8**, 345 (1960).

Knörr-Gärtner, H., Härle, I.: A modified method of culturing human amniotic fluid cells for prenatal detection of genetic disorders. Humangenetik **14**, 333–334 (1972).

Lee, C. L. Y., Gregson, N. M., Walker, S.: Eliminating red blood-cells from amniotic-fluid samples. Lancet **1970** II, 316–317.

Leeuwen, van, L., Jacoby, H., Charles, D.: Exfoliative cytology of amniotic fluid. Acta cytol. (Philad.) **9**, 442 (1965).

Lisgar, F., Gertner, M., Cherry, S., Hsu, L. Y., Hirschhorn, K.: Prenatal chromosome analysis. Nature (Lond.) **225**, 280–281 (1970).

Macintyre, M. N.: Chromosomal problems of intrauterine diagnosis. In: Intrauterine diagnosis. Birth defects, Orig. art. ser. VII/5 (D. Bergsma, ed.) The Nat. Fdn. – March of Dimes 1971.

Milunsky, A., Atkins, L., Littlefield, J. W.: Polyploidy in prenatal genetic diagnosis. J. Pediat. **79**, 303–305 (1971).

Milunsky, A., Atkins, L., Littlefield, J. W.: Amniocentesis for prenatal genetic studies. Obstet. and Gynec. **40**, 104–108 (1972).

Milunsky, A.: The prenatal diagnosis of hereditary disorders. Charles C. Thomas: Springfield, Ill. 1973.

Nadler, H. L.: Prenatal detection of genetic disorders. In: H. Harris, K. Hirschhorn eds., Advances in human genetics 3. New York-London: Plenum Press 1972.

Nadler, H. L., Gerbie, A. B.: Role of amniocentesis in the intrauterine detection of genetic disorders. New Engl. J. Med. **282**, 596–599 (1970).

Philip, J., Lebech, P., Niebuhr, E., Mikkelsen, M.: HMG-HCG stimulation, amniocentesis and prenatal chromosome analysis in a 14/21 translocation carrier with secondary amenorrhoea. Ugeskr. Laeg. **134**, 1850–1852 (1972).

Rook, A., Hsu, L. Y., Gertner, M., Hirschhorn, K.: Identification of Y and X chromosomes in amniotic fluid cells. Nature (Lond.) **230**, 53 (1971).

Rüdiger, H. W., Wolff, R., Wendel, U., Passarge, E.: Enhancement of amniotic fluid cell growth in culture. Humangenetik, in press.

Serr, D. M., Sachs, L., Danon, M.: Diagnosis of sex before birth using cells from the amniotic fluid. Bull. Res. Coun. Israel E **58**, 137–138 (1955).

Steele, M. W., Breg, W. R., Jr.: Chromosome analysis of human amniotic-fluid cells. Lancet **1966** I, 383–385.

Therkelsen, A. J., Petersen, G. B., Steenstrup, O. R., Jonasson, J., Lindsten, J., Zech, L.: Prenatal diagnosis of chromosome abnormalities. Acta pediat. scand. **61**, 397–404 (1972).

Thiede, H. A., Creasman, W. T., Metcalfe, S.: Antenatal analysis of human chromosomes. Amer. J. Obstet. Gynec. **94**, 569 (1966).

Valenti, C., Kehaty, T.: Culture of cells obtained by amniocentesis. J. Lab. clin. Med. **73**, 355–358 (1969).

Votta, R. A., Gagneten, C. B. de, Parada, O., Giulietti, M.: Cytologic study of amniotic fluid in pregnancy. Amer. J. Obstet. Gynec. **102**, 571 (1968).

Wahlström, J., Brosset, A., Bartsch, F.: Viability of amniotic cells at different stages of gestation. Lancet **1970** II, 1037.

CHAPTER IV

Preparation of Metaphase Chromosomes

Hans Georg Schwarzacher

1. Introduction

Chromosome preparations can principally be obtained from all tissues and suspensions in which mitoses occur. The initiation of cell cultures particularly suitable for chromosomal analysis owing to abundant mitoses, has been described in Chapters I and II. Tissues obtained directly from biopsies are also suitable for chromosome preparations.

The procedure aims at spreading the chromosomes of a cell so that they are clearly distinguishable. This requires preparing the cell as a whole and thereby ensuring that the chromosomes are individually laid out without overlapping. As much of the cytoplasm as possible should be removed so that it does not interfere with microscopic examination. This is best accomplished by pre-treatment of the cells in hypotonic solution, fixing in an acetic acid solution and subsequent squashing or air drying. The first reference to hypotonic pre-treatment of human chromosomes was made by Hsu in 1952, and, according to the postscript to this paper, this was an accidental finding. Its effects were described in detail by Hughes (1952).

To obtain an adequate number of suitable metaphases, the mitotic index must be as high as possible. Since chromosomes are most readily examined at the stages of late prophase and metaphase, substances that arrest mitosis in metaphase are useful. The choice of tissues with a sufficiently high number of mitoses is always important if an irksome search for suitable metaphase plates is to be avoided. Should there be a possibility, for example, of obtaining a higher yield of mitoses from the material on hand by setting up new cultures or continuing the original culture, this possibility should be fully exploited.

Chromosome preparations can be obtained either from cells in suspension or from monolayer cultures.

2. Preparation of Cell Suspensions

2.1. Material

Most suitable are suspension cultures from peripheral blood and bone marrow (see p. 3f.), suspensions from fibroblast or epithelial cultures (see p. 49f.) as well as cell suspensions from biopsy material containing a sufficient number of mitoses (e.g. bone marrow aspirates, tumor tissues, or neoplastic ascitic fluid).

Suspension cultures, bone marrow aspirates, and neoplastic effusions can be used directly, whereas cells from cultures growing on a surface must first be brought in suspension. This is effected, of course, in the same manner by which cells are removed from the surface when transferred for a subculture (e.g. by trypsinizing, p. 48). Biopsies from solid tissues should be cut into small pieces before cell suspensions with trypsin are made.

2.2. General Observations on Preparation Procedure

The preparative procedure is performed in three phases: 1. treatment with hypotonic solution, 2. fixation, and 3. spreading of chromosomes.

The hypotonic treatment of cells before fixation is indispensable for all types of preparation. Acetic acid (40–60%) or a mixture of acetic acid and alcohol are commonly used for fixation. The following two principal methods are available for spreading the chromosomes:

a) Air Drying. The cells are dropped onto a smooth surface (e.g. glass slide) on which they adhere and spread while drying as the film of fixative becomes thinner by evaporation. Only some residual cytoplasm remains, so that the preparation consists of a single layer of chromosomes from mitotic cells and interphase nuclei.

b) Squashing. The cells are spread and flattened by pressing them between two smooth surfaces (e.g. slide and cover slip, as described below).

Excellent results can be obtained with both methods, although in my own experience air-drying has proved more reliable and satisfactory than squashing. Squashing, however, is better adapted to certain tissues. A disadvantage of this method is that cells may be lost in making a permanent preparation.

2.3. Preparation by Air Drying

The different methods described in the literature are similar in principle and based on the work of Rothfels and Simonovich (1958), and Moorhead et al. (1960). Minor modifications are often necessary depending on type of materials and laboratory conditions. The procedure described below (see also Appendix p. 78f.) closely follows the basic method and has proved to be useful in our laboratory on many different cell suspensions.

After the suspension has been made, the subsequent steps of preparation are carried out in centrifuge tubes, those with 10–15 ml volume and a conical bottom being particularly suitable; the bottom should be accessible to a Pasteur pipette. These tubes are half-filled with cell suspension. It is better to use smaller centrifuge tubes (about 7 ml) of similar shape for smaller quantities.

The cells are repeatedly spun down at about 1,200–1,500 rpm (corresponding to 1,000–1,200 g) and resuspended in 5–8 ml liquid. A small laboratory centrifuge is quite adequate.

After each centrifugation the supernatant fluid is carefully removed by suction or simply by pouring off without disturbing the pellet. A drop of supernatant fluid is left in the tube and used to suspend the cell residue first. It is advisable to effect this resuspension by shaking rather than by pipetting. The new suspension fluid is then added with a pipette.

Hypotonic Treatment. After the first centrifugation (either from culture medium or from trypsin solution, depending on the type of culture), cells are suspended in a hypotonic solution for 10 min. The hypotonic solution should be about a quarter the molar concentration of a physiological saline solution. A 1:3 dilution of Hanks' solution or of the culture medium with distilled water, a 0.8–1.2% citrate solution, or 0.075 M potassium chloride may be used. It is recommended that the hypotonic solution be prewarmed to 37° C and that the suspension be maintained at this temperature. Slight changes of concentration or of the duration of hypotonic treatment may improve the preparation. Less time or higher concentrations reduce the swelling of the cells and the spreading of the chromosomes, more time or lower salt concentration may induce better spreading but eventually the chromosomes will appear fuzzy and become difficult to stain.

Fixation. Following hypotonic treatment, the cells are centrifuged and suspended in a fixative. By far the best fixative is acetic acid/alcohol in a ratio of 1 part concentrated acetic acid (glacial acetic acid) to 3 parts absolute ethanol or methanol. Other ratios can alter the appearance of the chromosomes. For example, by using 1 part glacial acetic acid + 1 part methyl alcohol, secondary constrictions are better visualized (Saksela and Moorhead, 1962). Other fixatives generally spread chromosomes poorly, with the possible exception of 50% acetic acid and a mixture of picric acid and alcohol.

The fixative should be changed at least twice by spinning and resuspending the cells in freshly prepared fixative. The cell suspension in acetic acid/alcohol fixative can be stored in the refrigerator at 4° C for several days, if well sealed.

Spreading of Chromosomes. When the cells have been centrifuged in the fixative for the last time, they are resuspended in a small volume of fresh fixative (about 5 times the volume of the cells) from which preparations can be made on cleaned and perfectly grease-free slides. A drop of suspension is placed on a wet (distilled water) and cool (4° C) slide held in the horizontal plane. Dry and prewarmed slides (60–70° C) can also be used. Insufficient spreading of the suspension can be avoided by tilting the slide and allowing the suspension to flow over the entire surface. Drying and adherence of the cells can be accelerated by passing the slide carefully through a flame.

This procedure results in a firm attachment of cell nuclei and chromosomes to the glass surface which persists throughout subsequent procedures, such as acid treatment in Feulgen stains, etc.

An initial test preparation should be made up and examined immediately under a phase contrast microscope at low magnification to determine whether the concentration of cells in the final suspension is correct, or whether it needs to be adjusted. Insufficient spreading of chromosomes can be improved by resuspending the cells in fresh fixative and recentrifuging .

The preparations are then processed further as desired (staining, autoradiography, etc.).

2.4. Preparation by Squashing

Hypotonic Treatment. This is performed in the same way as the air-drying method.

Fixation. This is done either as described for the air-drying method, i.e. fixation twice in acetic acid alcohol 1:3. Alternatively, cells may be fixed in 50% or 60% acetic acid. In this case the cells are *not* resuspended, but the fixative is introduced into the centrifuge tube without disturbing the pellet and left for 20 min before recentrifuging.

Spreading of Chromosomes. After centrifugation, the cell residue is resuspended carefully in about a five-fold volume of acetic acid/orcein (2% orcein in 50% acetic acid, see Appendix, p. 80). After a few minutes one drop of this suspension (whose concentration and staining intensity has been checked by microscopy) is placed on a clean slide and covered with a cover slip (not larger than 20×20 mm). Avoid air bubbles. Squashing is then performed under several layers of filter paper. It is best to place the thumb in the middle and roll it towards one side. The cover slip must not be moved on the slide under any circumstances. For the first squashing only light pressure should be applied, then the procedure is repeated 3–5 times with more pressure each time. After squashing, the cover slip should adhere firmly to the slide. Such a preparation can be preserved for several weeks by mounting with Krönig's cement, paraffin or nail varnish. For permanent preparations, the cover slip needs to be removed without loosening cells or ruining the results of the squashing. This can be very simply accomplished by placing the slide in absolute alcohol exactly up to the lower edge of the cover slip. After several hours, the alcohol will have been drawn up under the cover slip, and the squashed cells are then sufficiently fixed. Many cells remain on the slide upon lifting the cover slip. Better results are achieved by briefly freezing the preparation with dry ice and splitting off the cover slip with a knife. To obtain unstained permanent squash preparations from suspensions, these can be fixed in acetic acid-alcohol or 50% acetic acid following the hypotonic treatment and squashed immediately in fixative.

3. Direct Preparation from Monolayer Cultures

The first useful chromosome preparations from humans were obtained by this method (Hsu, 1952; Tjio and Levan, 1956; Lejeune et al., 1959) besides the squash preparations from testicular tissues (Ford and Hamerton, 1956). Direct preparations are still useful for routine work. The method is simple and enables primary cultures to be set up without trypsinization. Its disadvantage is that the chromosomes are not always successfully spread, at least not with the same degree of certainty as with suspension methods.

The cells are best grown on cover slips (see Chapter II), where they can be treated directly in the culture vessel, or by placing the cover slip with the cell layer facing upwards in a container that permits easy access (e.g. Petri dish). Care should be taken throughout the procedure, especially during hypotonic treatment and fixation, that the cover slips do not move. Changes of fluid must be done with special care, because cells in mitosis float off easily.

Hypotonic Treatment. The cover slip is covered by hypotonic solution and left for 20 min at $37°$ C.

Fixation. At first a few drops of fixative (glacial acetic acid 1 part + methyl alcohol 3 parts) are added to the hypotonic solution which is then pipetted off and entirely replaced by fixative. Fixation time is 30–45 min.

Spreading of Chromosomes. The fixative is drawn off and the cells are left to dry in the air.

Squash Preparations can also be produced from cover slips with monolayer cultures. In this case, the use of 50% acetic acid as fixative is recommended. After 30–45 min fixation, the cover glass with the cells facing downwards is placed on a slide covered with acetic acid/orcein. 5 min later, the squashing process is carried out as described above.

4. Direct Preparation of Embryonic Tissue

4.1. Field of Application

The cytogenetic examination of embryonic tissue of abortions has proved to be relevant to many problems in medical genetics. Direct chromosomal analysis of embryonic material can be carried out in about 50% of cases, obviating the need for setting up tissue cultures. If a chromosomal analysis of aborted material is planned, it is advisable to try the direct method described here and to store part of the material under aseptic conditions for a tissue culture in case the results of the direct preparation are not satisfactory.

4.2. Material

The embryonic material must be identified beyond doubt and separated from maternal tissue. This is sometimes difficult, but embryonic parts such as extremities can be recognized fairly easily in older specimens.

Good results can be obtained with 10–12 week old embryos, but useful preparations are also possible using material from both younger (from 7 weeks onwards) and older embryos (up to 19 weeks).

Preparations from liver, spleen or kidney are preferable when these organs can be well distinguished. Otherwise, more easily identifiable regions such as extremities or parts of the skull can be used. Brain generally yields good results, but it is seldom intact in embryos obtained surgically. Less satisfactory results are obtained with parts of the torso.

4.3. Technique

The following outline is based on a procedure developed by Evans *et al.* (1972); many details were kindly provided by Dr. E. P. Evans in a personal communication.

The piece of embryonic tissue (not too small – a few millimeters in longest diameter; in case of extremities, for example whole fingers) is placed immediately in a few ml of medium (TC 199 or McCoy 5 a) with Colcemid (0.05 µg per ml) or at least within 30 min of removal. Subsequently the material should be kept in the same medium containing Colcemid for several hours (or overnight) at 4° C.

Cell suspensions of inner organs (liver, spleen, kidney, lung) are made by mincing and shaking the tissue in the Colcemid medium, following which they are treated in the same way as other cell suspensions (see 2.3 and Appendix I).

Larger pieces of tissue, for example parts of extremities, should not be minced but placed *in toto* in hypotonic solution (1 % Na citrate, or distilled water 3 parts and Hanks' 1 part) in a small Petri dish for 10–15 min at room temperature. The hypotonic solution is then withdrawn by suction and replaced by fixative (freshly prepared acetic acid methyl alcohol 1:3).

The fixative should be changed twice every 5 min and the pieces of tissue then transferred to a small petri dish or a watch glass. Acetic acid 60% (roughly 4 times the volume of the tissue) is added. After 5 min the pieces are shaken for several minutes, which brings cells into suspension. This suspension is used directly for making air-dried preparations by dripping the suspended cells onto clean slides, or for squash preparations.

5. Staining

Various stains are available for preparations of cells and chromosomes adhering firmly to slides, as in air-dried or permanent squash preparations. The special staining methods to produce chromosome banding, as the quinacrine and the Giemsa banding techniques are described in detail in Chapters V and VI. To produce standard chromosome preparations, all the normal histological stains for cell nuclei can be used. However, since the preparations will have undergone hypotonic treatment and fixation with acetic acid or acetic acid/alcohol, the chromosomes are less dense than in directly fixed preparations, so staining may be difficult. In practice only a few methods have proved reliable (staining with Giemsa, aceto-orcein, pinacyanol, fuchsin or carmine), of which those most frequently employed are listed in the appendix (see p. 80). In general it is advisable to stick to one particular technique which has proved to be successful and convenient, and resort to others only if problems arise. The pinacyanol staining method (Klinger and Hammond, 1971) produces extremely dark-stained preparations. We have found it quite useful when preparations stain weaker than usual as a consequence of certain pre-treatments (e.g. fluorescence staining, autoradiography, long incubation in aqueous solutions).

In our experience, and contrary to that of other investigators, the Feulgen reaction has proved the least successful for chromosome preparations, even when carried out directly in the cell suspension.

Several other staining methods for chromosome preparations following hypotonic treatment have been published, but these do not yield better results than successful orcein or Giemsa stains and will therefore not be discussed further.

If chromosomes are not adequately stained for normal bright field microscopy, they can be better seen under phase contrast. However, we have not found that intentional weak staining of preparations in conjunction with phase contrast microscopy can be recommended as a routine procedure.

Appendix

Laboratory Guide

1. Chromosome Preparation from Cell Suspensions

Material: In principle all suspensions containing mitoses, for example blood and bone-marrow cultures, sternal aspirates, suspensions following trypsinization of fibroblast or epithelial cultures, ascites, or pleura aspirates.

1. Place suspension in a conical 15 ml centrifuge tube. Centrifuge for 8 min at about 1,200 rpm.
2. Carefully pour off the supernatant fluid. Resuspend the cell pellet with the last drop of supernatant by shaking. Do not use a pipette.
3. Add about 8 ml of prewarmed (37° C) hypotonic solution (1 part Hanks' solution + 3 parts distilled water, or similar) and let stand for 10 min at 37° C.
4. Centrifuge as in 1.
5. Pour off supernatant, stir up the cell pellet as in 2.
6. Add about 8 ml freshly prepared fixative fluid (1 part glacial acetic acid + 3 parts methanol).

7. Centrifuge as in 1.
8. Pour off supernatant, stir up residue as in 2.
9. Repeat steps 6–8 at least once.
10. Add fresh fixative to the cell pellet (about 5 times the volume of the cell pellet) and gently resuspend the cells, using a fine pipette so that a milky suspension is obtained.
11. Transfer a drop of the suspension with a fine pipette to a clean slide moistened with distilled water and distribute the suspension over the surface of the slide, then dry it by waving or by carefully drawing through a small flame. Check preparation under the microscope; if cell suspension is too dense, dilute with fixative. If chromosomes are poorly spread, repeat steps 6–8.
12. Stain and mount.

Cleaning and Preparation of Slides
1. Place new slides in pure ethyl alcohol (1:1) for at least 30 min.
2. Wipe off and dry with a clean cloth.
3. Place in about 0.1–0.5% solution of a detergent.
4. Place in distilled water and cool (refrigerator, 4° C).
5. The chilled slides are removed from the distilled water just before the suspension is transferred onto them. The surface must be completely wet; the formation of droplets indicates an unclean slide (e.g. a greasy surface) — such slides must be cleaned again. The excess water is quickly shaken off the slide and suspension is dripped onto it as above in step 11.

2. Direct Chromosome Preparations from Monolayer Cultures

Material: Cover slips (or slides) with a layer of cells (monolayer culture) are prepared in culture vessel or Petri dish.
1. Place a 3–5 mm layer of hypotonic solution (1 part Hanks' solution + 3 parts distilled water, or similar) on the cover slip and allow to stand for 20 min at 37° C.
2. Add a few drops of fixative (1 part glacial acetic acid + 3 parts methanol).
3. Draw off this fluid (hypotonic solution + a few drops of fixative) from the cover slip without moving it.
4. Overlay cover slip carefully with fixative and let stand for 30 min.
5. Draw off fixative.
6. Let stand uncovered until cells are air-dried.
7. Further treatment (staining, autoradiography, etc.) as desired.

3. Staining of Chromosome Preparations

Material: Preparations (slides, cover slips) of fixed and dried cells set up according to air-drying or squashing techniques (unstained permanent squash preparations). For special staining of chromosome banding see Chapters V and VI.

a) Standard Giemsa Stain
Staining solution: Commercial Giemsa solution is diluted 1:10 in phosphate buffer, pH 6.8.
1. Place slides in staining solution for 10 min.
2. Rinse in running tap water for 10–20 sec.
3. Air dry for 30 min at 37° C or for 1 hour at room temperature.
4. After xylene treatment, mount in DePeX.

b) Staining with Acetoorcein
1. Immerse preparations in orcein solution for 30 min at room temperature or for 10 min at 60° C.
2. Rinse off surplus stain solution by dipping in distilled water (30 sec to 1 min).
3. Differentiate for 1–2 min in 70% ethyl alcohol (check under the microscope).
4. 96% ethyl alcohol 1 min.
5. 100% ethyl alcohol, change twice, each 1 min.
6. Apply cover glass with Euparal or, after xylene, with DePeX, H.S.R., or similar.
   Orcein Solution. 2 g acetoorcein in 100 ml 50% acetic acid, boil for 30 min (caution! use a container with a narrow neck, add glass beads, heat slowly under a hood), then filter. The orcein solution can be used for several weeks. The solution should be boiled and filtered again each time before using.

c) Staining with Acetic Acid Carmine
Staining procedure as with orcein, except that 2 g carmine in 50% acetic acid is substituted for orcein. Carmine stain is as a rule weaker than orcein, but details of the finer structure of the chromosomes are well shown.

d) Staining with Pinacyanol (Klinger and Hammond, 1971).
This method produces very darkly stained chromosomes and cell nuclei. In most cases it is advisable to hydrolyse in HCl before staining. Only when a very strong staining is required should the HCl-treatment be omitted.
1. Place slides in 5 N HCl at room temperature (20⁰ C) for 2 min or in 1 N HCl at 60° C for 4 min.
2. Rinse in tap water for 2 min.
3. Stain in pinacyanol solution for 45 sec.

4. Differentiate in buffer solution (pH 6.4) for 45 sec.
5. Dip slides briefly (for a few sec) in tap water.
6. Dehydrate in two changes of absolute tertiary isopropanol or butanol.
7. Mount in Euparal or clear in xylene and mount in synthetic resin (DePeX).
   Staining Solution. Pinacyanol chloride 0.25% solution in 70% methanol.

4. Buffer Solutions

Phosphate-buffer after Sörensen
Solution A: 0.067 M KH_2PO_4 (= 9.08 g/1 liter H_2O)
Solution B: 0.067 M $Na_2HPO_4 \cdot 2H_2O$ (= 11.88 g/1 liter H_2O).

pH	Solution A	Solution B
6.4	73.2 ml	26.8 ml
6.8	50.8 ml	49.2 ml

References

Evans, E. P., Burtenshaw, M. D., Ford, C. E.: Chromosomes of mouse embryos and newborn young: preparations from membranes and tail tips. Stain Technol. **47**, 229–234 (1972).

Ford, C. E., Hamerton, J. L.: The chromosomes of man. Nature (Lond.) **178**, 1020–1023 (1956).

Hsu, T. C.: Mammalian chromosomes *in vitro*. I. The karyotype of man. J. Hered. **43**, 167–172 (1952).

Hughes, A.: Some effects of abnormal tonicity on dividing cells in chick tissue cultures. Quart. J. micr. Sci. **93**, 207–219 (1952).

Klinger, H. P., Hammond, D. O.: Rapid chromosome and sexchromatin staining with pinacyanol. Stain Technol. **46**, 43 (1971).

Lejeune, J., Gautier, M., Turpin, R.: Les chromosomes humains en culture de tissus. C.R. Acad. Sci. (Paris) **248**, 602–603 (1959).

Moorhead, P. S., Nowell, P. C., Mellman, W. J., Battips, D. M., Hungerford, D. A.: Chromosome preparations of leucocytes cultured from human peripheral blood. Exp. Cell Res. **20**, 613–616 (1960).

Rothfels, K. H., Siminovitch, L.: An air-drying technique for flattening chromosomes in mammalian cells grown *in vitro*. Stain Technol. **33**, 73–77 (1958).

Saksela, E., Moorhead, P. S.: Enhancement of secondary constrictions and the heterochromatic X in human cells. Cytogenetics **1**, 225–244 (1962).

Tjio, J. H., Levan, A.: The chromosome number of man. Hereditas (Lund) **42**, 1–6 (1956).

CHAPTER V

Fluorescence Microscopy of Chromosomes and Interphase Nuclei

HANS GEORG SCHWARZACHER

1. Introduction

With the fluorescence microscope, objects can be visualized that have the property of emitting light of longer wavelength (visible light) when excited by light of shorter wavelength (e.g. ultra-violet light). Unstained chromosomes show no suitable fluorescence of this type, but certain fluorescent dyes have a considerable specific affinity for nucleoproteins.

2. Fluorescent Dyes for Chromosomes and Cell Nuclei

2.1. Acridine Orange

Acridine orange (3,6-bis-dimethyl-amino-acridinum-chloride) was introduced by Bukatsch and Haitinger (1940), Strugger (1940) and Schümmelfeder (1950). With acridine orange it is possible to effect a differential staining of double stranded and single stranded DNA, and RNA. The differential binding to nucleic acids is due to their secondary structure: acridine orange binds to the highly ordered double helix of DNA in a monomer form, while with decreasing structural organization of the nucleic acids (as in single stranded denatured DNA or in RNA) an increasing amount of the acridine orange binds in an associated form (Rigler, 1966). The association of acridine orange molecules leads to a shift in the fluorescent wavelength (Zanker, 1962). The dye fluoresces green in the monomer form, and red in the associated form.

Therefore different functional structures of the DNA within the deoxyribonucleoprotein (DNP) complex can be differentiated, particularly euchromatin and heterochromatin (Rigler, 1966). Hence, the X-chromatin body, for example can be demonstrated with acridine orange, although acridine orange is not used for routine analysis of the X-chromatin.

After application of the special treatments which produce chromosome banding by a differential denaturation (see Chapter VI) acridine orange can be used instead of Giemsa (Chapelle *et al.*, 1973). Particularly clear pictures can be obtained with the R-banding technique (Dutrillaux and Lejeune, 1971). However, no more details are brought out by acridine orange than by the Giemsa staining.

2.1.1. Procedure of Staining with Acridine Orange

Air dried chromosome preparations (see Chapters IV and VI) or preparations of interphase nuclei (see Chapter IX) are immersed in buffer solution (phosphate buffer or citric acid phosphate buffer solution, pH 6) for a few minutes. The slides are stained for 10 min in a 0.01 % solution of acridine orange in buffer as above. After staining, the slides are rinsed in buffer twice for 2–3 min, then the cells are covered by a cover slip. The edges of the slip may be sealed with nail varnish, paraffin or any similar sealant. The intensity of the stain can be controlled by longer staining or prolonged rinsing after the staining.

Chromosomes and nuclear chromatin consequently fluoresce with a yellow-green color and the cytoplasm of intact cells contains more or less small particles fluorescing with a red color. The X-chromatin body may be visible as a green fluorescing spot. In preparations processed as for the R-banding, the bands fluoresce green and the interband regions dull red.

2.2. Quinacrine Derivatives

Some acridine derivatives, above all quinacrine mustard and quinacrine dihydrochloride, have been found to effect a differential fluorescence staining in certain parts of the chromosomes. This was first observed in chromosomes of different plant and animal species (Caspersson *et al.*, 1968 and 1969a), and later in human chromosomes as well (Zech, 1969; Caspersson *et al.*, 1970a). The chromosomes appear banded when stained with these dyes and today it is customary to refer to such bands as "*Q Bands*" (quinacrine bands).

2.2.1. Application

As detailed studies on the human karyotype have shown, *all human chromosomes* can be identified and characterized by staining with certain quinacrines. Caspersson and co-workers first showed that the characterization of individual chromosomes can be achieved by microphotometric measurement of the intensity of fluorescence along the length of the chromosomes. Later, Caspersson *et al.* (1971 b) gave the values of measurements on 5,000 chromosomes from 14 different persons, thus providing the standards for a visual identification of the entire chromosome complement of man.

Some chromosomes have such a characteristic fluorescence pattern that they can be immediately identified visually (see e.g. Zech, 1969; Pearson *et al.*, 1970; George, 1970 and 1971). Details of the identification of each individual chromosome will be described in Chapter VIII (see also Figs. in Chapter VIII).

In a good preparation processed by one of the *Giemsa-banding techniques* (see Chapter VI), chromosomes can in some cases perhaps be characterized more easily and accurately than in a quinacrine fluorescence preparation, but the Giemsa banding methods are somewhat tricky and do not produce consistently good results. The quinacrine fluorescence staining technique is, therefore, of great value because of its reliability. Once the fluorescence equipment is set up in a laboratory, it is performed easily and gives results quickly. It is particularly useful for chromosomes with a very characteristic quinacrine pattern such as the Y chromosome. Furthermore, it is possible to apply practically any other stain after quinacrine.

The characteristic fluorescence pattern obtained in mitotic chromosomes with quinacrine staining has also been observed in *meiotic chromosomes* (Pearson and Bobrow, 1970; Caspersson *et al.*, 1971 a), which affords a valuable aid for the study of meiosis.

The Y chromosome fluoresces so strongly, and retains its fluorescence during interphase, that this can also be perceived in *interphase nuclei* as a small luminous body (Caspersson *et al.*, 1970b; Pearson *et al.*, 1970; George, 1970). Since no other chromosome generally shows a fluorescent section of similar intensity and size following quinacrine staining, the presence of such a body in the cell nucleus (so called "Y-chromatin" or "Y body") can be taken to be a fairly certain sign of the presence of a Y chromosome in that particular cell (see Figs. 1 and 2, Chapter IX). Fluorescence staining with quinacrine is, therefore, an important method for the study of interphase cell nuclei (Chapter IX), which can be easily obtained from patients (e.g. from buccal smears, hair root cells, blood cells etc.). The Y-chromatin is also visible in Y carrying spermatozoa (Pearson and Bobrow, 1970).

The Giemsa banding methods do not allow clear distinction of the Y chromosome in interphase nuclei, although the Y chromosome is stained strongly by this method in metaphase. The quinacrine fluorescence stain is currently the only technique available to study Y chromosomes in interphase nuclei. The *X-chromatin body* (sex chromatin, Barr body) of female cells (see Chapter IX) may also show a somewhat stronger fluorescence after quinacrine staining in certain tissues (Mukherjee *et al.*, 1972). For a clear demonstration of the X-chromatin body, however, one of the non-fluorescent specific staining methods as described in Chapter IX should be applied.

2.2.2. Quinacrine Dyes

Quinacrine mustard (2-methoxy-6-chloro-9,4-bis(2-chlorethyl)amino-1-methylbutylamino acridine dihydrochloride) stains with particularly strong fluorescence. Some other quinacrine derivates (e.g. acranyl) likewise show a marked and specific affinity for certain chromosomal regions (Caspersson *et al.*, 1969b; Majewski *et al.*, 1971).

Quinacrine dihydrochloride, usually only referred to as "quinacrine" (2-methoxy-6-chloro-9,4-bis(2-ethyl)amino-1-methylbutylamino acridine dihydrochloride, "Atebrin", "Mepacrin"), gives also an excellent differential staining, though a little weaker than quinacrine mustard and which perhaps fades more rapidly on exposure to U.V.-light.

All quinacrine derivatives depend for their staining action on the intercalation of the acridine nucleus in the DNA molecule and on the ionic binding of the side chain to the phosphate groups of DNA. Weisblum and Haseth (1972) showed that the fluorescence of quinacrine is enhanced by AT-rich DNA and decreased by GC-rich DNA. Observations by Ellison and Barr (1972) confirmed the preference of AT-rich chromosome regions to stain with quinacrines.

In practice, both quinacrine mustard and quinacrine dihydrochloride have been found to give good results. Quinacrine mustard is especially recommended for quantitative studies. For qualitative investigation involving the differentiation of chromosomes and the visualization of the Y-chromatin, quinacrine dihydrochloride is very satisfactory. It may be used in preference to quinacrine mustard, because it is easier to obtain and cheaper.

2.2.3. Preparation of Specimens

2.2.3.1. Chromosome Preparations

All types of chromosome preparations done according to the methods described in Chapter IV (Mitotic chromosomes) and XI (Meiotic chromosomes) can be effectively stained with quinacrine. Air dried preparations can be stained immediately; in squash preparations the cover glass needs to be removed first (either by placing the mounted slide in 96% alcohol to reach the lower edge of the cover glass or by freezing and splitting off the cover slip; see Chapter IV).

The successful application of quinacrine fluorochromes depends to a large extent on the quality of the preparations. Samples with poorly spread chromosomes are usually not suitable for fluorescence staining, even if the number and main structures of chromosomes could be observed on normal stains.

2.2.3.2. Interphase Cells

Preparations of cell nuclei in interphase are mainly used to diagnose the Y-chromatin (Y chromosome). Most histological methods for demonstrating cell nuclei are suitable, but it should be noted that best results are obtained when the tissue has been fixed in acetic acid or in alcohol or in a mixture of both. Preparations containing intact nuclei (smears, teased tissue samples e.g. from the hair root, cell suspensions, cell cultures) are to be preferred over sections. The technical procedures for preparations of different tissues are described in detail in Chapter IX.

2.2.4. Staining Procedures

The procedure is simple: the preparations are placed in the staining solutions, rinsed with water or buffer solution, left for a short time in water or buffer for differentiation, covered by cover slip and examined under the fluorescence microscope.

Both quinacrine mustard and quinacrine dihydrochloride are used in aqueous solution. A 0.5% solution has been recommended in the literature, but should be regarded as the upper concentration limit, because it requires a relatively long period of differentiation. Good fluorescence and specific staining of nuclei and chromosomes can also be obtained at lower concentrations (down to 0.1%). According to Zech (1969) and Pearson et al. (1970), the staining solution should be adjusted to a pH of 5.5, whereas Caspersson et al. (1970c) recommend a pH of 7.0. Schwinger et al. (1971) point out that pure tap water is sufficient. Nevertheless, we recommend to use a freshly made up buffer solution. In our experience, the best results are obtained with a pH of 6.0 (e.g. phosphate buffer according to Sörensen or citric acid phosphate buffer according to McIlwain). The staining solution can be kept several days in the refrigerator. Storing for about one week may, in fact, improve the staining capacity of the solution. The addition of thymol (0.5%) prevents the growth of microorganisms in the solution.

Air dried preparations and also frozen sections should first be immersed in buffer solution. Paraffin sections are immersed in buffer solution after deparaffination and rehydration.

The slides are stained for 5 min. The excess dye is briefly washed out in buffer solution of pH 6.0. The slides are then transferred for differentiation to a fresh buffer solution for 5 min, and placed under a cover slip. The preparation should immediately be examined under the fluorescence microscope. Fluorescence may at first appear somewhat weak, but will usually intensify after waiting for a short period (10 min). If it remains weak, the slide must be stained again using a higher concentration, while the time of differentiation is reduced. Should the staining be too intense, the preparation can be rinsed again in buffer solution and a new cover slip mounted.

The intensity of the stain gradually diminshes on exposure to light. The preparations can be used for direct light microscopic observation for a period of a few hours, but any particular field kept in focus with full U.V. illumination will fade within a few minutes. Thus, usually only 2–3 photomicrographs can be made of a mitosis or of a cell nucleus. Precipitations of the stain may occur after some time. To prevent drying of the preparation, the cover slip can be sealed round the edge with nail varnish, paraffin or Krönig's cement. When using oil immersion, care must be taken that oil does not come in contact with the aqueous staining solution.

Faded preparations may recover some of their fluorescence within a few hours. The recovery is, however, hardly sufficient to demonstrate the fine chromosome banding but will sometimes enable a Y-chromatin diagnosis.

It is also possible to make a permanent preparation by dehydration in alcohol and mounting in an appropriate mounting medium (Klinger and Moser, 1972). After quinacrine staining, the slides are immersed in buffer solution for 5 min, then quickly dehydrated in 70%, 96%, 100% ethanol respectively, placed in Xylene and mounted in a neutral mounting medium (e.g. DePeX by G. T. Gurr). Chromosomes usually lose so much of the staining that the banding pattern is not very distinct. In preparations of interphase nuclei, the chromatin structure is, however, much clearer. Therefore, the Y-chromatin body as well as the X-chromatin body can be sometimes better diagnosed as in unmounted preparations. A further improvement for the staining of interphase nuclei is achieved by a short hydrolysis in HCl (e.g. $3N$ HCl at $22°$ C for 2 min, see Klinger and Moser, 1972).

2.2.5. Restaining with Other Dyes

Preparations stained with fluorochromes can subsequently be treated with any of the usual nuclear or chromosome stains. For this purpose, the slides together with cover slips are immersed for 30 min in 96% alcohol, which will remove the cover slip and dissolve the fluorochrome without impairing the cells. After brief air drying, any other staining method can be applied. For interphase nuclei, the Feulgen reaction for example (see Chapter IX) gives good results. A subsequent chromosome stain may turn out somewhat weak; here the use of intense stains, such as Giemsa or pinacyanol (see Chapter IV) is recommended.

Blood smears can be restained with standard methods, but here it is advisable to make several smears and stain one of them for the analysis of drumsticks (see Chapter X). On the other hand, the stain of regular blood smears can be removed by methanol and a fluorochrome stain with quinacrine be attempted. Furthermore, chromosome preparations stained by one of the standard methods (Giemsa, Orcein) can be de-stained with alcohol (70%) or acetic acid (50%) or both, and sub-sequently treated with quinacrine.

3. Other Fluorochromes

Fluorescent dyes other than quinacrine mustard and quinacrine dihydro-chloride have thus far not gained any importance for practical use, but they are of theoretical interest (see e.g. Caspersson *et al.*, 1969 b; Majewski *et al.*, 1971; Hillwig and Gropp, 1972).

4. Technique of Fluorescence Microscopy

4.1. Principle

When using fluorochromes, the visualization of the stained structures results from stimulation of the stain to fluoresce by means of U.V.- or blue light. This secondary fluorescence is of longer wavelength and is therefore in the visible spectrum and can be studied. A fluorescence microscope must be fitted with 1) a U.V.-and/or blue light source, 2) an excitor filter between the light source and the object, filtering out the visible light of longer wavelengths as well as the heat radiation, and 3) a barrier filter between object and eye (or the camera) to filter out the exciting U.V. and blue light. The first filter prevents light of longer wavelength from superimposing on the fluorescence; the second filter is necessary for enhancing the contrast and also for protecting the eyes and shielding the film when taking microphotographs.

4.2. Microscope Equipment

High-pressure mercury lamps (HBO 100 or HBO 200) are usually used as light source. This type of lamp provides a high intensity within a few narrow bands of the spectrum (the Hg lines). Xenon lamps, which are also used for fluorescence microscopy, have the advantage of an almost continuous spectrum, but their intensity is not as high as that of the Hg lamps in the violet and blue regions which are necessary to excite the quinacrine fluorescence.

The ideal excitor filter should extinguish all light with a wavelength of over 470 nm; the ideal barrier filter should extinguish all light below 470 nm. In practice, the Schott "BG 12" can be used as excitor filter and a combination of the Schott "OG-515" and "GG9" as barrier filter. The thickness of the filterglass depends on the light source and the kind of illumination used. With direct illumination, the 3 mm BG 12 is usually the best, while with epi-illumination, 5 mm might be necessary. Pearson (1973) recommends the use of the Schott AL 436 as excitor filter in combination with a 490 nm barrier filter.

In many instances, it is sufficient to use a fluorescence microscope with direct illumination. All unnecessary glass filters and accessory lenses of the illuminating system should be removed to maintain maximum light intensity, but the numerical aperture of the condensor should be kept as high as possible. Therefore, immersion objectives and immersion of the condensor are necessary to bring out fine details and to give maximum contrast. The non-fluorescent immersion oils provided by most optical companies have to be used.

The fluorescence intensity is further enhanced by epi-illumination systems because the loss of light energy is low.

The use of a dark-field illumination condensor offers another possibility to increase fluorescent light intensity and contrast with very satisfactory results, and it is less expensive than an epi-illumination system.

4.3. Microphotography

The low intensity of light inevitably causes long exposure times which demands that the camera is not susceptible to vibration. It is recommended, therefore, that the microscope be placed on a separate sturdy table and that no other work be done on it or in its vicinity during exposure. With regard to film material, highly sensitive black and white films (e.g. Kodak 3-X; Ilford HP 4) are often recommended for routine use.

Finer grained but less sensitive films may require prolonged exposure which could cause fading or artefacts due to vibration or shifting of the object, although their higher resolution is advantageous. Using dark field illumination condensor or epi-illumination, the exposure time can be kept as low as 5–10 sec with highly sensitive films at highest magnifications. Fine grained films (e.g. Kodak-Panatomic-X, Agfa-Gevaert Isopan IF and similar types) need long exposure times (1–2 min) and cause so much fading that it is not possible to take more than one or two photographs of the same cell. In order to reveal finest details, the use of a fine grained film would, of course, be advantageous. Breg (1972) particularly recommends the H & W Control VTE Panchromatic film for such purposes. A good compromise is e.g. the Ilford Pan F which needs about 30–60 sec with maximal illumination. When working with a microscope fitted to an automatic camera, in order to save light the automat regulating the exposure time should not be used (e.g. in the Zeiss "Photomicroscope", one third of the light passes through the photometer).

5. Analysis of Preparations

5.1. General Comments

Since fluorescent preparations do not keep long and fade on exposure to light, only a rapid evaluation is possible directly under the microscope. A more detailed analysis of chromosomes can usually be carried out only on photographs, because smaller differences in fluorescence intensity which may not be perceived by the eye will be enhanced by the photographic process. Hence, for chromosome preparations, it is advisable to photograph a sufficient number of metaphases rather than waste time trying to evaluate the preparations directly under the microscope.

For examination of the Y-chromatin in interphase nuclei, one will usually find direct observation and counting under the microscope to be sufficient.

5.2. Chromosome Preparations

Full details concerning the characteristic fluorescence pattern of individual chromosomes are given in Chapter VIII. Since fluorochrome preparations do not uniformly give good results, and any one metaphase is rarely fully informative, a sufficient number of cells should be photographed. Apart from anomalies of the Y chromosome, which can be rather well recognized even from relatively few cells, the fluorescence method helps in the elucidation of more complex chromosomal anomalies, which will usually require a detailed study of a larger number of cells.

5.3. Interphase Cells and Y-Chromatin

The Y-chromatin appears in normal male cell nuclei as a bright dot of about $0.3–1.0 \mu$ in diameter. When one or more Y chromosomes are present, a corresponding number of intensively fluorescent bodies (Y-chromatin bodies) can be observed in the cell nucleus. Full details of the evaluation of interphase nuclei are given in Chapter IX.

Appendix

1. Staining with Quinacrine-Dihydrochloride and Quinacrine-Mustard

Material: Fixed and air dried chromosome preparations and interphase cell nuclei preparations.
1. Immerse slides for 5 min in buffer solution at pH 6.0.
2. Place slides for 5 min in solution of quinacrine dihydrochloride (or quinacrine mustard) 0.2% in phosphate buffer at pH 6.0.
3. Place slides for 5 min in buffer solution at pH 6.0.
4. Cover the cells on the wet slides with a cover glass. Avoid air bubbles.
5. Check under the fluorescence microscope: if staining is too weak, wait for 10 min; if still too weak, remove cover slip and restain in the dye solution for again 5 min, shorten the time of differentiation in buffer (point 3), put on a new cover slip; if stained too strongly (no banding pattern of chromosomes, no distinct Y-chromatin), remove cover slip and place slides again in fresh buffer for a 5 min, and mount. If preparation is stained satisfactorily, seal coverglass with nail varnish.

2. Suppliers of Quinacrine Stains

Quinacrine mustard: *Polyscience*, Inc., Paul Valley Industrial Park, Warrington, Pa. 18976, U.S.A.
Quinacrine dihydrochloride: *SIGMA Chemical Company*
3500 DeKalb-St., St. Louis, Missouri 63118
U.S.A.

3. Phosphate Buffer Solution (Sörensen)

Solution A: 0.067 M KH_2PO_4 (= 9.08 g/l H_2O)
Solution B: 0.067 M $Na_2HPO_4 \cdot 2H_2O$ (= 11.88 g/l H_2O).
Ready-made buffer salts to be diluted in distilled water are supplied by E. Gurr, Ltd., London S.W. 14, England.

Solution A (ml)	Solution B (ml)	pH
98.8	1.2	5.0
87.7	12.3	6.0
73.2	26.8	6.4
50.8	49.2	6.8
39.2	60.8	7.0
28.5	71.5	7.2

References

Breg, W. R.: Quinacrine fluorescence for identifying metaphase chromosomes, with special reference to photomicrography. Stain Technol. **47**, 87–93 (1972).

Bukatsch, F., Haitinger, M.: Beiträge zur fluoreszenzmikroskopischen Darstellung des Zellinhaltes, insbesondere des Cytoplasmas und des Zellkernes. Protoplasma (Wien) **34**, 515–523 (1940).

Caspersson, T., Farber, S., Foley, G. E., Kudynowski, J., Modest, E. J., Simonsson, E., Wagh, U., Zech, L.: Chemical differentiation along metaphase chromosomes. Exp. Cell Res. **49**, 219–222 (1968).

Caspersson, T., Hultén, M., Lindsten, J., Zech, L.: Identification of chromosome bivalents in human male meiosis by quinacrine mustard fluorescence analysis. Hereditas (Lund) **67**, 147–149 (1971a).

Caspersson, T., Lomakka, G., Zech, L.: The 24 fluorescence patterns of the human metaphase chromosomes — distinguishing characters and variability. Hereditas (Lund) **67**, 89–102 (1971b).

Caspersson, T., Zech, L., Johansson, C.: Differential binding of alkylating fluorochromes in human chromosomes. Exp. Cell Res. **60**, 315–319 (1970a).

Caspersson, T., Zech, L., Johansson, C., Lindsten, J., Hultén, M.: Fluorescent staining of heteropycnotic chromosome regions in human interphase nuclei. Exp. Cell Res. **61**, 472–474 (1970b).

Caspersson, T., Zech, L., Johansson, C., Modest, E. J.: Identification of human chromosomes by DNA-binding fluorescent agents. Chromosoma (Berl.) **30**, 215–227 (1970c).

Caspersson, T., Zech, L., Modest, E. J., Foley, G. E., Wagh, K., Simonsson, E.: Chemical differentiation with fluorescent alkylating agents in *Vicia faba* metaphase chromosomes. Exp. Cell Res. **58**, 128–140 (1969a).

Caspersson, T., Zech, L., Wagh, U., Modest, E., Simonsson, E.: DNA-binding fluorochromes for the study of the organization of the metaphase nucleus. Exp. Cell Res. **58**, 141–152 (1969b).

Chapelle, de la A., Schröder, I., Selander, R. K., Stenstrand, K.: Differences in DNA composition along mammalian chromosomes. chromosoma (Berl.) **42**, 365–382 (1973).

Dutrillaux, B., Lejeune, J.: Sur une nouvelle technique d'analyse du caryotype humani. C. R. Acad. Sci. (Paris), Série D **272**, 2638–2640 (1971).

Ellison, J. R., Barr, H. J.: Quinacrine fluorescence of specific chromosome regions. Chromosoma (Berl.) **36**, 375–390 (1972).

George, K. P.: Cytochemical differentiation along human chromosomes. Nature (Lond.) **226**, 80–81 (1970).

George, K. P.: Quinacrine mustard — a selective fluorescent stain for the Y-chromosome in human tissue for routine cytogenetic screening. Stain Technol. **46**, 34–36 (1971).

Hilwig, I., Gropp, A.: Staining of constitutive heterochromatin in mammalian chromosomes with a new fluorochrome. Exp. Cell Res. **75**, 122–126 (1972).

Klinger, H. P., Moser, G. C.: Improved chromatin-fluorescence technique. Lancet **1972** II, 1366.

Majewski, F., Bier, L., Pfeiffer, R. A.: Fluoreszenzmikroskopischer Nachweis des menschlichen Y Chromosomes in Interphasekernen durch Acridinderivate (Atebrin, Acranil). Klin. Wschr. **49**, 814–818 (1971).

Mukherjee, A. B., Moser, G., Nitowsky, H. M.: Fluorescence of X and Y-chromatin in human interphase cells. Cytogenetics **11**, 216–227 (1972).

Pearson, P. L.: The identification of mammalian chromosomes by differential staining techniques. In: New Techniques in Biophysics and Cell Biology, ed. R. H. Pain and B. J. Smith. London: John Wiley & Sons 1973.

Pearson, P. L., Bobrow, M.: Fluorescent staining of the Y-chromosome in meiotic stages of the human male. J. Reprod. Fertil. **22**, 177–179 (1970).

Pearson, P. L., Bobrow, M., Vosa, C. G.: Technique for identifying Y chromosomes in human interphase nuclei. Nature (Lond.) **226**, 78–80 (1970).

Rigler, R.: Microfluorometric characterization of intracellular nucleic acids and nucleoproteins by acridine orange. Acta physiol. scand. **67**, Suppl. 267, 1–122 (1966).

Schümmelfeder, N.: Die Fluorochromierung des lebenden überlebenden und toten Protoplasmas mit dem basischen Farbstoff Acridin-Orange und ihre Beziehung zur Stoffwechselaktivität der Zelle. Virchows Arch. path. Anat. **318**, 119–154 (1950).

Schwinger, E., Rakebrand, E., Müller, H. J., Bühler, E., Tettenborn, U.: Y-body in hair roots. Humangenetik **12**, 79–80 (1971).

Strugger, S.: Fluoreszenzmikroskopische Untersuchungen über die Aufnahme und Speicherung des Akridinorange durch lebende und tote Pflanzenzellen. Jena. Z. Med. Naturw. **73**, 97 (1940).

Weisblum, B., Haseth, P. L.: Quinacrine, a chromosome stain specific for deoxyadenylate-deoxythymidylate-rich regions in DNA. Proc. nat. Acad. Sci. (Wash.) **60**, 629–632 (1972).

Zanker, V.: Über den Nachweis definierter reversibler Assoziate („reversibler Polymerisate") des Acridinorange durch Absorptions- und Fluoreszenzmessungen in wäßriger Lösung. Hoppe-Seylers Z. physiol. Chem. **199**, 225–258 (1952).

Zech, L.: Investigation of metaphase chromosomes with DNA-binding fluorochromes. Exp. Cell Res. **58**, 463 (1969).

CHAPTER VI

Banding Patterns in Human Chromosomes Visualized by Giemsa Staining after Various Pretreatments

WOLFGANG SCHNEDL

With 5 Figures

1. Introduction

Many methods appropriate for demonstrating banding patterns within chromosomes now exist. These banding patterns allow the differentiation of chromosomes which previously could not be identified by standard staining methods. Since different areas within a single chromosome can thus be precisely localized, studies of structural abnormalities of chromosomes are greatly facilitated.

The first demonstration of differential staining of chromosomes was achieved by quinacrine fluorescence analysis introduced by Caspersson (Caspersson *et al.*, 1970; see Chapter V). In this chapter only those methods that demonstrate banding patterns within chromosomes by transmitted light microscopy will be described. They consist of various procedures carried out on chromosome preparations and subsequent staining with Giemsa solution.

The first of these methods was derived from *in situ* hybridization procedures. They served to localize certain DNA fractions within chromosomes in cytological preparations. In principle, the *in situ* hybridization methods utilize the fact that only those single strands of DNA will hybridize which have a complementary base sequence. The procedures are as follows: first DNA within cells is dissociated (for instance by a NaOH treatment) in a cytological preparation; subsequently the preparations are incubated in a solution of single-stranded nucleic acid molecules which are labeled with tritium. The regions of the chromosomes that contain the respective base sequences hybridize with the added molecules and can be localized by autoradiography.

However, only DNA fractions made up of many *repetitive* DNA sequences can be localized by *in situ* hybridization methods (repetitive DNA is composed of a great many of identical or near identical DNA sequences; in some cases one single diploid nucleus may contain more than 1 million copies of the same DNA sequence). The methods are not sensitive enough for localizing non-repetitive DNA sequences.

Working with such methods, Arrighi and Hsu (1970) noticed that simple treatment of chromosome preparations with NaOH and subsequent incubation with $2 \times$ SSC (a solution containing 0.3 M NaCl and 0.03 M trisodium citrate at pH 7.0) followed by staining in a Giemsa solution was sufficient to effect differential staining of the chromosomes.

In human chromosomes, the centromeric regions in particular stain more intensely than other regions. In mouse the situation is even more striking, because most centromeric regions stain very strongly. Earlier *in situ* hybridization studies of mouse chromosomes had shown that the centromeric regions contain the highly repetitive DNA fraction of this species (Pardue and Gall, 1970; Jones, 1970).

Similar methods were developed subsequently which also showed differentially stained regions along the entire length of the chromatids, forming a very characteristic banding pattern in each chromosome.

It appears to be of particular interest that treatment of chromosomes with enzymes (pronase or trypsin) and subsequent Giemsa staining results in similar banding patterns of the chromosomes (Dutrillaux *et al.*, 1971). Later on it was shown that still other agents can produce banding patterns which can be visualized by the Giemsa stain. Utakoji (1972) demonstrated banding patterns in human chromosomes pretreated with potassium permanganate. Urea can also be used for demonstrating bands (for instance Shiraishi and Yosida, 1972; Kato and Yosida, 1972). There are still other agents which may produce banding under appropriate conditions. Yosida and Sagai (1972) used a solution of sodium dodecyl sulfate. Kato and Moriwaki (1972), testing several reagents, found that many salts with alkaline pH, strong bases, and surface active agents are potent band inducing agents.

These results indicate that solubilization of some chromosomal proteins may be involved in producing the bands. It could be shown, however, that DNA is also attacked by some of the Giemsa banding procedures (Schnedl, 1973). Large amounts of DNA are removed, when these methods are applied.

The actual mechanisms working in the various methods are still unknown, nevertheless, in general they all yield the same results, with some differences in the quality of the banding patterns.

It should be stressed that most of the Giemsa bands coincide with the bands visualized by the quinacrine fluorescence method. In addition, regions that can be seen by the Arrighi and Hsu method (1970) are shown. Possibly two or more mechanisms are superimposed in the production of these banding patterns.

2. General Comments on the Different Methods

Arrighi and Hsu (1970) and Yunis *et al.* (1971) described the first methods. Many procedures demonstrating banding patterns in chromosomes have been described since then, most of them derived from these original methods. Similarly, the enzyme method originally described by Dutrillaux *et al.* (1971) was modified later by other authors.

At least one reason for the development of so many, slightly different methods is the fact that preparations from different laboratories will require a different treatment. Even different supplies of slides prepared under identical conditions may require modification of a procedure. Therefore, each particular method will have to be adapted to the prevailing conditions. The steps of the Giemsa banding procedures which can be altered for obtaining optimal results will be especially pointed out in this chapter.

It should be emphasized that chromosome preparations for the Giemsa banding techniques should not be stored longer than 2–3 months. On the other hand, fresh slides should be used 2–3 days after preparation at the earliest. It is possible to "age" preparations artificially (for instance by incubating at 60° C for several hours).

Preparations stored for as long as one year or even longer may give satisfactory results with the enzyme methods.

The final step in nearly all procedures consists of staining with Giemsa solution. The Giemsa solutions used by different authors are not the same. Nevertheless, it is sufficient to use any kind of Giemsa stain solution in a buffer at pH of 6.8–7.0 (for instance 10 ml Giemsa stock solution of various sources and 90 ml phosphate buffer of pH 6.8 or 7.0 with a low molarity, e.g. 0.01 M).

Most Giemsa banding methods described in this chapter can be performed on preparations which were previously analysed by quinacrine fluorescence. This allows study of the same metaphase cell with both methods. A preparation already stained by a Giemsa method cannot be subsequently analysed by quinacrine fluorescence.

For staining a preparation by the Giemsa banding technique following quinacrine fluorescence, the coverslip is carefully removed, the slide washed for 2 hours in 95% ethanol, and dried. The Giemsa staining method desired can then be carried out immediately.

3. Methods for Demonstrating the Centromeric Heterochromatin

In man and in most species studied so far, the methods described below are able to demonstrate strongly stained regions at the centromeres. Therefore, the Paris Conference (1971) proposed the term "C bands" for these regions. The term "C-bands" is, however, not appropriate for all mammals: in cattle, for instance, the centromeric regions remain nearly unstained when using these procedures (Schnedl, 1972).

In human chromosomes, the following regions will appear strongly stained (Fig. 1):

1. the centromeres of most chromosomes (or, in fact, the regions adjacent to the centromeres). The precise position with regard to the centromere cannot be decided with certainty in each case. Some chromosomes show only low amounts of this material (for instance, chromosome No. 2);

2. the secondary constrictions of the chromosomes Nos. 1, 9, and 16;

3. in the Y chromosome the distal part of the long arm (the same region is brilliantly fluorescent when analysed by quinacrine fluorescence).

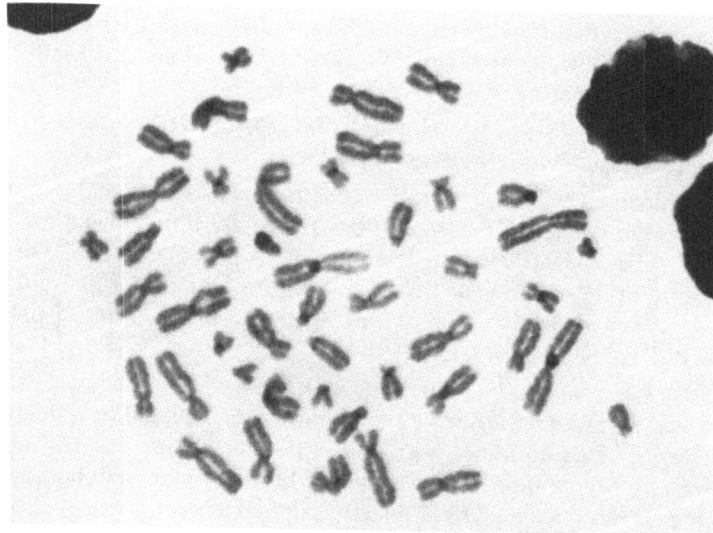

Fig. 1. Human male metaphase cell, stained by the Arrighi and Hsu method (1970)

3.1. The Method of Arrighi and Hsu (1970)

This method is directly derived from *in situ* hybridization procedures. The original method consists of treating slides first with 0.2 N HCl (to remove the histones and other non-acid proteins) and then with RNase, followed by 0.07 N NaOH for 2 min. After rinsing in ethanol, they are incubated for more than 12 hours in 2×SSC at 65° C, and stained with Giemsa. Not all of these steps are necessary; the HCl and the RNase treatment may be omitted, because chromosome preparations do not contain large amounts of RNA after hypotonic treatment.

The critical factor in all procedures involving NaOH treatment is its concentration and its duration. In many cases, 0.07 N NaOH for 2 min is too severe, whereas 0.007 N or even 0.002 N NaOH, respectively, and shorter duration of treatment (30–90 sec) may be satisfactory.

Flame dried and air dried preparations can be directly subjected to this procedure. Squash preparations should be made on slides which were coated with a layer of 0.1 % gelatine in 0.01 % chrome alum solution to prevent the cells from floating away during the procedure. Another possibility consists of coating the preparations with 0.5 % agar.

3.2. The Method of Yunis et al. (1971)

Chromosome preparations are first heated at 85–100° C for 10 min in 0.06 M phosphate buffer (pH 6.8), incubated in the same buffer at 65° C for several hours, and then stained with Giemsa.

This method is easier to perform than the one by Arrighi and Hsu, but the results are less reproducible.

The most critical point in this method is heating of the slides in buffer. We use a temperature of 90° C for 5 min, which is enough in most cases. For the second step, the authors recommend an incubation time of more than 30 min at 65° C. An incubation time of more than 10 hours was necessary to obtain good results.

For Giemsa stain, one of the routine methods may be used, as in other procedures described in this chapter.

Although this method was originally only described to show spots at the centromeres of human and mouse chromosomes, banding patterns of chromatids in cattle chromosomes could be demonstrated by this method (Schnedl, 1972). The results in man are practically identical with those obtained by the method of Arrighi and Hsu.

3.3. Meiotic Chromosomes

In meiotic preparations from human males, the Giemsa banding techniques thus far have failed to demonstrate banding patterns in the

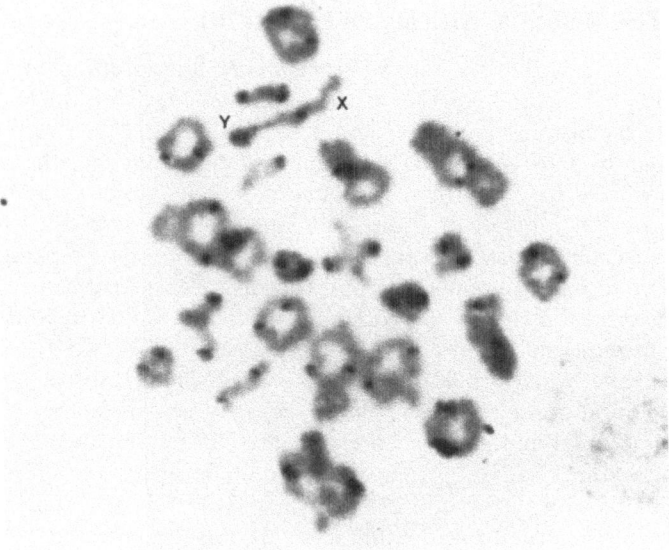

Fig. 2. Human male meiosis (diakinesis). Staining of centromeric regions.
Courtesy of Dr. A. Chandley

diakinetic or metaphase I cells. However, the centromeric regions of the
chromosomes and the distal part of the Y are stained as in the "C band"
techniques. Of course, the original method of Arrighi and Hsu (1970)
gives the same result. A modification using barium hydrochloride
(Sumner, 1973) has been particularly recommended by Chandley (1973)
(Fig. 2).

Location of the centromeres and the distal part of the Y may be
desirable in meiotic cells for studying nondisjunction and other problems.

3.4. Methods for Demonstrating the Secondary Constriction of Chromosome No. 9 (Bobrow et al., 1972; Gagné and Laberge, 1972)

Staining of chromosome preparations with a Giemsa solution adjusted
to pH 11 may produce a preferential staining of the secondary constriction of chromosome No. 9. This region appears red-purple, whereas the
other chromosomes only show a pale blue color. Centromeric regions
of other chromosomes may also be stained.

With these methods the secondary constriction of the chromosome No. 9 may be also visualized in interphase nuclei, e.g. blood cells or spermatozoa.

The staining time required for specific staining of the chromosome No. 9 is longer at higher pH values.

4. Methods for Demonstrating Banding Patterns over the Entire Chromosome

The bands made visible by these methods were termed "G bands" ("Giemsa bands") by the Paris Conference (1971). They correspond well with those visualized by the quinacrine fluorescence method with some exceptions.

The results of differential staining of human chromosomes obtained by the three basic methods can be summarized as follows:

Preferential staining of	Giemsa band-ing method	Quinacrine fluorescence	Arrighi and Hsu method
Centromeric regions	+	−	+
Sec. constrictions in Nos. 1 and 16	+	−	+
Sec. constriction of No. 9	−	−	+
Y. distal part	+	+ +	+
Y. proximal part	±	−	−
Fine bands in the chromatid	+	+	−

Many different methods producing banding patterns along chromosomes are now available. The following sections provide a detailed description of the original methods that were published independently by different laboratories. Later descriptions of minor modifications are discussed when necessary. Procedural schemes for the methods are listed in the appendix for laboratory use.

4.1. Methods Using NaOH Treatment Followed by an Incubation in Buffer

4.1.1. The Method of Schnedl (1971)

In view of extensive personal experience, this method will be described in greater detail. In principle, the procedure consists of treating chromosome preparations with NaOH for 1–2 min, followed by incubation in

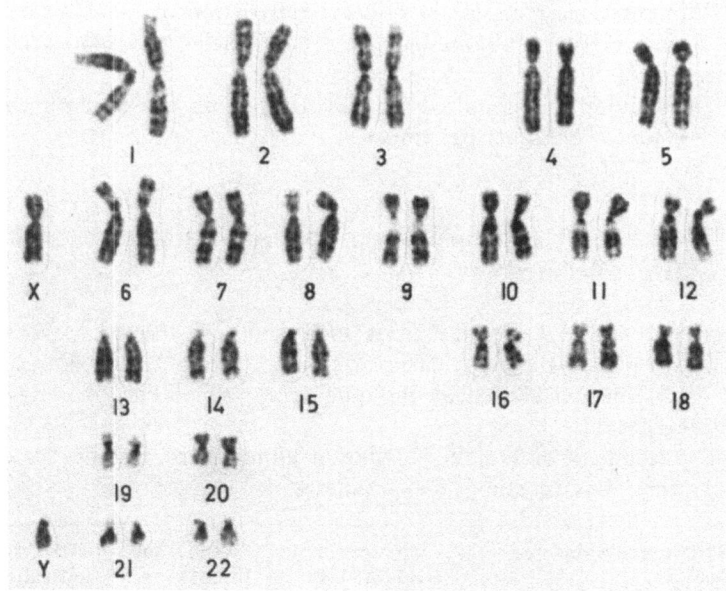

Fig. 3. Human male metaphase cell, stained by the method of Schnedl (1971), karyotype. Banding patterns characteristic of each chromosome pair are visible

phosphate buffer (pH 6.8) at 59° C for 16 hours or longer. After staining with Giemsa solution, the chromosomes show a very clear and detailed banding pattern (Fig. 3).

In addition, regions differentiated by the Arrighi and Hsu method (1970) are strongly stained, except for the secondary constriction of chromosome No. 9, which stains rather weakly.

Particularly detailed banding patterns can be produced in the elongated chromosomes of prometaphases (Fig. 4). These numerous fine bands fuse in the course of mitosis and in this way the bands of the contracted metaphase chromosomes are formed (Fig. 3).

As in most Giemsa banding methods, preparations made under different conditions may require modified techniques. Two steps of the present method may be altered to adapt the method to particular conditions: 1. the pretreatment of the preparations, e.g. use of agar-coated or uncoated slides, and 2. the concentration of the NaOH solution and the time of treatment.

If chromosomes are washed off by the NaOH treatment or if their structure is grossly altered, agar coating of the slides will be helpful. On the other hand, if the concentration of the NaOH is too high and/or the time of treatment too long, no banding patterns will be found in the

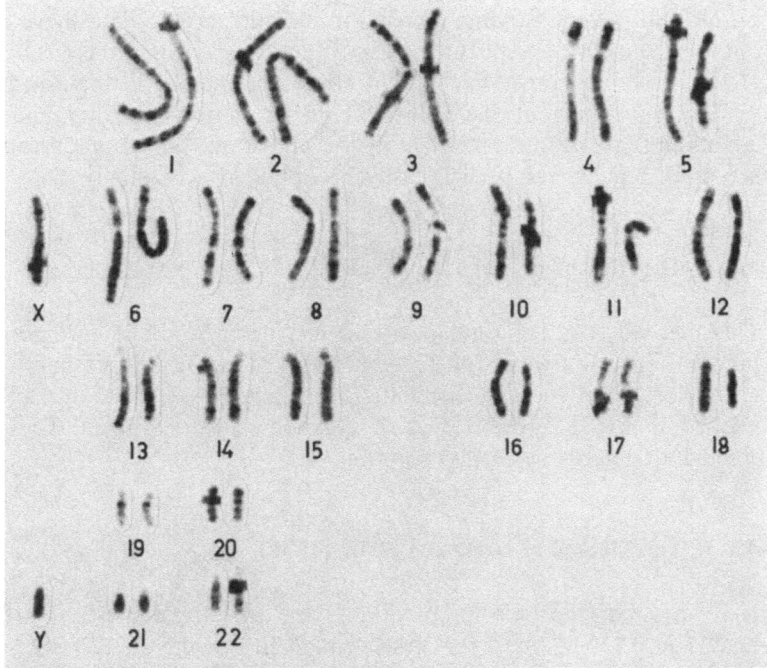

Fig. 4. Human male prometaphase cell, stained by the method of Schnedl (1971), karyotype. Very complex banding patterns can be observed in the elongated prophase chromosomes

chromosomes and the result will be similar to that of the Arrighi and Hsu method.

In our laboratory agar coated preparations treated for 60–90 sec in 0.07 N NaOH usually give the best results. Preparations from other laboratories may require agar coating and treatment in 0.007 N or even 0.002 N NaOH.

Chromosome preparations obtained from fibroblast cultures are more resistant to NaOH treatment than preparations made of blood cultures or direct bone marrow preparations. The method of preparing slides will also influence results. Too high concentrations of colchicine should not be used. Also the hypotonic medium has some effect: in our laboratory we use Gey's solution diluted with 3 parts of distilled water, others use 0.075 M KCl, which renders preparations less resistant to the NaOH treatment.

The method can be modified by prolonging the time of NaOH treatment and/or increasing the concentration of the NaOH solution, which will give the same results as the Arrighi and Hsu method (1970).

No satisfactory banding patterns in meiotic chromosomes have been obtained thus far, except for the centromeric regions and the distal part of the Y chromosome which could always be readily distinguished.

The advantage of the method consists in its versatility and the adaptability to various materials. If the above mentioned modifications are used, reliable results will easily be obtained accordingly.

4.1.2. The Method of Gagné et al. (1971)

This method is rather similar to that just described. The chromosome preparations are treated for 20 sec in 0.014 N NaOH, rinsed in ethanol and dried. Then they are incubated for 18 hours at 66° C in 6 × SSC, washed in 6 × SSC, transferred to ethanol, and dried again. They are stained afterwards in Giemsa solution.

4.1.3. The Method of Drets and Shaw (1971)

Here the chromosome preparations are first treated with 0.07 N NaOH in 0.112 M NaCl for 30 sec, rinsed thoroughly in 12 × SSC, and incubated with 12 × SSC at 65° C for 60–72 hours. The slides are passed through 70% and 95% ethanol, dried and stained with a buffered Giemsa solution (pH 6.6).

Pretreatment of slides with 0.02 N HCl for 15 min at room temperature prior to exposure to NaOH may result in more distinct bands.

The NaOH treatment, as in the related techniques, is the critical point of the procedure. The time of treatment may be shortened or the concentration of the NaOH be lowered to 0.014 or 0.007 N if necessary.

Since the incubation in 12 × SSC as originally described is very time consuming (60–72 hours) the authors also tried shorter incubation periods and found that 18–24 hours will suffice in most cases.

4.2. Methods Using Various Incubation Procedures without NaOH

4.2.1. The Method of Sumner et al. (1971)

(Termed "ASG"-Technique, Derived from "Acetic/Saline/Giemsa")
In the original description, this is a very short and simple method.

The chromosome preparations are simply incubated in 2xSSC at 60° C for various intervals (45 min to 16 hours) and afterwards stained with a buffered Giemsa solution.

The chromosomes progressively loose their ability to take up stain under the 2xSSC treatment, certain regions prior to others. The banding pattern produced in this way is in general the same as that obtained by other methods. The critical factor in this method is the incubation time which should be tested for every new series of chromosome preparations.

The chromosome preparations to be stained by this method should be made in a room with a relative humidity of 40–50%.

No banding will be obtained if the slides are prepared at high humidity.

If the chromosomes stain too intensely, simple reduction of the staining time may be helpful in some cases.

The most effective modification of the technique consists of treatment with 0.01 N or 0.007 N NaOH for 10–30 sec prior to the incubation in 2xSSC, which makes the method similar to those described above.

4.2.2. The Method of Patil et al. (1971) "Giemsa-9-Technique"

In this method the alkali treatment and the Giemsa staining is combined to form one single step by changing the pH of the usual Giemsa solutions from neutral to pH 9.0. This results in differential staining of chromosomes. Different staining times should be tried (5–15 min). This method is very reliable for routine use, but other methods are preferable for more advanced studies.

Finally, several other methods involving various incubation procedures have been described (for instance Utakoji, 1972; Shiraishi and Yosida, 1972; Kato and Yosida, 1972; Kato and Moriwaki, 1972).

4.3. Methods Using Enzymatic Digestion (Pronase, Trypsin)

These methods have the advantage that preparations can be obtained within a few min and that good results can be obtained from slides which were stored for 1 year or longer. However, since the activities of the enzymes used are rather variable, it is difficult to specify standard procedures.

Chromosomes treated with enzymes (pronase, trypsin) tend to be swollen and distorted, but the bands stand out very clearly in successful preparations: the bands are sharply delineated because they appear to be compressed by the less stained, swollen areas (Fig. 5).

4.3.1. The Method of Dutrillaux et al. (1971)

Air dried chromosome preparations, made by a special procedure, are treated with 0.005% pronase solution for 3–6 min at 37° C, rinsed in distilled water and stained with Giemsa solution.

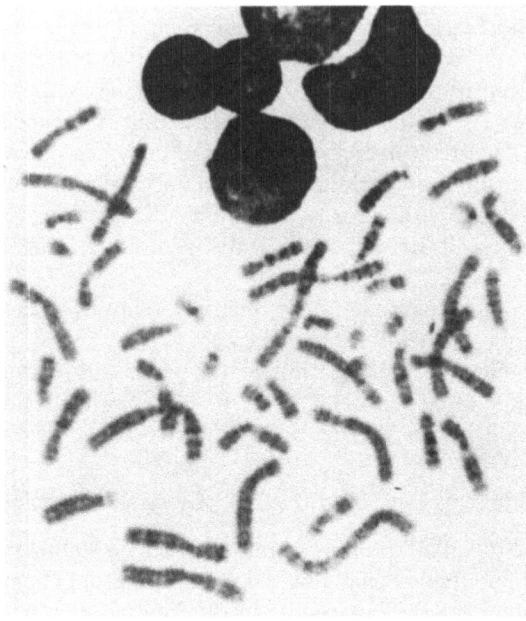

Fig. 5. Human male metaphase cell, stained after trypsin treatment (Dutrillaux
et al., 1971; Seabright, 1971)

Different incubation times should be tried for every new batch of
preparations.

4.3.2. The Method of Seabright (1971)

Chromosome preparations made by conventional air drying, are treated
with 0.25% trypsin solution (Difco) for 10–60 sec and stained with
buffered Leishman stain, which is quite similar to the usual May-
Grünwald or Wright stain.

Trypsin preparations obtained from other sources than Difco give
the same results if used in appropriate concentration.

It is important to work at a constant pH such as pH 6.8; the pH
optimum of trypsin lies at a higher pH value, but at pH 8.0 or even
pH 9.0 some unspecific effects on the chromosomes will be observed.
Different incubation temperatures can be used (between 20 and 40° C).
However, a given temperature should be used throughout one batch to
obtain standard conditions. The results are better reproducible at lower
temperatures.

Each new batch of material for preparation should be tried at several different incubation times (10–60 sec). It is preferable to use a less active trypsin solution which requires a longer incubation time, rather than more active trypsin solutions with short incubation times which give less consistent results.

4.3.3. The Method of Wang and Fedoroff (1972)

This method uses a less concentrated trypsin solution than in the technique of Seabright, though the above modifications of the latter, may be applied to this method.

4.3.4. The Method by Sun, Chu and Chang (1973)

A method involving an alcoholic Leishman stain solution containing trypsin as a combined enzyme-stain solution has been described by Sperling and Wiesner (1972). In our hands a similar method resting on the same principle but using a special pretreatment turned out to be highly reliable not only in man but also in all species tried so far (Sun, Chu and Chang, 1973).

4.4. The Reverse Bands

4.4.1. Reverse Bands Using Giemsa (Dutrillaux and Lejeune, 1971)

These bands were the first type of multiple Giemsa bands made visible in human chromosomes (communicated on April 26th, 1971). This kind of bands was later termed "reverse bands" because they are precisely localized at chromosomal regions where other Giemsa banding techniques produced light areas. At the Paris Conference (1971), they were designated accordingly as "R-bands".

This method simply consists of incubating chromosome preparations in heated phosphate buffer (pH 6.5 at 87° C) for 10 min and staining with Giemsa. However, the "reverse bands" are less reproducible than the other kinds of Giemsa bands.

4.4.2. The Acridine Orange Reverse Bands (for instance Bobrow et al., 1972b)

The acridine reverse bands, observed by fluorescence microscopy, show exactly the same picture as the R bands, however, the bands appear greenish-yellow on red background. The pretreatment is very similar to that one used for obtaining Giemsa R bands. Older preparations require longer treatment times. Afterwards the slides are stained with acridine orange instead of Giemsa.

Appendix

Laboratory Procedures

I. Methods for Demonstrating Centromeric Heterochromatin

1. The Method of Arrighi and Hsu (1970)

Preparations. Flame dried or air dried chromosome preparations; squash preparations; slides for squash preparations should be coated with a solution containing 0.1 % gelatine and 0.01 % chrome alum (chromium potassium sulfate). The fixed cells are squashed on the dried slides without stain; the coverslip is removed by the dry-ice method and the preparation is rinsed twice with 95 % ethanol, and dried.

Procedure. 1. Treat the slides with 0.2 N HCl at 20° C for 30 min.
  2. Wash thouroughly in distilled water and air dry.
  3. Treat with pancreatic RNase (100 µg/ml in 2 × SSC) at 37° C in a moist chamber for 60 min. The solution should be heated in a boiling water bath for 5 to 10 min before use to inactivate DNase contained in the commercially available RNase.
  4. Rinse 3 times in 2 × SSC, 70 % ethanol and afterwards in 95 % ethanol and dry.
  5. Treat the slides with 0.07 N NaOH (2 min). This is the most critical point in the procedure. Variations of the NaOH treatment (see below) may be necessary to obtain good results.
  6. Rinse 3 times in 70 % ethanol, transfer to 95 % ethanol and dry. It is necessary to very quickly remove all residual NaOH.
  7. Incubate in 2 × SSC (or 6 × SSC) at 65° C for a minimum of 12 hours (better 24 hours).
  8. Rinse in 70 % and 95 % ethanol.
  9. Stain for 15–30 min in one of the Giemsa solutions described below.

The *Giemsa solutions* used by us are composed as follows:
Solution a)   50   ml distilled water
                   1.5 ml 0.1 M citric acid
               adjust pH with 0.2 M Na$_2$HPO$_4$ to 6.8–7.2
                   1.5 ml absolute methanol
                   5   ml stock Giemsa solution (W. H. Curtin).

Solution b) 100   ml phosphate buffer (pH 7.0; 0.01 M)
          10   ml stock Giemsa solution (W. H. Curtin).

10. Rinse quickly in distilled water and dry; mount in Permount.

Steps Nos. 1–4, aimed at removing histones and RNA, can be omitted. However, if cells have not been subjected to hypotonic treatment, it is necessary to remove the RNA, which might interfere with the procedure. In normal chromosome preparations involving hypotonic treatment very little RNA is left.

If the chromosomes are swollen and distorted after Giemsa staining, the NaOH treatment was too severe. In this case, the NaOH concentration or the duration of treatment or both can be reduced (for instance 0.014 or 0.007 N NaOH may give better results; duration of treatment is between 30 and 90 sec).

Step No. 8 (rinsing of preparations after the SSC-incubation) can be omitted, but the slides should be cooled before placing into Giemsa stain to avoid precipitation.

2. *The Method of Yunis et al. (1971)*

   Procedure. 1. Place the slides in 0.06 M phosphate buffer, for instance, Sörensen's buffer: 0.06 M KH_2PO_4/Na_2HPO_4) at pH 6.8 at 85–100° C. According to our experience, 90° C will give good results. Incubate for 5–10 min (5 min will be enough in most cases).

   2. Transfer to 0.06 M phosphate buffer (pH 6.8) at 65° C and incubate for more than 10 hours.

   3. After cooling the slides, stain with one of the Giemsa solutions.

3. a) *The Method of Bobrow et al. (1972)*

   This procedure and a similar one by Gagné and Laberge (1972) specifically stain the secondary constriction of chromosome No. 9.

   Procedure. The chromosome preparations (3–4 days old) are stained for about 10–20 min in a freshly prepared Giemsa solution consisting of 2 ml Giemsa solution (G. Gurr) in 98 ml distilled water;

   adjust this mixture to pH 11 with NaOH.

   After staining, the slides are rinsed in distilled water, dried, soaked in xylene and mounted. Less precipitate is formed on the slide if it is stained at 37° C.

   b) *The Method of Gagné and Laberge (1972)*

   Preparations. The authors recommend fixing the cells in methanol/glacial acetic acid (3:1) and suspension in 45% acetic acid for 3–4 min before drying with warm air.

   Procedure. The slides are stained for 5 min with a 2% Giemsa solution (Harleco, Azure Blend type) in 0.1% $Na_2HPO_4 \cdot 12H_2O$

adjusted to pH 11.6 with NaOH. They are washed in a jet of tap water for 1–2 min, transferred to distilled water, dried, soaked in xylene, and mounted.

In our laboratory, longer staining times were necessary to give specific staining of chromosome No. 9 by either method (up to 2 hours).

II. Methods for Demonstrating Multiple Banding Patterns

A. Methods Using NaOH Treatment Followed by an Incubation in Buffer

1. The Method of Schnedl (1971)

Preparation (tested thus far). Conventional *air dried chromosome preparations* from blood or fibroblast culture cells. Hypotonic treatment: (10 min) Gey's solution diluted with 3 parts of distilled water. Fixative: a freshly prepared mixture of 1 part glacial acetic acid and 3 parts methanol.

Bone marrow preparations: as chromosomes from this material tend to become too contracted for obtaining good banding patterns, don't colchicinize for too long.

Male meiotic preparations: Hypotonic treatment (15 min): 0.9% Na citrate. Fixative: same as for somatic metaphase preparations.

Procedure. 1. Coat slides with 0.5% agar (necessary only if cells are washed away by the NaOH treatment). Dip the slides into the agar solution (liquified by boiling, then cooled to 60° C before use) and allow to dry completely.

2. Treatment with 0.07 N NaOH at 20° C: The slides (agar coated or without agar layer) are immersed in the NaOH solution for 90 sec, and directly transferred to 70% ethanol; after 2 further changes of 70% ethanol, the slides are transferred to 96% and then absolute ethanol and dried.

3. Incubation in phosphate buffer ($M/15, KH_2PO_4/Na_2HPO_4$) pH 6.8 at 59° C: The slides are immersed in the prewarmed buffer solution. They should be incubated at least for 15, or better 24 hours.

4. Staining in buffered Giemsa solution: (1 part Giemsa solution (Merck) and 9 parts phosphate buffer (1.14 g $Na_2HPO_4 \cdot 2H_2O$ and 0.49 g KH_2PO_4 in 1 l distilled water). The wet preparations are cooled or rinsed with distilled water, transferred to this solution and allowed to stain for approximately 20 min; prolonged staining may sometimes be necessary. The stained slides are washed briefly in water (in order to remove precipitates of the stain), immersed in distilled water and dried thoroughly. Then the slides

are dipped into xylene and mounted (for instance, in DePeX, Gurr).

This Giemsa staining procedure can also be used for the other methods described in this chapter and for conventional chromosome staining. The critical stage in this method is again the treatment with NaOH. In several cases (see text), the 0.07 N NaOH turned out to be too strong, 0.007 N NaOH or even 0.002 N NaOH might have given the optimal result.

For every new batch of chromosome preparations, we try 6 modifications of the technique:

Uncoated and agar coated slides, respectively

0.07 N NaOH   90 sec
0.007 N NaOH   90 sec
0.002 N NaOH   90 sec.

Sometimes time of treatment also has to be reduced (to 30–60 sec). The agar coat does not interfere with the Giemsa stain, and in most instances it is removed during incubation in Sörensen's buffer. If desired, it can be removed with a jet of water after the Giemsa staining.

2. The Method of Gagné et al. (1971)

Preparations. Conventional air-dried chromosome preparations.

Procedure. 1. Treatment with 0.014 N NaOH: immerse the slides the NaOH-solution for 15–20 sec, transfer quickly to 70% ethanol; after 2 further changes of 70% ethanol, they are placed into 95% ethanol and dried.

2. Incubation in $6 \times$ SSC at 66° C: transfer the slides to the warmed $6 \times$ SSC solution. After 18 hours, wash twice in $6 \times$ SSC, transfer to 70%, then 95% ethanol, and dry.

3. Staining in buffered Giemsa solution: the authors recommend the following mixture [nearly the same as solution A used by Arrighi and Hsu (1970)]:

50 ml   distilled water
1.5 ml citric acid at pH 6.9
1.5 ml methanol
5 ml   stock Giemsa solution (Harleco)

stain in this solution for 30 min, rinse in distilled water and dry, differentiate under microscopic control in 25% ethanol, transfer to acetone, then to toluene, and mount.

3. The Method of Drets and Shaw (1971)

Preparations. Conventional flame dried chromosome preparations.
Hypotonic solution: 1% sodium citrate.

Procedure. 1. Treatment with 0.07 N NaOH in 0.112 M NaCl (pH 12.0): the slides are immersed into this solution for 30 sec and rinsed 3 times (10 min each) in 12 × SSC (pH 7.0).

2. Incubation in 12 × SSC at 65° C: the wet slides are covered with a coverslip to maintain a layer of 12 × SSC between slide and coverslip, and transferred to a chamber (Petri dish) moistened with 12 × SSC. The moist chamber containing the slide is incubated at 65° C for 60–72 hours. The coverslip is then removed carefully without scratching the cell layer and the slide is rinsed 3 times in 70% ethanol for a total of 10 min. After rinsing in 95% ethanol, the slides are air dried.

3. Staining in buffered Giemsa solution: stain for 5 min in the same Giemsa solution as used for the Arrighi and Hsu procedure; rinse in distilled water, air dry and mount.

Exposing the slides to 0.02 N HCl for 15 min before NaOH treatment will sometimes enhance the banding pattern. After HCl treatment, the slides have to be washed thoroughly with ethanol and dried prior to the NaOH treatment.

If chromosomes turn out to be swollen and distorted, the time of NaOH treatment must be reduced.

The incubation in 12 × SSC at 65° C can also be performed in a coplin jar (using a moist chamber is rather troublesome). The incubation time can be reduced to 18–24 hours without great loss in quality of the banding pattern obtained.

B. Methods Using Various Incubation Procedures without NaOH

1. *The Method of Sumner et al. (1971), "ASG-Technique"*

Preparations. Conventional air dried preparations from blood cultures or mouse bone marrow cells (without colchicinizing the animal), hypotonic solution 0.075 M KCl,
the relative humidity during preparations may influence the banding patterns. Preparations made at a high relative humidity will show no banding patterns. Best results are obtained at 40–50% R.H.

Procedure. 1. Incubation in 2 × SSC (1 hour at 60° C), and afterwards rinse in demineralized water.

2. Staining with a buffered Giemsa solution.
Stain for 1–1.5 hours in the following mixture:
50 ml buffer (pH 6.8) made with Gurr's buffer tablets
1 ml Giemsa stock solution (Gurr's Giemsa R 60).
Afterwards rinse briefly in demineralized water, dry, soak in xylene and mount.

If the staining is too strong and good banding is not obtained, some modifications can be tried.

The slides may be treated with 0.01 M NaOH for 5–30 sec, rinsed in distilled water and then incubated in the usual way in 2 × SSC. In other cases, simply a prolonged incubation with 2 × SSC may improve the banding pattern. In most cases a combination of both modifications (NaOH-treatment and prolonged incubation in 2 × SSC) will have the best effect.

For every new batch of preparations several different incubation times should be tried (30, 60, and 90 min; 2, 6, and 18 hours).

Sometimes the banding patterns can be differentiated better after a shorter period of staining in Giemsa.

2. The Method of Patil et al. (1971); "Giemsa-9 Technique"

Preparations. Hypotonic solution 0.075 M KCl. Cells are spread on cooled wet slides which are dried by blowing and heating on a warmed plate at exactly 60° C for 90–120 sec (to avoid inconsistent effects of flame drying).

Procedure. The chromosome preparations are stained in the following mixture:

96 ml distilled water
  2 ml 0.14 M Na_2HPO_4
  2 ml Harleco Giemsa blood stain (original Azure blend type; Gurr's Giemsa R 66 can also be used).

The mixture is adjusted to pH 9.0 if necessary. The slides are stained for 5 min; since the time of staining is critical, different times should be tried (1–30 min).

The stained preparations are briefly rinsed in distilled water, dried, soaked with xylene and mounted.

3. The Method of Utakoji (1972)

Procedure. The slides are incubated in the following solution at 0° C (ice bath):

10 mM Potassium permanganate in 33 mM phosphate buffer at pH 7.0 containing 5 mM magnesium sulphate.

The author recommends an incubation time of 20–40 min, though longer times of treatment will sometimes be necessary.

Afterwards the slides are washed in running tap water for about 30 sec, soaked in absolute ethanol for 5 min, and dried.

Giemsa staining for about 5 min in a mixture of

  1 ml Giemsa stock solution
99 ml phosphate buffer at pH 7.0.

After staining, wash in distilled water, dry, soak in xylene, and mount.

C. Methods Using Enzymatic Digestion

1. The Method of Dutrillaux et al. (1971)

Preparations. Hypotonic treatment (15 min) in the following mixture:
15 parts serum, diluted with the 5-fold volume of distilled water
1 part 3.39 % $MgCl_2$, Hyaluronidase 2.5 U/ml.
Fixation: First fixation (35 min) in a mixture of:
3 parts chloroform,
1 part glacial acetic acid,
6 parts 100 % ethanol.
Second fixation (15 min) in a mixture of:
1 part glacial acetic acid,
3 parts 100 % ethanol.
Slides are prepared according to the conventional air drying procedure
(on cooled wet slides).
Procedure. Pronase treatment: 3–6 min at 37 °C in a solution of 5 mg
pronase in 100 ml distilled water.
Giemsa staining: for instance with the Giemsa solution used by
Schnedl.

2. The Method of Seabright (1971)

Preparations. Air dried chromosome preparations (flame drying is less
suitable), hypotonic treatment with 0.075 M KCl 3–4 min at 37° C.
Procedure. Trypsin treatment: for 10–60 sec in 0.25 % Trypsin (Difco)
in an isotonic salt solution (or simply in distilled water at room
temperature or higher temperatures up to 37° C).
Staining: 3–4 min in Leishman stain (BDH) diluted with buffer pH
6.8 (G. T. Gurr) 1:4; subsequently wash rapidly in distilled water,
air dry; soak in xylene, mount.

3. The Method of Wang and Fedoroff (1972)

Preparations. Flame drying technique.
Procedure. Trypsin treatment: the slides are immersed into 0.025 to
0.05 % Trypsin (in Ca- and Mg-free balanced salt solution) at 25–30° C
for 10–15 min. Alternatively, this solution can be diluted with 0.02 %
EDTA 1:1. The pH of the enzymatic solutions is adjusted to pH 7.0.
The slides are afterwards rinsed in several changes of 70 %, 96 %,
absolute ethanol, and dried.
Giemsa staining: stain 1–2 min in the mixture described in the Arrighi
and Hsu method. Wang and Fedoroff used the Giemsa stock solution
by Fisher Scientific Co.
The slides are rinsed twice in distilled water, dried, soaked with xylene,
and mounted.

4. *The Method of Sun, Chu and Chang (1973)*
Preparations. Air drying technique.
Procedure. 1. Dry slides at 60° C for 16–24 hours (or even longer).
2. Incubate in 0.025 M phosphate buffer at pH 6.8 and 56° C for 10 min.
3. Stain with a freshly prepared buffered trypsin-Giemsa solution for 15–30 min:
73 ml 0.025 phosphate buffer, pH 6.8
27 ml methanol
50 mg Giemsa powder (E. Gurr)
0.3–0.4 ml 0.25% trypsin (Difco)
4. Wash thoroughly with tap water, rinse in distilled water, dry, soak in xylene and mount in DPX (E. Gurr).
0.025 M phosphate buffer (pH 6.8): 3.4 g KH_2PO_4 in 1 liter distilled water, adjust to pH 6.8 with concentrated NaOH.

III. The Reverse Bands Methods

1. *R-Bands Using Giemsa Stain (Dutrillaux and Lejeune, 1971).*

Preparations. Chromosome preparations are produced as for the enzymatic method of Dutrillaux *et al.* (1971) (see p. 114).
Procedure. Incubation in phosphate buffer at pH 6.5 (20 mM) at 87° C; the slides are incubated for 10–12 min, and rinsed with tap water for a few sec.
Giemsa staining: stain for about 10 min in the following mixture:
92 parts distilled water,
4 parts phosphate buffer at pH 6.7,
4 parts Giemsa stock solution (for instance Merck)
rinse with distilled water, dry, soak with xylene, and mount.
2. *Acridine Orange Reverse Bands (for instance Bobrow et al., 1972b).*
Preparations. Air drying technique.
Procedure. 1. Incubate in Sörensen's phosphate buffer (pH 6.5) at 85° C for 5–30 min (old slides require longer treatment times).
2. Stain with 0.01% acridine orange in Sörensen's buffer (pH 6.5) for 5 min.
3. Differentiate in Sörensen's buffer (pH 6.5) for 2 min, mount in the same buffer, analyze with a fluorescence microscope equipped as for quinacrine fluorescence studies. The exposure times usually are a quart to half of the time necessary for quinacrine fluorescence.

IV. Solutions

10 × SSC: 1.5 M NaCl (87.7 g/l)
0.15 M trisodium citrate (44.1 g/l)

adjust to pH 7.0 with 1 N NaOH, dilute to any strength with distilled water.

0.067 M phosphate buffer (Sörensen's buffer):
Solution A: 9.078 KH_2PO_4 g/l
Solution B: 11.876 $Na_2HPO_4 \cdot 2 H_2O$ g/l.

pH	ml solution B (+ solution A to make 100 ml)	pH	ml solution B (+ solution A to make 100 ml)
6.5	31.3	7.1	67.0
6.6	37.2	7.2	72.6
6.7	43.0	7.3	77.7
6.8	49.2	7.4	81.8
6.9	55.2	7.5	85.2
7.0	61.2		

References

Arrighi, F. E., Hsu, T. C.: Localization of heterochromatin in human chromosomes. Cytogenetics **10**, 81–86 (1970).

Bobrow, M., Madan, K., Pearson, P. L.: Staining of some specific regions of human chromosomes, particularly the secondary constriction of No. 9. Nature New Biol. (Lond.) **238**, 122–124 (1972a).

Bobrow, M., Collacott, H. E. A. C., Madan, K.: Chromosome banding with acridine orange. Lancet **1972**b **II**, 1311.

Caspersson, T., Zech, L., Johansson, C., Modest, E. J.: Identification of human chromosomes by DNA-binding fluorescent agents. Chromosoma (Berl.) **30**, 215–227 (1970).

Chandley, A.: Personal communication.

Drets, M. E., Shaw, M. W.: Specific banding patterns of human chromosomes. Proc. nat. Acad. Sci. (Wash.) **68**, 2073–2077 (1971).

Dutrillaux, B., de Grouchy, J., Finaz, C., Lejeune, J.: Mise en evidence de la structure fine des chromosomes humains per digestion enzymatique (pronase en particulier). C.R. Acad. Sci. (Paris), Série D **273**, 587–588 (1971).

Dutrillaux, B., Lejeune, J.: Sur une nouvelle technique d'analyse du caryotype humain. C. R. Acad. Sci. (Paris), Série D **272**, 2638–2640 (1971).

Gagné, R., Laberge, C.: Specific cytological recognition of the heterochromatic segment of number 9 chromosome in man. Exp. Cell Res. **73**, 239–242 (1972).

Gagné, R., Tangnay, R., Laberge, C.: Differential staining patterns of heterochromatin in man. Nature New Biol. (Lond.) **232**, 29 (1971).

Jones, K. W.: Chromosomal and nuclear location of mouse satellite DNA in individual cells. Nature (Lond.) **225**, 912–915 (1970).

Kato, H., Moriwaki, K.: Factors involved in the production of banded structures in mammalian chromosomes. Chromosoma (Berl.) **38**, 105–120 (1972).

Kato, H., Yosida, T. H.: Banding patterns of Chinese hamster chromosomes revealed by new techniques. Chromosoma (Berl.) **36**, 272–280 (1972).

Pardue, M. L., Gall, J. G.: Chromosomal localization of mouse satellite DNA. Science **168**, 1356–1358 (1970).

Patil, S. R., Merrick, S., Lubs, H. A.: Identification of each human chromosome with a modified Giemsa stain. Science **173**, 821–822 (1971).

Schnedl, W.: Banding pattern of human chromosomes. Nature New Biol. (Lond.) **233**, 93–94 (1971).

Schnedl, W.: Analysis of the human karyotype using a reassociation technique. Chromosoma (Berl.) **34**, 448–454 (1971).

Schnedl, W.: Giemsa banding, Quinacrine fluorescence and DNA replication in chromosomes of cattle (Bos taurus). Chromosoma (Berl.) **38**, 319–328 (1972).

Schnedl, W.: Observations on the mechanisms of Giemsa staining methods, p. 342–345. In: Chromosome identification, T. Caspersson and L. Zech, eds. 23. Nobel Symposium New York and London: Academic Press 1973.

Seabright, M.: A rapid banding technique for human chromosomes. Lancet **1971 II**, 971–972.

Shiraishi, Y., Yosida, T. H.: Banding pattern analysis of human chromosomes by use of a urea treatment technique. Chromosoma (Berl.) **37**, 75–83 (1972).

Sperling, K., Wiesner, R.: A rapid banding technique for routine use in human and comparative cytogenetics. Humangenetik **15**, 349–353 (1972).

Sumner, A. T., Evans, H. J., Buckland, R. A.: New technique for distinguishing between human chromosomes. Nature New Biol. (Lond.) **232**, 31–32 (1971).

Sun, N. C., Chu, E. H. Y., Chang, C. C.: Staining method for the banding patterns of human mitotic chromosomes. Mammalian Chromosome Newsletter **14**, 26–28 (1973).

Utakoji, T.: Differential staining patterns of human chromosomes treated with potassium permanganate. Nature (Lond.) **239**, 168–170 (1972).

Wang, H. C., Fedoroff, S.: Banding in human chromosomes treated with trypsin. Nature New Biol. (Lond.) **235**, 52–53 (1972).

Yosida, T. H., Sagai, T.: Banding pattern analysis of polymorphic karyotypes in the black rat by a new differential staining technique. Chromosoma (Berl.) **37**, 387–394 (1972).

Yunis, J. J., Roldan, L., Yasmineh, W. G., Lee, J. C.: Staining of satellite DNA in metaphase chromosomes. Nature (Lond.) **231**, 532–533 (1971).

CHAPTER VII

Autoradiography of Human Chromosomes with ³H-Thymidine

WOLFGANG GEY

With 4 Figures

1. Introduction

The autoradiographic method introduced by Bélanger and Leblond (1946) and Pelc (1947) permits the localization of chemical compounds in tissues, cells, or the individual components of cells. These chemical compounds are revealed with the aid of radioactive isotopes which have previously been introduced in appropriate precursors. Suitable photographic emulsions spread out over the labeled material then register the radiation emitted, which is subsequently made visible in a microscopic preparation after appropriate photographic development.

Deoxyribonucleic acid (DNA) is usually employed for autoradiography of chromosomes, because it is the fundamental constituent of chromosomes and is known to be metabolically stable in keeping with its function as the carrier of genetic information. A specific precursor of DNA is the nucleoside thymidine, which is incorporated into chromosomal DNA by a direct metabolic pathway thus making radioactive labeling possible. Thymidine is normally labeled by a radioactive isotope in its pyrimidine ring. Tritium (³H) is the best label for autoradiographic studies of chromosomes, because it emits electrons of low energy which permits high resolution since a single disintegration causes a visible track, in contrast to electrons of higher energy. Taylor *et al.* first produced tritiated thymidine in 1957 and incorporated it into the DNA of plant chromosomes. Numerous similar experiments on bacteria, plants, animals, and man have since been carried out. These have made important contributions to the understanding of the time sequence of DNA synthesis within single chromosomes, and moreover demonstrated how the newly synthesized DNA is distributed to the daughter cells during cell division. This method was also responsible for considerable progress in the identification of chromosomes (for recent reviews see Giannelli, 1970 and Miller, 1970).

2. Theoretical Basis

2.1. The DNA Synthesis Cycle

When cells divide mitotically, the DNA of the chromosomes is transmitted unchanged from cell to cell after identical reduplication. Our present knowledge of DNA synthesis in chromosomes is largely based on experiments with cells with a very short mitotic cycle.

During the interval between two mitoses, the interphase, the chromosomes are not contracted and therefore not visible as discrete structures. The cell nuclei then permit the identification of the chromosomes merely in the form of chromatin, sometimes with individual condensations (called chromocenters).

If radioactive precursors of DNA (e.g. ³H-thymidine) are introduced to a tissue or cell culture, then all cells in which DNA synthesis takes place will take up these labeled precursors and incorporate them into the chromosomes. Autoradiographs of the cells produced immediately or shortly after introduction of ³H-thymidine show radioactivity in the interphase nucleus of cells which are carrying out DNA synthesis. If a longer period of time is permitted between introduction of ³H-thymidine, and preparation and autoradiography, radioactive labeling can also be found in cells which are just in division or have already divided. Cells taken at intervals from cell cultures exposed to radioactive DNA precursors can, therefore, be used to determine the time required for labeled mitoses to appear first, the proportion of labeled and unlabeled mitoses, and the degree of uptake in interphase nuclei. With experiments of this type, Howard and Pelc (1953) first demonstrated in *Vicia faba* that the DNA synthesis of chromosomes takes place during a precisely defined period in the interphase (synthesis period = S-period). Prior to the onset of mitosis and after its conclusion, there are two periods (gaps), G_2 and G_1, in which no DNA is synthesized. Continuously growing cells pass through a cycle from M (mitosis) through G_1, S, G_2 to the next mitosis (Fig. 1).

Investigations on mammalian cells in tissue culture showed that the duration of the S-period is specific for a particular species, as it is for a particular cell type. According to Bender and Prescott (1962), human lymphocytes in short term cultures from peripheral blood require for

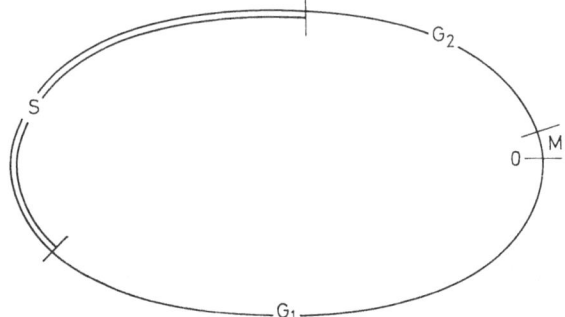

Fig. 1. Relationship of the DNA synthesis cycle to cell division cycle. S = DNA synthesis period, M = mitosis, G_1 and G_2 = periods in which no DNA is synthesized

at least 12 hours DNA synthesis before the first division. The duration of the S-period for fibroblasts from human biopsies was ascertained by Moorhead and Defendi (1963) to be $7-7^1/_2$ hours, by Schwarzacher and Schnedl (1965), 9 hours. The G_2 period varies between 2 and 5 hours in somatic cells of the numerous mammals investigated up to now (e.g. Taylor, 1960; German, 1962; Bender and Prescott, 1962; Grumbach et al., 1963; Galton and Holt, 1964; Schwarzacher and Schnedl, 1965; Bianchi and Bianchi, 1965), while in most cases the mitosis only required less than 1 hour (Odartchenko et al., 1964). The differences in the duration of the G_1 period are particularly great. Values of from 4 up to 20 hours were observed in vitro for cells in the logarithmic growth phase (Schwarzacher and Schnedl, 1965). In cells with a very long growth cycle (weeks to months), as is the case with many cells in vivo, this is due to a long G_1 period, while S, G_2 and M still lie within the above limits. Temperature, radiation, and the particular composition of the nutrient medium can influence the duration of the cell cycle (survey by Lark, 1963).

The final stage of the S-period (approximately the last 2 hours) in human cells as well as in cells of other species exhibits several peculiarities. During this phase, larger chromosome segments have completed the process of replication, while in other chromosome segments replication continues. The autoradiographic pictures of chromosomes previously labeled with tritiated thymidine reveal characteristic patterns resulting from more or less elongated adjacent labeled and unlabeled chromosome segments. These patterns may be used for further characterization of non-homologous morphologically similar chromosomes (Schmid, 1963; see Chapter VIII). It has been demonstrated that homologous chromosomes generally pass through the same labeling pattern in the same sequence. Biological and technical factors certainly influence the consistency in appearence of autoradiographic pictures (Gey, 1966; Back et al,. 1967; Büchner et al., 1968).

2.2. Incorporation of Thymidine into DNA

From the experiments of Lehman et al. (1958) and Bessman et al. (1958) on cell-free extracts of exponentially grown cultures of E. coli, it is known that ^{14}C-thymidine is incorporated into DNA only in the presence of ATP and Mg^{++}. Numerous additional experiments confirm this metabolic requirement, which also occurs in cultured cells of higher organisms (Feinendegen, 1966).

The following considerations should be borne in mind:

1. Not all cells are capable of utilizing thymidine (Adelstein et al., 1964).

2. Some cells are capable of demethylating thymidine so that tritium in the methyl group is lost during incorporation (Fink et al., 1962).

3. The quantity of stored acid-soluble thymidine per cell (thymidine pool) varies in different cell types (Potter et al., 1963).

4. The thymidine pool is renewed only during the DNA synthesis period (Feinendegen and Bond, 1962; Stone et al., 1965).

The incorporation of thymidine into DNA takes place in two phases. The thymidine (TdR) is first phosphorylated by specific kinases, and thymidine monophosphate (TMP), thymidine diphosphate (TDP), and finally thymidine triphosphate (TTP) are produced sequentially. In the second step, DNA polymerase causes polymerization of the TTP together with the three other deoxyribonucleoside triphosphates to DNA with the concomitant elimination of pyrophosphate. The phosphorylation of thymidine is the basis of feed-back controls which inhibit thymidine kinase, particularly TTP (Potter, 1963). Similar feed-back controls of phosphorylated thymidine concentration also influence RNA synthesis (Feinendegen et al., 1961; Gentry et al., 1965). Even in small quantities, thymidine has an inhibitory effect on continuous cell growth.

2.3. Tritiated Thymidine

Tritiated thymidine consists of the nucleoside thymidine labeled with tritium ("heavy hydrogen", radioactive hydrogen isotope with mass number 3). The isotope disintegrates with a half-life of 12.26 years by

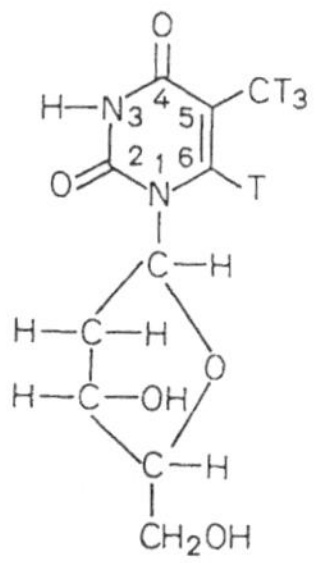

Fig. 2. Thymidine molecule (Chem. Abstr. numbering system). The two possible positions of tritium are indicated by T

emitting exclusively very soft β radiation. Owing to its low average energy of 5.7 keV the β radiation penetrates only 1–2 μ, a small part of it up to 6 μ, into the tissue, while its average in air is 0.5 mm. Small variations in distance between object and photographic emulsion, therefore, lead to considerable differences in the radiation dose absorbed (Robertson *et al.*, 1959). This precludes any quantitative assessment of the distribution of tritium in labeled material from the number of silver grains exposed per unit area. The low range of the β radiation results in an excellent resolution of labeled adjacent structures, because practically no exposed silver grains lie further than 1 μ from the radiation source (Hughes, 1957).

The nucleoside thymidine, composed of the heterocyclic base thymidine and the sugar deoxyribose, is mainly labeled with tritium in position 6 (Fig. 2), though, analyses have shown that about 15% of the tritium occurs in the 5-methyl group. Thymidine can also be labeled only in the methyl group. The radioactive preparations so obtained possess a certain specific activity, that is, the activity in Ci per mMol of preparation. Recently the specific activites of tritiated thymidine have almost reached their theoretical limit of 29 Ci/mMol (The Radiochemical Centre, Amersham, England). The molecule disintegrates to a certain extent under the influence of its own radiation. The disintegration rate depends on specific activity, storage temperature, specific radiolabel concentration, presence of traps for radicals (e.g. ethanol), and traces of chemical impurities. With a storage temperature of about $+2°$ C, the disintegration rate is generally less than 2% per month.

The extremely weak β radiation of tritium represents no external radiation danger, even with the use of higher specific activities. Yet, the possible danger of absorption by personnel in the course of work with this isotope, and the resulting danger of internal irradiation must be considered. In addition, contamination with a β emittent cannot be discovered by the usual Geiger counter methods and may pass unnoticed for a long period of time.

2.4. Autoradiographic Film Material

In recent years, a series of photographic emulsions has been developed which registers the effect of charged atom particles of different energy, such as electrons, protons, and α-particles. These emulsions consist of silver bromide crystals or grains (about 0.2 μ in diameter) embedded in gelatine. Although the autoradiographic emulsions have grains of different sizes, their size fluctuates very little in any given emulsion, in contrast to the usual photoemulsions. Another peculiarity of autoradiographic emulsion is that the grains are very tightly packed, i.e. the emulsion contains very little gelatine. The smaller the grains, and generally the thinner the emulsion layer, the greater is the resolving power. Through exposure to radiation from radioactive isotopes, latent images are produced in the affected grains; by treatment with suitable photographic developers they reduce the silver bromide of the grains to metallic silver.

Light, heat, and contact with various chemical compounds cause background radiation effects which may reduce the clarity of the autoradiographic picture. With appropriate precautions, however, this can be avoided in most cases. Detailed descriptions of diagnosis and prevention of background fogging by various methods have been provided by Prescott (1964) and Rogers (1973).

Autoradiographic film material is commercially available as gel or in the form of a thin layer on a glass plate.

2.4.1. Emulsions

At room temperature, the emulsions are in gel form and must be liquefied by heating to about 40° C before spreading on the labeled material. Owing to rapid development of background fogging, it is recommended that emulsions be stored at $+4°$ C no longer than 2–3 months. Kodak supplies suitable products, e.g. Nuclear track emulsion, type NTB2 and NTB3. NTB3 has a higher sensitivity and a smaller grain size than NTB2. Thus, NTB3 may be preferred if the radioactivity in the sample is low, but it tends to develop more background than NTB2. For use in the dark room, a lamp with a special filter (e.g. Kodak Wratten Safelight,

Series 1) and a 15 watt bulb are permissible; these should not be brought nearer than 30 cm to the film.

2.4.2. Stripping Films

The films consist of a 10 µ layer of gelatine as carrier under a 5 µ radiation sensitive reactive layer mounted together on a glass plate. We use a Kodak film (fine grain autoradiographic stripping plate AR10). The film can be cut into the required sizes and be subsequently easily removed from the glass. The films can be kept at +4° C for at least ¹/₂ year without developing any significant background. The same restrictions as to the fluid emulsions apply to stripping films with respect to darkroom illumination.

3. Techniques

In principle, the incorporation of tritium thymidine into DNA can be attempted in all tissues and suspensions containing dividing cells. Blood and bone marrow suspension cultures as well as fibroblast or epithelial cells cultured on glass surfaces are particularly suitable for studying mitotic chromosomes.

3.1. Labeling with Tritiated Thymidine

We use tritiated thymidine which is supplied in standardized solutions with specific activity from 1.9 Ci to 3 Ci/mMol. Some radioactive fluid may be lost when the solution is withdrawn from the ampules, and care has to be taken to remove traces of fluid in order to avoid possible ingestion of the isotope.

Tritiated thymidine can be applied to the cells for a continuous period, or for a limited period as a so-called "pulse".

3.1.1. Continuous Labeling

In this method, a given quantity of tritiated thymidine is added to the cell culture and left until termination. The following table indicates the quantities of isotope and duration of treatment which have proved satisfactory for routine investigations of the last phases in the S-period in different cell types.

	Quantity of ^{3}H-thymidine per ml of medium	Duration of treatment (hours)
Peripheral blood	1 µCi	4–7
Bone marrow	1 µCi	2–6
Fibroblats	1 µCi	5–8$^1/_2$

The time of isotope treatment must be adjusted according to the length of the cell cycle in the particular culture; the length of the G_2 period varies considerably from case to case. Moreover, depending on the point of time at which the isotope is administered, mitoses are obtained from more or less advanced stages of the DNA synthesis. Thus for example, most mitoses from a peripheral blood culture show only a few isolated labeled chromosome segments if tritiated thymidine is available for the final 4–5$^1/_2$ hours only. When tritiated thymidine is available for 7 hours, numerous labeled chromosome segments are visible in almost all mitoses.

3.1.2. Pulse Labeling

Cells cultured in suspension as well as in monolayers are first incubated with 1 µCi tritiated thymidine per ml medium for 10–15 min. The incorporation of the isotope into the cells is then interrupted by centrifugation of suspensions for 5 min at 1,000 rpm, discarding the radioactive supernatant, and placing the cells until preparation in a prewarmed medium containing a 100–1,000 fold quantity of non-radioactive thymidine. Centrifugation and addition of non-radioactive thymidine are superfluous in monolayer cultures: the culture fluid is poured off and collected in a sterile vessel, tritiated thymidine is added to the cells in prewarmed medium for 10–15 min. Following this, the cells are washed again, and the original medium is restored.

3.2. Harvest of Cultures and Chromosome Preparations

1. After the cell cultures have been labeled continuously or pulse labeled, Colcemid is added for the final 2–3 hours of culture. We usually administer 1 µg Colcemid/ml medium.
2. Termination of the culture now takes place in the usual manner (see Chapter IV). The preparations are stained best with acetic acid orcein (see p. 80). Before other stains are used, it should be checked whether they subsequently interact with the autoradiographic emulsion by either reducing its radiosensitivity or causing increased background.

3. The chromosome preparations are screened for well spread and complete mitoses, which are photographed and the location is recorded.

4. Before coating squash preparations with film, we remove the cover slips according to a method described by Conger and Fairchild (1953): The slide is placed with its cellfree side on the flat surface of a block of dry ice for at least 10 min. The cover slip is then lifted with a sharp safety razor blade. Carefully clean off the Krönig's wax with the razor blade and rinse the preparation twice for 2 min in absolute alcohol. The preparation can now be stored for several weeks.

Semipermanent air dried preparations can be treated in the same way.

3.3. The Autoradiographic Technique

3.3.1. Covering the Preparation with Autoradiographic Film Material

Until fixation, all work with film must be carried out in a specially lit dark room (see also autoradiographic film material, p. 124f.).

Emulsion Technique

The emulsion types which we use, NTB2 and NTB3 (Kodak, see p. 124) must first be liquefied. For this purpose a suitable amount of gel (112 ml) is taken from the supply bottle, placed in a glass beaker and diluted 1:1 with distilled water. The mixture is warmed in a water bath at 42–45° C for 15 min. For coating the slide with film, we use a special vessel filled with liquefied emulsion which is placed in a water bath. The vessel should be glass of such a shape that two slides can be easily dipped into it. Air bubbles may be present in the emulsion, but can be removed by dipping a clean slide into it several times. In order to use the emulsion economically, two pre-warmed slides with their cell-free sides facing each other are simultaneously dipped into the special vessel for 3–5 sec. The slides are then immediately separated and clamped horizontally with the clips of a special preparation holder (Fig. 3). The emulsion can also be dripped onto the slide and spread by tilting the slide in all directions, or by rolling a glass rod along the surface. Care has to be taken throughout the procedure to avoid mechanical stress in the emulsion. Following this, the preparation is dried for 2–3 hours in a dust-free part of the dark room or in a lightproof drying cupboard, through which cool air is blown with an electric fan. A detailed description of the emulsion technique was presented by Prescott (1964).

Stripping Film Technique

Before applying the film, we spread chrome alum on the back side of the slide and the cell free part of the front side so that the film later adheres firmly to the glass.

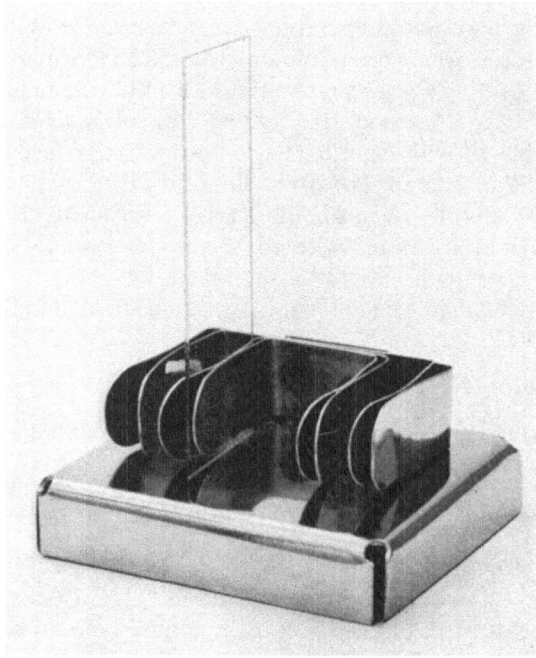

Fig. 3. Holder for drying slides covered with film emulsion

Preparation of chrome alum gelatine:

10 ml K-chrome alum solution (0.5 g K-chrome alum Merck No. 1034 dissolved in 100 ml distilled water) is added to 90 ml gelatine solution (0.5 g gelatine Merck No. 4078 dissolved in 100 ml distilled water).

The film is first divided into 4×4 cm pieces on a cutting apparatus (Schmid 1965; Fig. 4) with the aid of a scalpel. The air in the dark room should be humid enough to prevent static electricity during cutting. The glass plate with the film facing upwards is now placed in a dish containing 70% ethanol for 2–3 min and then transferred to another dish containing absolute ethanol. After this pretreatment the individual squares of the film can be quite easily removed from the glass plate with pointed forceps. These squares are now dropped, emulsion side downwards, into a dish containing distilled water. They unfold themselves on the surface of the water and swell. A slide is now brought underneath the floating square and lifted out of the water with a sudden jerk, together with the square. The overhanging ends of the film thereby adhere to the reverse side of the slide. We then lay the slide on a glass plate, drip distilled water onto the film and do 5 more preparations the same way.

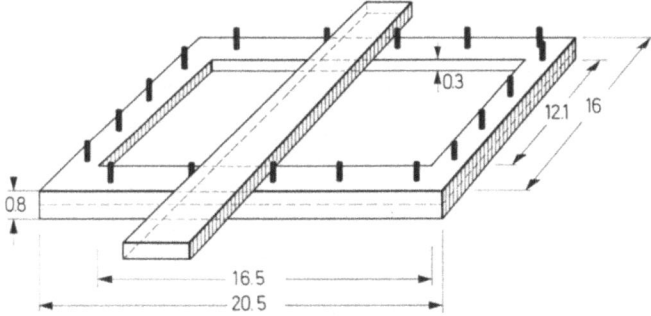

Fig. 4. Apparatus for cutting A.R. 10 films

Meanwhile, the film which has already been stripped has swollen and is finally smoothed with a soft, wet brush. The preparations are dried under the same conditions as the emulsion technique.

3.3.2. Exposure

When the film is completely dry, slides are placed in lightproof plastic boxes which are sealed with adhesive tape and stored in the refrigerator at $+4°$ C during exposure. We place a small quantity of a drying agent ($CaCl_2$) in the box. Pilot preparations are developed at different time intervals according to the labeling employed in order to obtain optimal exposure. This test of labeling intensity is necessary from the 5 th day of exposure onwards with the quantities of tritiated thymidine and specific activity we use. As the emission of the β rays from the tritium may be regarded as constant, it is possible to calculate the point in time of doubling, tripling, etc. in the number of the expected silver grains from the first pilot preparation.

3.3. Development and Fixation of the Autoradiographic Film

1. Place preparation in a developer (e.g. Kodak D-19b) for 2 min at 20° C.
2. Neutralize for a few seconds in tap water at 19–21° C.
. Leave 3–8 min in acid fixative at 19–21° C.
   Rinse 10–20 min.

5. Dry 2–3 hours in a dust-free part of the dark-room or in a drying box with a fan.

The developed and fixed preparations can now either be stored for several days or stained immediately.

3.3.4. Staining

The stain for application *after* exposure of the autoradiographic film which has proved best in our laboratory is buffered Giemsa solution (Gude *et al.*, 1955). It stains chromosomes light blue and provides good contrast to the black silver grains of the autoradiographic film.

1. Place squash preparation for 3–5 min in buffered Giemsa solution. Chromosomes of air dried preparations will often be sufficiently stained from the initial Orcein so that further staining with Giemsa solution is superfluous or requires only 2–3 min.

2. Remove residual stain by dipping twice in distilled water.

3. Dry for 2–3 hours.

4. Remove the film from the reverse side of the preparation with a sharp safety razor blade.

5. Mount the preparation in Euparal, Eukitt or similar.

3.3.5. Evaluation of Chromosomes in Autoradiographs

Following the staining process, the mitoses recorded previously are located again and photographed according to the quality of the auto-radiographic pattern.

3.3.6. Preparation of Control Pictures by Removal of Silver Grains and Dissolving Autoradiographic Film

For most routine investigations, it is not absolutely necessary to photo-graph mitoses before coating the preparation with film. Control pictures of good quality can also be obtained by removing the metallic silver from the film emulsion (Bianchi *et al.*, 1964):

1. Place the preparation for several hours in xylene. The cover slips can then be separated easily from the film with a lancet.

2. Remove the remaining material used for embedding the preparation (in the case of Euparal, by means of repeated dipping in absolute alcohol; with Eukitt, DePeX, H.S.R. etc., by repeated dipping in xylene).

3. Descending series of alcohols.

4. 10 min distilled water.

5. 5 min in 1.5% K ferric cyanide solution.

6. 5 min in 20% Na thiosulphate solution.

7. 1 min distilled water.

8. Dry for 2–3 hours.

9. Mount in Euparal or similar.

After removal of the silver grains it is possible to dissolve the entire radiographic film by hydrolysis. Following this, the chromosomes are stained once more (e.g. with Schiff's reagent). A film can now be mounted again. This procedure prevents the loss of valuable material if excessive background is present or if the intensity of labeling should be influenced by changing the exposure time.

1. Hydrolysis in 1 N HCl at 60° C, 8 min.

2. Remove HCl residue by dipping in distilled water.

3. Stain for 2 hours in Schiff's reagent.

4. Remove residual Schiff's reagent by dipping in distilled water.

5. Ascending series of alcohols and embedding in synthetic resin.

This procedure, however, sometimes results in the loss of part or all of the cells, which come off along with the film; it may also occur that small pieces of emulsion remain sticking to the glass surface and cover parts of prephotographed mitosis. It is advisable, therefore, to overlay the stained slide with a plastic film before applying the autoradiographic emulsion. Bishop and Bishop (1963) use Vinalak, Stubblefield (1965) applies polyvinyl Formal (Formvar) for this purpose. A Formvar membrane on the slide is obtained in the following way (Hamerton, 1971): A solution of 0.4% Formvar in 1,2 dichloroethane is prepared and a drop of it placed in a dish of convenient size containing distilled water. A thin film of plastic forms on the water surface. The slide is now brought under the membrane and withdrawn so that the membrane folds around the back. A more cautious way to place the Formvar membrane on the preparation was described by Stubblefield (1965) who first covers a clean slide with the membrane and then transfers it to the preparation in a water bath. The plastic layer is usually thin enough that no appreciable amount of radiation is absorbed. In preparations protected in this way the silver grains need not be oxidized and the autoradiographic emulsion can easily be peeled or washed off. This procedure is of special use for repeated autoradiography (grain counting, two emulsion technique with ³H and ¹⁴C, etc.).

Appendix

Summary of the Procedure

A. Labeling with Tritiated Thymidine
Continuous or pulse labeling.

B. Processing of Cultures and Making up Chromosome Preparations
Application of the usual methods.

C. Autoradiographic Techniques

1. Place autoradiographic film material on the preparation.
   Emulsion Technique. Take gel from supply flask in the dark room. Dilute 1:1 with distilled water. Heat the mixture in a water bath. Dip slide into liquid emulsion. Dry preparation.
   Stripping Film Technique. Before applying the film, spread chrome alum gelatine on the reverse side and the cell-free area of the front side of the slide. Cut up the film in the dark room. After cutting film, place glass plate upwards in 70% alcohol, then in absolute alcohol. Loosen squares of film from the glass plate and drop onto distilled water, emulsion side facing downwards. Floating squares are lifted together with slide out of the water, allowed to swell for a few minutes, then smoothed with a soft brush. Dry preparation under lightproof conditions.
2. *Exposure.* Slide with completely dry film in light proof plastic box. After a few days remove test preparations to check exposure.
3. Develop and fix the autoradiographic film.
4. Stain preparation with buffered Giemsa solution. Then dry preparation, remove film from the reverse side and mount.
5. Photograph suitable cells on autoradiographs.
6. Prepare control pictures by removing silver grains and dissolving the autoradiographic film.
7. Stain preparations.
8. Photograph control pictures.

References

Adelstein, S. J., Lyman, C. P., O'Brien, R. O.: Variations in the incorporation of thymidine into the DNA of some rodent species. Comp. Biochem. Physiol. **12**, 223–231 (1964).

Back, F., Dörmer, P., Baumann, P., Olbrich, E.: Zur Problematik der Chromosomenautoradiographie. Humangenetik **4**, 305–319 (1967).

Bélanger, L. F., Leblond, C. P.: A method for locating radioactive elements in tissues by covering histological sections with a photographic emulsion. Endocrinology **39**, 8–13 (1946).

Bender, M. A., Prescott, D. M.: DNA synthesis and mitosis in cultures of human peripheral leucocytes. Exp. Cell Res. **27**, 221–229 (1962).

Bessman, M. J., Lehman, I. R., Simms, E. S., Kornberg, A.: Enzymatic synthesis of deoxyribonucleic acid. II. General properties of the reaction. J. biol. Chem. **233**, 171–177 (1958).

Bianchi, N., Lima-de-Faria, A., Jaworska, H.: A technique for removing silver grains and gelatin from tritium autoradiographs of human chromosomes. Hereditas (Lund) **51**, 207–211 (1964).

Bianchi, N. O., de Bianchi, M. S. A.: DNA replication sequence of human chromosomes in blood cultures. Chromosoma (Berl.) **17**, 273–290 (1965).

Bishop, A., Bishop, O. N.: Analysis of tritium-labelled human chromosomes and sex chromatin. Nature (Lond.) **199**, 930–932 (1963).

Büchner, Th., Wilkens, A., Pfeiffer, R. A.: Autoradiographische Markierungsmuster der Chromosomen 1, 2, 3, 4, 5, 13–15, 16 und Grad der Übereinstimmung der Homologen nach Einbau von H³-Thymidin während der späten S-Phase. Quantitative Untersuchungen an Zellen der Blutkultur. Klin. Wschr. **46**, 187–194 (1968).

Conger, A. D., Fairchild, L. M.: A quickfreeze method for making smear slides permanent. Stain Technol. **28**, 281–283 (1953).

Feinendegen, L. E.: Tritium labeled molecules in biology and medicine. New York-London: Academic Press Inc. 1966.

Feinendegen, L. E., Bond, V. P.: Differential uptake of H³-thymidine into the soluble fraction of simple bone marrow cells, determined by autoradiography. Exp. Cell Res. **27**, 474–484 (1962).

Feinendegen, L. E., Bond, V. P., Hughes, W. L.: RNA mediation in the DNA synthesis in Hela cells studied with tritium labeled cytidine and thymidine. Exp. Cell Res. **25**, 627–647 (1961).

Fink, R. M., Fink, K.: Relative retention of H³ and C¹⁴ labels of nucleosides incorporated into nucleic acids of Neurospora. J. biol. Chem. **237**, 2889–2891 (1962).

Galton, M., Holt, S. F.: DNA replication patterns of the sex chromosomes in somatic cells of the Syrian Hamster. Cytogenetics **3**, 97–111 (1964).

Gentry, G. A., Morse, P. A., Potter, R. van: Pyrimidine metabolism in tissue culture cells derived from rat hepatomas. Cancer Res. **25**, 517–524 (1965).

German, J. L.: DNA synthesis in human chromosomes. Trans. N.Y. Acad. Sci. **24**, 395–407 (1962).

Gey, W.: Untersuchungen über die DNA-Replikationsmuster der Chromosomengruppen 4–5, 13–15 und 21–22 an in vitro gezüchteten menschlichen Lymphocyten. Humangenetik **2**, 246–261 (1966).

Giannelli, F.: Human chromosomes DNA synthesis. Basel: Karger 1970.

Grumbach, M. M., Morishima, A., Taylor, J. H.: Human sex chromosome abnormalities in relation to DNA replication and heterochromatinization. Proc. nat. Acad. Sci. (Wash.) **49**, 581–589 (1963).

Gude, W. D., Upton, A. C., Odell, T. T.: Giemsa staining of autoradiograms prepared with stripping film. Stain Technol. **30**, 161–162 (1955).

Hamerton, J. L.: Human cytogenetics, vol. I. New York-London: Academic Press 1971.

Howard, A., Pelc, S. R.: Synthesis of deoxyribonucleic acid in normal irradiated cells and its relation to chromosome breakage. Hereditas (Lund), Suppl. **6**, 261–273 (1953).

Hughes, W. L.: In: Proceedings of the Symp. on Tritium in Tracer Applications, New York 1957.

Lark, K. G.: Cellular control of DNA biosynthesis. In: Molecular genetics, part I (H. J. Taylor, ed.), p. 153–266. New York-London: Academic Press 1963.

Lehman, I. R., Bessman, M. J., Simms, E. S., Kornberg, A.: Enzymatic synthesis of deoxyribonucleic acid. I. Preparation of substrates and partial purification of an enzyme from Escherichia coli. J. biol. Chem. **233**, 163–170 (1958).

Miller, O. J.: Autoradiography in human cytogenetics. In: Advances in human genetics, vol. I (H. Harris and K. Hirschhorn, eds.), p. 35–130. New York-London: Plenum Press 1970.

Moorhead, P. S., Defendi, V.: Asynchrony of DNA synthesis in chromosomes of human diploid cells. J. Cell Biol. **16**, 202–209 (1963).

Odartchenko, N., Cottier, H., Feinendegen, L. E., Bond, V. P.: Evaluation of mitotic time in vivo, using tritiated thymidine as a cell marker: successive labeling with time of separate mitotic phases. Exp. Cell Res. **35**, 402–411 (1964).

Pelc, S. R.: Autoradiograph technique. Nature (Lond.) **160**, 749–750 (1947).

Potter, R.: Feedback inhibition of thymidine kinase by thymidine triphosphate. Exp. Cell Res., Suppl. **9**, 259–261 (1963).

Potter, R. L., Nygaard, O. F.: The conversion of thymidine to thymine nucleosides and deoxyribonucleic acid in vivo. J. biol. Chem. **238**, 2150–2155 (1963).

Prescott, D. M.: Autoradiography with liquid emulsion. In: Methods in cell physiology (D. M. Prescott, ed.), vol. I, p. 365–370. New York: Academic Press 1964.

Robertson, J. S., Bond, V. P., Cronkite, E. P.: Resolution and image spread in autoradiographs of tritium-labeled cells. Int. J. appl. Radiat. **7**, 33–37 (1959).

Rogers, A. W.: Techniques of autoradiography, 2nd ed. Amsterdam: Elsevier 1973.

Schmid, W.: DNA replication patterns of human chromosomes. Cytogenetics **2**, 175–193 (1963).

Schmid, W.: Autoradiography of human chromosomes. In: Human chromosome methodology (J. J. Yunis, ed.). New York: Academic Press 1965.

Schwarzacher, H. G., Schnedl, W.: Der Zellzyklus in Fibroblastenkulturen von Menschen. Z. Zellforsch. **67**, 165–173 (1965).

Stone, G. E., Miller, O. L., Prescott, D. M.: H^3-thymidine derivative pools in relation to macronuclear DNA synthesis in Tetrahymena pyriformis. J. Cell Biol. **25**, 171–177 (1965).

Stubblefield, E.: Quantitative tritium autoradiography of mammalian chromosomes. I. The basic method. J. Cell Biol. **25**, 137–147 (1965).

Taylor, J. H.: Asynchronous duplication of chromosomes in cultured cells of Chinese hamster. J. biophys. biochem. Cytol. **7**, 455–463 (1960).

Taylor, J. H., Woods, P. S., Hughes, W. L.: The organization and duplication of chromosomes as revealed by autoradiographic studies using tritium-labeled thymidine. Proc. nat. Acad. Sci. (Wash.) **43**, 122–128 (1957).

CHAPTER VIII

The Human Karyotype

Analysis of Chromosomes in Mitosis and Evaluation of Cytogenetic Data

EBERHARD PASSARGE

With 20 Figures

1. Introduction

This chapter deals with methods for analysis of human chromosomal preparations and evaluation of cytogenetic findings. It describes the human karyotype determined by standard staining procedures, the characteristics of each individual chromosome as defined by recent techniques, and the evaluation of cytogenetic data for diagnosis. In addition, the normal variants of the karyotype are described in detail. A description of the results of cytogenetic studies or specific problems related to chromosomal aberrations is beyond the scope of this chapter and the reader is referred to recent surveys, such as those by Hamerton (1971), Jacobs (1969 and 1972), Wright *et al.* (1972), Ford (1973), Hirschhorn (1973), Ferguson-Smith (1973), or Caspersson and Zech (1973), and the literature cited there.

The writing of this text was aided by earlier reviews of the subject by Ferguson-Smith (1964), German (1964b), Patau (1965), Turpin and Lejeune (1965), Court Brown *et al.* (1966, 1967, 1969), Jacobs *et al.* (1970), but it draws from numerous other sources which are not always individually cited here, including personal experience. The texts of the Third and Fourth Standardization Conferences on Human Cytogenetics (Chicago Conference, 1966; Paris Conference, 1971) were very helpful.

The references are selected according to their usefulness as an introduction to the literature without regard to completeness or scientific priorities. An appendix contains some tables which are thought to be useful in cytogenetic diagnosis.

2. Chromosomes at Normal Metaphase

This chapter is concerned with the analysis of metaphase mitotic chromosomes as seen through the light microscope. Chromosomes in other phases of the mitotic cycle and electron microscopic preparations have thus far eluded an exact analysis of the chromosomal complement for diagnostic purposes and will not be considered here: relevant information can be found in the proceedings of a recent cytogenetic conference (Wright, Crandall, Boyer, 1972).

2.1. The Material for Analysis

Metaphases suitable for cytogenetic analysis consist of well spread chromosomes with little or no overlap. Such cells are located under low magnification (e.g. through a 10× objective) as seen in Fig. 1 and then analysed at about 1,000×. At this magnification the 46 chromosomes of a normal human metaphase are visible as elongated structures which are divided longitudinally into the two *chromatids*. Each of these would have been separated at the end of cell division and distributed to each of the two daughter cells, had this process not been stopped during preparation. Homologous chromatids of each chromosome in metaphase lie parallel and are held together by a constriction, the *centromere* or primary constriction, to which the mitotic spindle fibers are attached in normal cell division. The position of the centromere is constant for each chromosome and divides it into *chromosome arms*. The position of the centromere and the total length are the most important parameters for the arrangement into pairs and groups of the karyotype as described below. Additional important features characterize a chromosome, but some require special techniques (see Section 4 of this chapter).

As only well spread metaphases can be reliably analysed, a selection of suitable cells must be made. In routine analysis one usually selects according to quality, but it is important to realize that this may introduce bias in the results obtained. For investigative purposes it is essential to establish precise criteria for selection prior to analysis.

Cells in early metaphases are most suitable for analysis, because chromatids are then neither contracted too strongly, as can be observed after long application of colchicine, nor is analysis hindered by insufficient spread of chromosomes, as seen in prophase. There is no reliable preparative procedure to control quality and characteristics of metaphases. Each metaphase exhibits its individual degree of contraction and spreading, and usually permits one to recognize an individual cell as well as to distinguish it from neighboring metaphases.

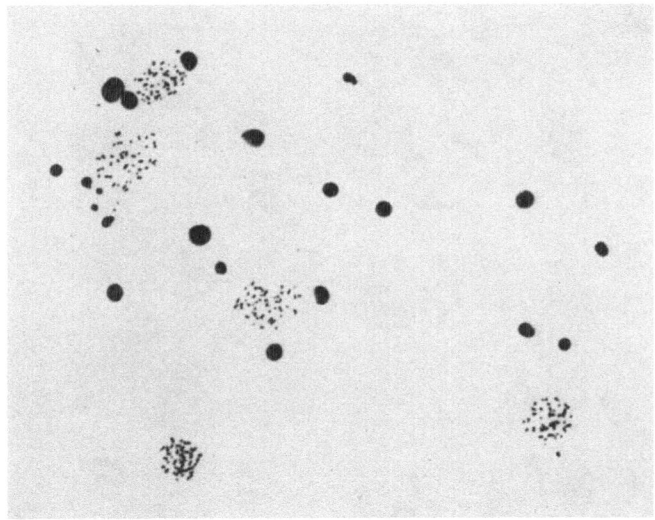

Fig. 1. Chromosomal preparation from lymphocytes at low magnification (× 150) as seen when screening for metaphases. Five metaphases are clearly visible

Artificial variations in chromosome number occur in about 1–2% of cells as a result of spreading, even with careful handling. Loss of one or more chromosomes during preparation is easier to understand than the presence of additional chromosomes and is actually more common. Statistically such artifacts are observed more often in groups with high numbers of chromosomes, e.g. in the C (6-X-12) group which accounts for 16/46 or 33% of the chromosomal complement. However, if aberrant chromosome counts are obtained in technically good preparations, even in a few cells, it might be advisable to repeat the culture. This is recommended as a general rule to verify any unusual result.

2.2. Location of Individual Chromosomes

There is non-random position of some chromosomes at metaphase. The short arms of two or more of the satellite-carrying acrocentric chromosomes (D group and G group, see Section 3) tend to lie together (Fig. 2). The association of acrocentric chromosomes is thought to be caused by the nucleolus organizing regions of these chromosomes when they combine to form a common nucleolus (Ferguson-Smith, 1964; Schmid, 1969).

Zang and Back (1968) have defined useful criteria for acrocentric chromosome association: two or more acrocentric chromosomes may

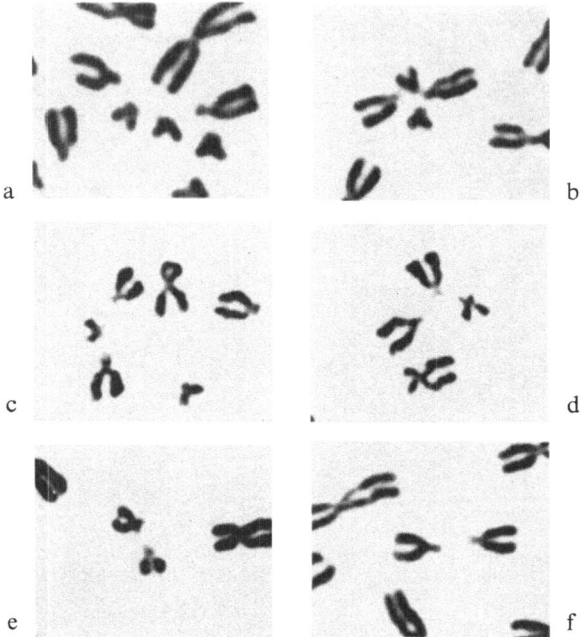

Fig. 2a–f. Association of acrocentric chromosomes in different arrangements. (a) Association of five acrocentrics in chain formation. (b) Two D and two G group chromosomes. (c and d) Two D and one G group chromosome. (e and f) Two G and D group chromosomes, respectively

be considered to be 'associated' if they fulfill the following conditions: (i) their distance should not exceed the length of the long arm of the largest G group chromosome of the same metaphase; (ii) larger distances are accepted if "associated" acrocentric chromosomes are connected by clearly visible thread-like structures; (iii) along the same longitudinal axis larger distances are accepted, i.e. up to the length of the long arm of a D group chromosome; (iv) the short arms of all additional "associated" chromosomes must lie on or above the "centromere line" of the first associated chromosome. The centromere line is defined as the line that crosses the centromere perpendicular to the chromosomal longitudinal axis (see Fig. 1 of Zang and Back, 1968).

The method and, perhaps, origin of the culture influence the frequency of acrocentric associations, which seem to occur less often in micro methods. On the average about 0.8–3.0 associations can empirically be expected per cell, regardless of sex or age of the individual. Most associations involve only 2 or 3 acrocentric chromosomes, but occasionally larger groups are seen (Fig. 2a–f).

Previous studies of the distribution of chromosomes involved in acrocentric association were based on autoradiographically identified chromosomes (Cooke, 1972), pointing to interindividual variability of association frequencies. A recent study by Patil and Lubs (1971)[1], based on reliable identification of each acrocentric chromosome by fluorescence technique, appears to have clarified the matter. The authors studied 202 metaphases from 11 individuals and found 234 associations in 149 cells; 39 associations involved three or more chromosomes. The distribution was non-random, as follows: chromosome 13 = 108 times, chromosome 14 = 124 times, chromosome 15 = 72 times, chromosome 21 = 123 times, chromosome 22 = 101 times (expected were 106.2 per chromosome).

The individual chromosomes were involved in the following frequencies per association:
0.54 = chromosome 14, 0.52 = chromosome 21, 0.45 = chromosome 13, 0.43 = chromosome 22, 0.30 = chromosome 15, except in one of the ten individuals where chromosome 22 was more often involved than chromosome 21.

In 195 associations involving only two chromosomes, 30 were between homologous (21–21 in 11, 22–22 in 8, 14–14 in 7, 13–13 in 3, 15–15 in 1) and 165 between non-homologous chromosomes (14–21 in 23 as the most common and 15–22 in 11 as the least common).

The Y chromosome does not participate in association of acrocentrics, but satellited chromosomes also associate with other secondary constrictions of non-satellited chromosomes, in particular with the paracentric constrictions of chromosome 1, 9 and 16 (Ferguson-Smith and Handmaker, 1963). Other chromosomes have been reported by some to be non-randomly distributed at metaphase, i.e. to take a position at the periphery more often than expected, e.g. the sex chromosomes, chromosomes 17–18 (Miller *et al.*, 1963; Barton *et al.*, 1965), but not by others, at least for the D and G groups (German, 1964b), and the Y chromosome (Spence *et al.*, 1973). A marker chromosome 17[2] was found to be peripherally located in 95% of metaphases (Schmid and Bauchinger, 1969). Autoradiographic studies by Ockey (1969) indicated that the later replicating member of two homologs and the longer chromosomes tend to lie more peripherally. Galperin (1969) studied the non-random location of chromosomes in detail.

---

[1] The data obtained by these authors are the consequence of pooling the findings from different individuals. As was shown recently (Schmid-Dorn, M., Vogel, W., Krone, W.: On the relationship between the frequency of association and the nucleolar constriction of individual acrocentric chromosomes, Humangenetik, in press), the association frequency is a constant feature of the individual acrocentrics, and is correlated with the structure of the nucleolus organizer region.

[2] A marker chromosome is a morphologically conspicuous chromosome that can regularly be distinguished from its homologous partner.

The significance of presumptive non-random location remains unclear at present and cannot yet be used for the analysis of a single cell. It was assumed by Comings (1968) that chromosomes have a non-random arrangement at interphase owing to specific sites of attachment to the nuclear membrane.

2.3. Effect of Ageing on the Karyotype

Several studies have correlated an increasing number of aneuploid cells in metaphases from lymphocyte cultures with increasing age of an individual although contradictory data exist as summarized elsewhere (Passarge, 1968). Court Brown et al. (1966, 1967) in particular have described in detail the effect of age. In blood cultures of women over 60 years of age, aneuploid cells with 45 chromosomes were found in 5% of metaphases, and this proportion rose to 12% in women over 75. In elderly men, from 65 years on, there was a similar, but less marked tendency to aneuploidy up to about 1% of cells, while in younger men, such cells only occurred in less than 0.5%.These hypomodal cells with 45 chromosomes were assumed to lack an X chromosome in women and could recently be shown to lack an Y chromosome in men (Pierre and Hoagland, 1971).

These observations may have practical consequences, because in older persons they would require to distinguish an age-dependent XO/XX or XO/XY mosaicism from a congenital aberration of this type, which could be difficult when the deviant cell is in the minority. Table 1 in the Appendix provides some criteria for the diagnosis of true mosaicism (see also Section 6.2 of this chapter).

2.4. Influence of Culture Conditions on the Number and Structure of Chromosomes

Certain numerical or structural chromosomal deviations occur in normal cultures regardless of the age of the individual examined. They may be artifacts, but some are actual results of culture conditions in vitro and must be interpreted with caution for the purposes of medical diagnosis. About 1–2% of metaphases from lymphocyte or fibroblast cultures and a variable proportion from cultured amnion cells (see Chapter III) are polyploid, in most cases tetraploid. Tetraploid cells may result e.g. from fusion of cell nuclei or from endomitotic reduplication. Data on the frequency and origin of polyploid and endoreduplicated cells have been presented by several authors (Schwarzacher and Schnedl, 1965; Powsner, 1966; Pawlowitzki and Cenani, 1967; Milunsky et al., 1970). The usual frequency of endoreduplication (tetraploid cells with diplo-chromosomes) varies from below 0.1 % to about 1 %.

Various structural changes which may lead to misinterpretation can be observed in normal cultures. Some of these may be secondary constrictions which are part of the normal karyotype as described in Section 5 of this chapter, and are subject to considerable individual variation as discussed in Section 5. *A secondary constriction* denotes a constriction in both chromatids, possibly due to formation of an achromatic chromatid segment, but without a clear break in the continuity of the chromatid (Fig. 13c and d, 16). It is possible to enhance the frequency of secondary constrictions by certain chemicals, e.g. 5-bromodeoxy-uridine (BUdR), hydroxylamine hydrochloride, or mitomycin C (for reviews see Brøgger and Johansen, 1972; Brown *et al.*, 1972).

Breaks of chromosomes are usually oberved in a few percent of metaphases. There are no established frequency figures because of the different conditions in different laboratories, the use of different nutrient media, and numerous other factors which usually are impossible to define, but which probably cause the considerable differences and the variability observed between laboratories or within a laboratory over a certain time span. Undisclosed radiations, unnoticed exposure to ultra-violet light, or virus infections may be responsible for chromosomal breaks that are from time to time observed in seemingly increased frequency without explanation. Depletion of arginine in nutrient media due to mycoplasma infections can cause chromosomal breaks (Aula and Nichols, 1968) and may be rather common causes of unexplained chromosomal breakage, especially in fibroblast cultures.

Two simple types of breaks are most frequently found in normal cultures and are usually observed in about 3–4 % of analysed metaphase cells: 1. chromatid gaps or breaks (only one chromatid either barely or clearly interrupted; the lesion having occurred in the G_2 phase following DNA synthesis); 2. isochromatid breaks (also called "chromosome breaks") with both chromatids clearly interrupted at the same position due to a lesion that must have occurred in the G_1 phase before DNA synthesis. Other types of structural aberrations generally occur less frequently and their observation may indicate an anomalous high chromosomal instability.

Littlefield and Goh (1973) reported breakages in the range of 4–7 % (excluding single chromatid breaks) in a study of 27,709 metaphases from 305 lymphocyte cultures of 31 control persons (21 female, 10 male) over a three-year period. They noted considerable variability in breakage frequencies in three-day cultures from the same person and among cultures on the same day from different persons. In addition, they found a slightly higher variability and incidence of breaks in women (1,146 of

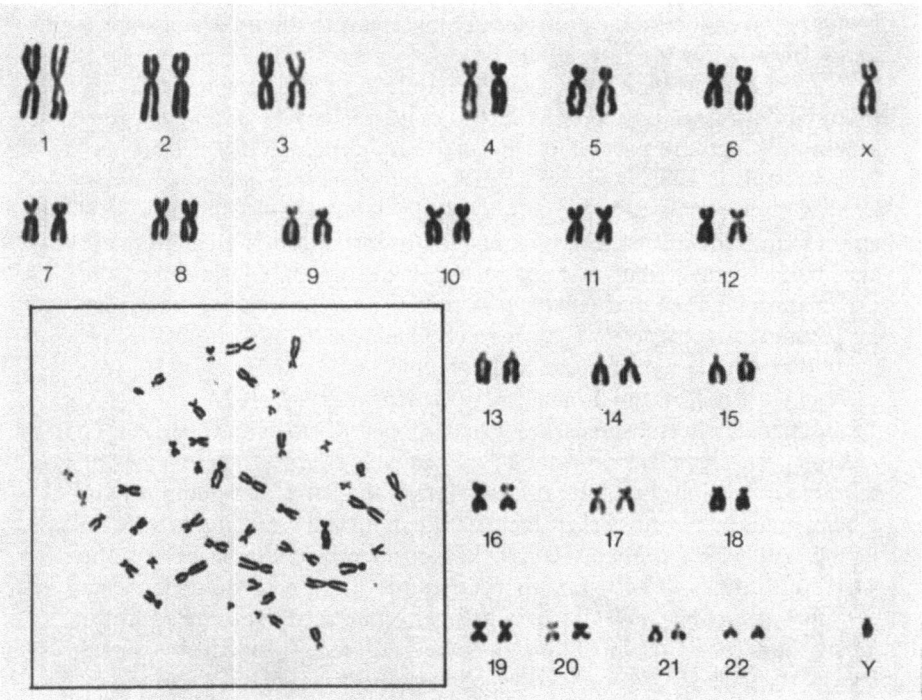

Fig. 3. Normal male karyotype on standard stain (Giemsa), arranged in pairs according to the Denver-London system. Individual identification is possible only for pairs 1–3 and 16–18, and the Y chromosome. Karyotype reproduced at about 1,500 ×, the metaphase plate from which it was prepared (left lower corner) at about × 750

17,759 or 6.5% abnormal cells) than in men (648 of 11,950 or 5.6%). The frequency of individual types was as follows: isolocus breaks (referred to as isochromatid breaks above) 4.25% (average of the female 4.8% and the male values of 3.7%), deletions 1.7% (1.8 female, 1.6 male), dicentric chromosomes 0.12% (0.08 female, 0.18 male), translocations 0.28% (0.3 and 0.27), inversions 0.075% (0.07 and 0.08), homologous crossing-over figures 0.02% (only in females observed), partial endoreduplication 0.01% in females (none in males), pulverized chromosomes and metaphases 0.11% (0.1 in females, 0.12 in males).

Court Brown *et al.* (1966) found a total of 405 (3.26%) chromatid aberrations in 12,420 metaphases from lymphocytes cultured for 2 or 3 days in individuals 15–75 years of age. Of these, 206 aberrations occurred in persons over 65 years. Dicentric chromosomes were observed in Court Brown's study in 0.055% (1:1,800 cells) and crossing-over

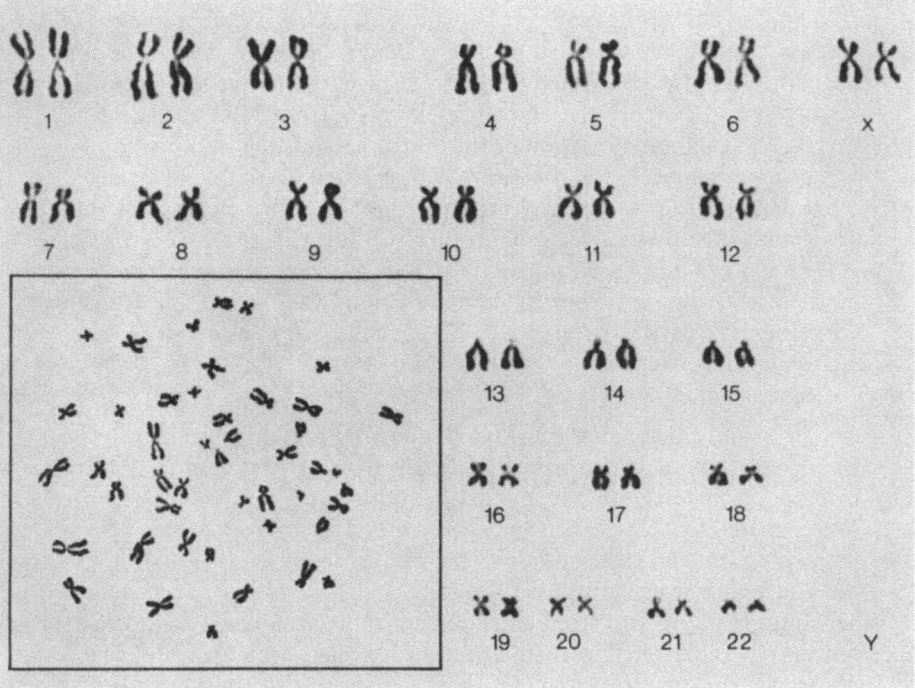

Fig. 4. Normal female karyotype, arranged as in Fig. 3. Note prominent secondary constriction of one member of chromosome 1 (also in Fig. 3)

reunion figures in 0.056% (7 among 12,420 metaphases), which corresponds to my own experience. I suggest that, as a rule, further search for chromosomal instability is indicated if one of the rare types of chromosomal rearrangement is encountered. Methods to analyse chromosomal breakage are described in section 6.3 of this chapter.

3. The Standard Karyotype

The karyotype is the systematic arrangement of chromosomes in groups of homologous pairs according to features that are important for characterizing the chromosomal complement of a given cell. This includes the morphology as well as distinct patterns produced by certain staining reactions or isotopes. The term "karyotype" may refer to an individual or to a single cell, an ambiguity which may have to be observed. However, other terms such as "idiogram" have not come into general use.

The basic classification of human metaphase chromosomes was derived from conferences in Denver (Denver Report, 1960), London (London Report, 1963), and Chicago (Chicago Conference, 1966). It was agreed to number the autosomes in pairs from 1–22 according to decreasing length and position of the centromere as well as other criteria (see below). The sex chromosomes were designated X and Y and arranged separately where possible, and in the vicinity of autosomal groups that are similar in size and structure (group 6-X-12 or C, and 21–22-Y or G). Additional criteria, based on special staining properties, were set forth more recently at the Fourth Standardization Conference on Human Cytogenetics (Paris Conference, 1971). These are considered in later sections of this chapter.

The 22 pairs of autosomes are divided into 7 groups (A–G) which can be readily distinguished (Figs. 3 and 4) on routine staining. The following types of chromosomes occur in man: 1. *metacentric* (centromere approximately in the middle of both arms); 2. *submetacentric* (centromere divides the chromosome into a long and a short arm, designated in shorthand notation as q and p, respectively, according to the Chicago nomenclature, see Section 7.2); 3. *acrocentric* (centromere separates the long arm from an extremely short arm which may or may not carry visible satellites). As mentioned earlier, each arm of a metaphase chromosome consists of two chromatids (sister chromatids). In practice, the singular form is often used although "the arm" consists of two sister chromatids. Three parameters are important for the basic morphological characterization of a chromosome: 1. the relative *length*; 2. the chromosome *arm index*, defined as the relationship of the longer to the shorter arm; 3. the *centromere index*, defined as the relationship of the shorter arm to the total length of the chromosome, which determines the relative position of the centromere. In addition, chromosomes can be characterized by other means, such as autoradiography, special stains, or measurements, as discussed in Section 4 of this chapter. An important morphological feature of acrocentric chromosomes (D group and G group) are the *satellites*, although they are not always visible. They consist of a pair of deeply stained small round bodies at the end of the short arm from which they are separated by a region that takes up little or no stain. As mentioned earlier, Section 2.2, this region is thought to be the nucleolar organizing region (see Hamerton, 1971).

The human karyotype consists of the following groups which are described in more detail below:

Group A (chromosomes 1–3): large, metacentric and submetacentric chromosomes; *group B* (chromosomes 4–5): large, submetacentric chromosomes; *group C* (chromosomes 6–12): medium-sized, submetacentric chromosomes which include the *X chromosome(s)* although they cannot be morphologically distinguished with certainity from this group; *group D* (chromosomes 13–15): medium sized, acrocentric chromosomes with satellites on the short arm; *group E* (chromosomes 16–18): relatively short metacentric to submetacentric chromosomes; *group F* (chromosomes 19–20): small, metacentric chromosomes; *group G* (chromosomes 21–22): small, acrocentric chromosomes with satellites on the short arm. The *Y chromosome* is similar to this group in shape and size. Figs. 3 and 4 show a male and a female karyotype, respectively.

The arrangement is depicted according to the internationally used Denver and London system. It should be noted, however, that even though every chromosome pair is placed in a numbered position, in man only 6 of the 22 autosomal pairs (chromosomes Nos. 1, 2, 3, 16, 17, 18) and (usually) the Y chromosome can be individually identified by shape (arm index, centromere index) and relative size (length) alone and have a clearly defined position in the karyotype. Although each pair can be clearly identified by special staining procedures (see Section 4), it was felt that a detailed description of the morphological features of the karyotype as revealed by standard stains would still be desirable, until the special staining techniques have become routine everywhere.

3.1. The Individual Chromosomes on Routine Stain

A Group
Chromosome 1 is the largest metacentric chromosome of the karyotype. The centromere lies almost exactly in the middle. The chromosome arm index is 1.1 and the centromere index 48–49. A secondary constriction (for definition see Section 2.4) is often present near the centromere in the proximal section of the arm that is regarded as the longer one, and may cause a frequently observed variation in length of the long arm (see Section 5).

Chromosome 2 is the largest submetacentric chromosome, with an arm index of 1.5–1.6 and a centromere index of 38–40.

Chromosome 3 is the second largest metacentric chromosome, with an arm index of 1.2 and a centromere index of 45–46. This chromosome is about 20% shorter than chromosome 1 and can easily be distinguished from it. Arm index and centromere index in numerous cells indicate that a short arm can be distinguished from a long arm. All chromosomes of the A group can be readily identified.

B Group
Chromosomes 4–5 cannot be differentiated on morphological grounds alone. They are the largest very distinctly submetacentric chromosomes

and can easily be identified as a group. Their arm index is 2.6–3.2 and
the centromere index is 24–30. No chromosome of the C group has such
a low centromere index and consequently they can hardly be confused
with this group. Chromosome 4 has been said to be about 5–8 % longer
than No. 5 when measured. Since this falls within the normal variation
of length, Patau (1965) considers length measurements not to be useful
in differentiating the two pairs. The issue has been settled by distinct
autoradiographic and fluorescent differences which unambiguously
distinguish the two numbers of this group, as described in Section 4.

C Group and X Chromosome
No individual chromosome of the C and X group can be clearly distinguished. According to the London convention (London Report 1963)
pair Nos. 6, 7, 8, and 11 are relatively metacentric with a centromere
index of about 35–40, whereas Nos. 9, 10, and 12 are relatively submetacentric with a lower centromere index of about 27–35. These differences can be used to pair presumptively homologous chromosomes.
Some authors consider these efforts to be of little use (Patau, 1965),
while others (Turpin and Lejeune, 1965) feel that a morphological
differentiation within the C group is possible. The fluorochrome and
Giemsa banding patterns described in Section 4 have resolved the
problem. Chromosome 6 is the largest chromosome of this group, but
only a little larger than the presumptive X chromosome and chromosome 7. The length of the X chromosome will be discussed in Section 4.
As mentioned later, a secondary constriction is often found in the
proximal long arm of a pair of this group which is designated pair 9.

D Group
Chromosomes 13–15 are easily recognized as a group of large acrocentric
chromosomes. The centromere index is only 15 which is the lowest in
the human karyotype. These three pairs carry satellites which, probably
for technical reasons and also because of individual variation in size,
are rarely all visible at once in a single cell. The D group chromosomes
differ slightly in length (up to about 10%), which will usually permit
their arrangement in descending order of length. The lack of clear
morphological differences can be overcome with special identification
methods, as described in Section 4.

E Group
Chromosomes 16–18 of this group are fairly short and have a median
or submedian centromere. Chromosome 16 can regularly be distinguished from the two other pairs of this group. It is metacentric, or
slightly submetacentric, with an arm index of 1.4–1.8 and a centromere
index of about 40. Generally its length equals about one third of the
length of a chromosome 1. Chromosome 16 possesses considerable

individual variations in size owing to a frequently observed secondary constriction in the proximal long arm, as described in more detail in Section 5. Chromosomes 17 and 18 can be distinguished in good preparations by their length and the position of the centromere. Chromosome 18 is on the average about 5–10% shorter and has distinctly shorter *short* arms with an arm index of 2.4–4.2 and a centromere index of about 26 (21–29). The centromere index of chromosome 17 is 31 (a range of 23–36) and the arm index is about 1.8–3.1, indicating a more proximal position of the centromere. Obvious autoradiographic and special staining differences also exist in this group (see Section 4).

F Group
Chromosomes 19 and 20 are small metacentric chromosomes which are not distinguishable from each other by any means except by fluorochrome banding and special Giemsa staining patterns. The arm index is given as 1.2–1.9 and the centromere index as approximately 40 with a range of 34–46 (Chicago Conference, 1966). As a group these chromosomes are easy to recognize.

G Group
Chromosomes 21–22 are very short acrocentric chromosomes with satellites at the end of the short arm, although these usually are not all apparent in one cell. Both pairs often show slight differences in lengths. The smaller one is now designated as No. 21 (see Section 4), because definite differences in the fluorescent and special Giemsa staining properties have permitted distinction of the two pairs.

4. Identification of Individual Chromosomes by Special Methods

Two technical advances, the fluorescent banding and the Giemsa banding techniques, introduced in 1970 and 1971, respectively, provide important methods for identification of each individual chromosome. The technical aspects have been described in Chapters V and VI. In essence, the identification of each chromosome rests on a distinct pattern of bands in specific regions along the metaphase or prometaphase chromosome. These bands are produced either by quinacrine fluorescence (Caspersson *et al.*, 1970a, 1970b, 1971; Manolov *et al.*, 1971), now called *Q bands* (Fig. 5), or by various modifications of Giemsa stain after certain denaturation procedures (Sumner, Evans, Buckland, 1971; Patil, Merrick, Lubs, 1971; Drets and Shaw, 1971; Schnedl, 1971), now called *G bands* (Figs. 6–8). Originally the Giemsa band studies were preceeded by methods to stain selectively the "constitutive heterochromatin" by

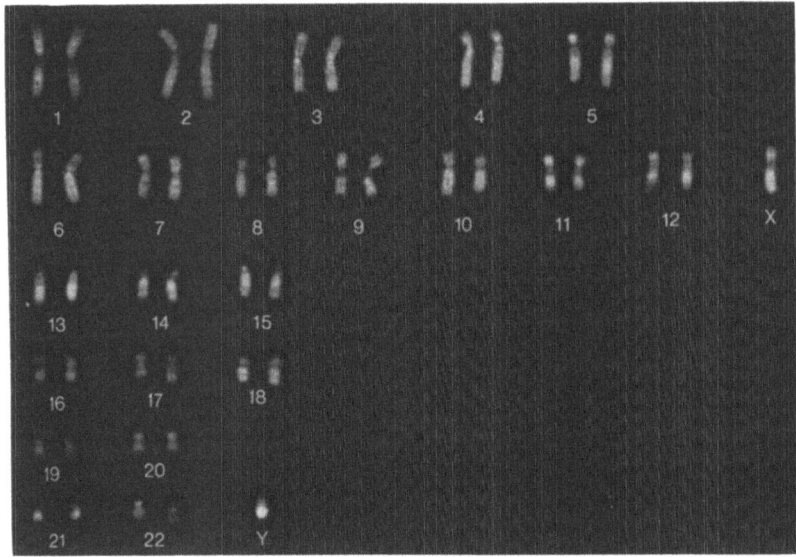

Fig. 5. Q banding pattern of a normal male karyotype on quinacrine mustard fluorescent stain (courtesy of Dr. Lore Zech). Note variant fluorescent spot at centromere of chromosome 3

in situ DNA–RNA hybridization (Pardue and Gall, 1970; Arrighi and Hsu, 1971), now called *C bands* (Fig. 10). A special use of some pretreatment (incubation at 87° C in buffer solution) produced patterns which were the reverse of the other Giemsa staining methods (Dutrillaux and Lejeune, 1971), now called *R bands*. Other modifications have produced slightly different patterns or accentuated the established ones such as the use of Giemsa at pH 11 (Bobrow, Madan, and Pearson, 1972; Gagné and Laberge, 1972). The use of proteolytic enzymes followed by Leishman or Giemsa stain also produced *G bands* (Dutrillaux *et al.*, 1971; Seabright, 1971 and 1972; Wang and Fedoroff, 1972; Fig. 8). Recently a special type of terminal bands *(T bands)* has been described that is useful for the determination of juxta-telomeric break points (Bobrow *et al.*, 1972; Dutrillaux, 1973; Dutrillaux *et al.*, 1973).

In this context, a *chromosome band* is defined as part of the chromosome that is clearly distinguishable from its adjacent segments by appearing darker or lighter in either of the techniques (Paris Conference, 1971). Each band is designated by its midline rather than its margins. Since all areas of a chromosome are specified as belonging to a band, there are no interband regions. A *chromosome region* is defined as any area on a chromosome that lies between the midlines of two adjacent

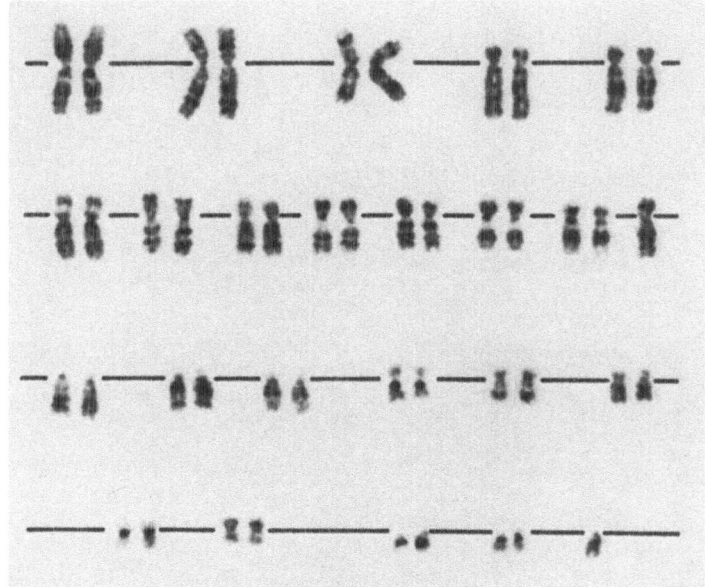

Fig. 6. G banding pattern on acid-saline-Giemsa stain of a normal male karyotype (courtesy of Dr. H. John Evans and The National Foundation, New York). First row: group A and B, second row: group C and the X chromosome, third row: group D and E, fourth row: group F and G, and the Y chromosome

landmarks. A *chromosome landmark* is defined as a consistent and distinct cytological feature that is an important aid in identifying a chromosome and includes bands, centromeres, and tips (cited from the Paris Conference, 1971). Chromosome bands, regions, and landmarks are shown in Fig. 9 and listed in Table 2 of the Appendix. Variable chromosome bands are not to be used as landmarks. The Paris Conference (1971) recommends the following terms to indicate the intensity of fluorescence:

negative = no or almost no fluorescence;
pale   = faint fluorescence as on distal 1p;
medium = as the two broad bands on 9q;
intense = as the distal half of 13q;
brilliant = as on distal Yq.

Although the banding patterns produced by the different staining reactions are probably analogous in one way or another, the description of individual chromosome patterns given below is subdivided according to the main technique employed, except for the patterns produced by

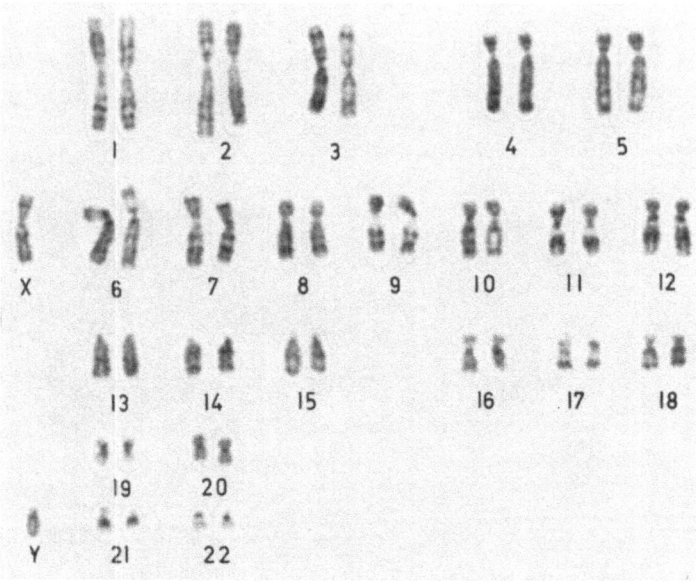

Fig. 7. G banding pattern after alkaline pretreatment followed by Giemsa
stain of a normal male karyotype (courtesy of Dr. W. Schnedl)

tritium autoradiography which is a basically different method. The
following section is intended to provide a diagnostic aid in the recognition and characterization of an individual chromosome as observed
through the microscope or in a photomicrograph. An absolutely complete but technically unrealistic definition has not been attempted, and
special photometric and densitometric methods have been omitted when
they have confirmed the visual identification (for recent reviews see
Caspersson and Zech, 1972 and 1973). The number and width of most
bands is to some extent correlated with the degree of contraction. There
tend to be more and narrowed bands in prometaphase than in metaphase (see Schnedl, 1971) as illustrated by Fig. 7.

The following patterns are based on the references given above and
own experience: *Q bands* revealed by differential binding of fluorochromes (quinacrine) shown in Fig. 5 and *G bands* revealed by different
modifications of the Giemsa techniques shown in Figs. 6–8 are described
below. The *C bands* are described separately (Section 4.2). The *R bands*
are not described separately, because in principle they are the reverse of
the *G bands* (like a negative).

Recent reviews are now available for detailed discussions of the
application and the theoretical background of the new staining techniques

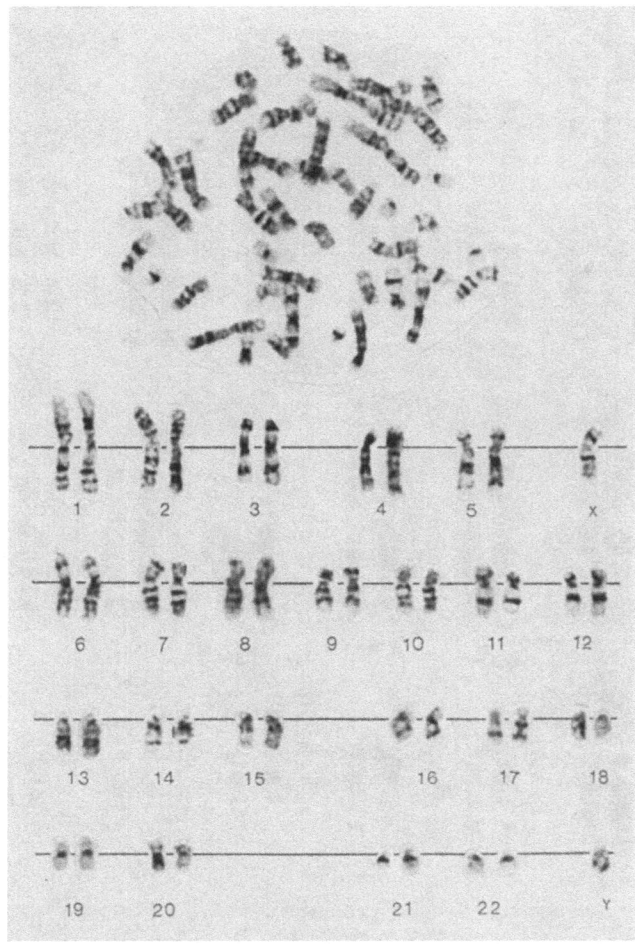

Fig. 8. G banding pattern after pretreatment with trypsin followed by Giemsa stain of a normal male karyotype

(Caspersson and Zech, 1973; Miller *et al.*, 1973; Comings, 1973; de la Chapelle *et al.*, 1973).

4.1. The Individual Banding Patterns (Q and G Bands)

Both patterns are shown in a diagram (Fig. 9) and listed in Table 2.

Chromosome 1

Q bands: The short arm has a medium fluorescent band over its proximal region towards the non-fluorescent centromere and a medium band

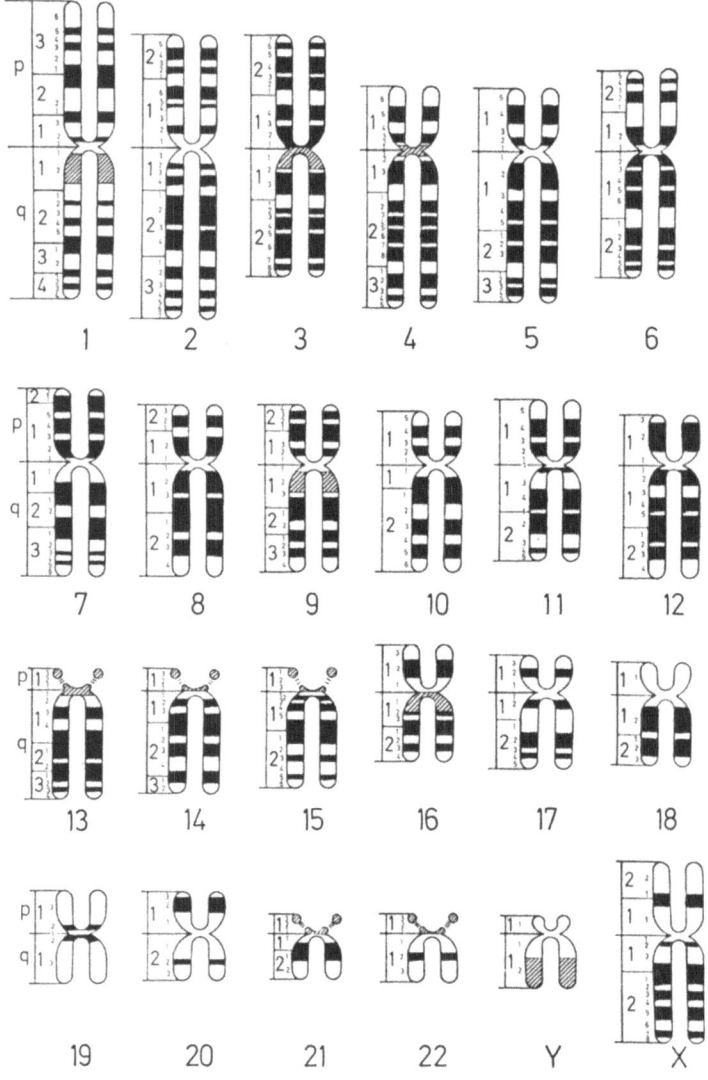

Fig. 9. Diagrammatic representation according to the Paris nomenclature of chromosome bands and regions as observed with the Q-, G-, or R-staining methods; centromere representative of Q-staining only. Positive Q and G bands, and negative R bands are shown in black, variable regions are shown hatched. Arm designations (p for short arm, q for long arm), region numbers, and band number are indicated to the left of each row, as provided by the Paris nomenclature (see Section 7). (From the Paris Conference, 1971, courtesy of Dr. Daniel Bergsma and The National Foundation, New York)

over the central regions 2–3. The long arm has little or no fluorescence over the middle part of the long arm (junction of regions 2 and 3).

G bands: The short arm has two broad bands over its proximal half and four bands over the middle and distal part of the long arm. The secondary constriction stains heavily and is highly polymorphic (see Section 5).

Chromosome 2

Q bands: There is fairly indistinct medium fluorescence over the short arm and the proximal as well as the distal long arm, while the middle part of the long arm may show two relatively bright fluorescent bands which may be fused into one broad segment (region 2).

G bands: Four to seven bands may be present over the entire long arm and three to four over the short arm. The centromere is only lightly stained.

Chromosome 3

Q bands: An almost symmetrical pattern results from a pale band in the middle section of each arm, which separates segments of medium fluorescence distal and proximal to it, the distal segment being slightly longer in the long arm. The centromere shows polymorphic fluorescence at the proximal end of the long arm as described in Section 5.

G bands: The centromere and the paracentric region is fairly densely stained. The distal parts of both arms stain intensely while the central parts of both arms remain light and appear as a white band (junction region 1 and 2).

Chromosome 4

Q bands: Diffuse medium fluorescence of the entire chromosome, in particular of the long arm. There may be a bright band at the proximal part of the long arm near the centromere (region 1). A pericentric narrow bright band occurs as a polymorphic character (see Section 5).

G bands: Generally more evenly and densely stained than No. 5. Four evenly spaced wide bands are present over the long arm, one or two over the short arm.

Chromosome 5

Q bands: The middle of the long arm contains a broad, medium to intense fluorescent band or number of bands while its distal region (3) and the centromere display no or little fluorescence. The short arm of chromosome 5 cannot be well distinguished from that of chromosome 4, except that its central band (region 1, band 4) tends to be brighter and shorter than on 4p.

G bands: A heavily stained broad band covers the middle third of the long arm (distal region 1 and proximal region 2). It is separated from a telomeric band by absence of staining over the distal part of the long arm.

Group 6–12 (C) and X

Identification of individual chromosomes has now become possible for this group. Chromosomes 11 and 12 have a rather similar fluorescence pattern, but all other pairs are quite readily identified in technically good preparations.

Chromosome 6

Q bands: Two distinct bands of medium brightness in the distal short arm are separated by a faint band. The long arm may show up to four bands of medium intensity. The centromere is evenly bright.

G bands: The short arm has a wide proximal area of rather faint staining which sets off a heavily stained distal quarter of the short arms resembling satellites. The paracentric region of both arms and the entire long arm, except its telomeric region, are heavily stained.

Chromosome 7

Q bands: The short arm terminates with a narrow medium-intense band. In good preparations, a proximal band of medium intensity may be present over the short arm. The centromere is faint. Two bright bands are located over the proximal and central part of the long arm.

G bands: Three distinct bands are found over the long arm. The centromere may be stained. The short arm has 2–3 bands, one band being telomeric.

Chromosome 8

Q bands: This chromosome shows fairly even medium fluorescence and has a fairly indistinct pattern. The slightly more fluorescent long arm may show three broad bands in good preparations.

G bands: A broad band, sometimes split into 3–5 individual bands, is located over the middle part of the long arm. Its proximal region stains lightly. The short arm has 1–2 bands.

Chromosome 9

Q bands: Two evenly spaced bands of medium brightness over the long arm allow easy recognition of this chromosome. The secondary constriction at the proximal long arm remains dark and is highly polymorphic as described in Section 5. The short arm contains one central broad band of medium intensity.

G bands: The middle and the distal part of the long arm have two wide bands which are sometimes split in two bands each. The secondary constriction remains negative with both Giemsa and quinacrine stain, but is highly polymorphic (see Section 5). It stains almost selectively with the Giemsa-11 technique (Giemsa stain at pH 11, Bobrow, Madan, and Pearson, 1972; Gagné and Laberge, 1972).

Chromosome 10

Q bands: Three evenly spaced bands, which may appear as a broad band, of medium to bright intensity are located over the long arm. The centromere is dark. The short arm is evenly fluorescent, two bands (2 and 4) being visible in good preparations.

G bands: The long arm regularly has three bands with band 5 of region 2 sometimes split. The most proximal band stains most heavily. The short arm has one broad or two narrow bands over the middle part (bands 2 and 4).

Chromosome 11

Q bands: The centromere is of medium to faint intensity and separated from both arms by symmetrical dark bands. Broad bands of medium intensity are located over the central parts of both arms. Dark distal bands delimit the proximal and the distal end of the long arm.

G bands: Two bands are located over the middle part of the long arm. They may be separated by a narrow negative band. The centromere may be stained and separated from the short arm by a narrow negative band from a broad positive band.

Chromosome 12

Q bands: The faint fluorescence and the localization of fluorescent bands resemble chromosome 11, but the short arm is shorter reflecting the lower centromere index of this chromosome. The long arm is brighter than the short arm.

G bands: The G pattern is similar to chromosome 11 as in the quinacrine stain, but the central band over the proximal part in the long arm is wider than in chromosome 11.

X Chromosome

Q bands: Paracentric dark segments separate evenly bright bands over most of the distal short arm. Over the long arm one broad band is present, which is composed of three distinct and almost even bands in good preparations. The size of the X chromosome lies between chromosome 7 and 8. Pattern or size for the X chromosomes in females do not differ.

G bands: There are 2–4 well marked bands over the long arm, the proximal one being most prominent, and one over the middle part of the short arm. No difference exists between the X chromosomes in female individuals.

Group 13–15 (D)

Unequivocal differentiation of the three pairs of this group is possible by all special methods. The satellite regions and the short arms are variable and polymorphic as described in Section 5. The satellites may be very bright as polymorphic marker in any member of this group.

Chromosome 13

Q bands: The distal half of the long arm has a broad bright band which may be divided into two bands. The short arms and satellites may be very bright.

G bands: The long arm stains positively over most of its length except for a narrow telomeric segment. Up to four individual bands may be present. The satellite region stains positively.

Chromosome 14

Q bands: The proximal half of the long arm is bright while the distal is faint save for a bright distal band.

G bands: A broad proximally located band which may be split into 2–3 bands and a negatively stained band over the distal third (band 4, region 2) characterize the long arm. There is a narrow distal band of medium intensity at the end of the long arm.

Chromosome 15

Q bands: The overall fluorescence is less than in Nos. 13 and 14. The proximal half of the long arm is faintly fluorescent.

G bands: The distal half of the long arm is lightly stained. The proximal half of this arm has 2–3 bands.

Group 16–18 (E)

Owing to its distinct morphological features, chromosomes of this group are easy to distinguish in good preparations. Fluorescence is useful to distinguish Nos. 17 and 18.

Chromosome 16

Q bands: The secondary constriction at the proximal part of the long arm is dark and quite variable in length as a heritable polymorphic trait. Distal to the region is a broad band of medium intensity. The short arm tends to be slightly less fluorescent.

G bands: The paracentric secondary constriction at the proximal long arm is heavily stained by the special Giemsa technique in contrast to its negative fluorescence. It is highly variable as a polymorphic trait. There is a lightly stained band over the junction of the distal to the middle third of the long arm. The short arm shows generally light staining, but 1–2 bands may be present.

Chromosome 17

Q bands: This chromosome shows low overall density and is one of the darkest of the complement. The long arm has distal one faint broad or two narrow bands which are separated by a narrow dark band.

G bands: The entire chromosome takes little stain and remains rather pale. There is one usually narrow band over the short arm and one over the distal long arm.

Chromosome 18

Q bands: The fluorescence is more even and brighter than in No. 17. Two medium intense bands or one broad band are visible over the long arm, the proximal one being brighter.

G bands: This is a densely stained chromosome. Two broad bands appear over the long arm, the proximal one being brighter and broader than the other. A medium narrow band may be present over the short arm.

Group 19–20 (F)

Previously it had been impossible to differentiate homologous pairs with any degree of certainty. Fluorescence microscopy has not solved all problems, but differences can now be recognized. The special Giemsa technique seems to be the most useful with respect to this group.

Chromosome 19

Q bands: This is the darkest fluorescent chromosome of the karyotype. In good preparations, a paracentric faint band can be seen on each side of the dark centromere. There is no distal brightness.

G bands: There is a fairly positively stained area around the centromere.

Chromosome 20

Q bands: The overall brightness clearly exceeds chromosome 18 and the most fluorescent bands of medium intensity are located over the distal part of the short arm, while the middle part of the long arm is slightly less intense.

G bands: The distal ends of both arms stain positively while the centromere remains less stained than in chromosome 19.

Group 21–22 (G)

The distinct fluorescence of these chromosomes now permits unambiguous differentiation and has resolved the problem of the so-called mongolism chromosome.

Chromosome 21

Q bands: This is the smaller, but distinctly brighter fluorescent pair which occurs as trisomy in Down's syndrome. The long arm shows proximal a bright and distal a faint segment. The satellites and short arms show variable, probably polymorphic, intense fluorescence.

G bands: Chromosome 21 is darker and smaller than No. 22. In particular, the long arm is densely stained, with 1–2 proximal bands discernible.

Chromosome 22

Q bands: Weak fluorescence which tends to fade towards the distal part of the long arm. The centromere remains negative.

G bands: There is stain mainly around the centromere, but not in the distal two-thirds of the long arm.

Y Chromosome

Q bands: The Y chromosome has a characteristic brilliant fluorescent segment on the distal half of the long arm. This is the most intense region of the karyotype and allows unequivocal identification of the Y in most cells. It may be subdivided into two or more bands. This region may vary in length, which causes the well known polymorphism in the length of the Y chromosome (Schnedl, 1971; Bobrow *et al.*, 1971), as described in Section 5. The short arm and the proximal long arm show faint and even fluorescence.

G bands: The distal half of the long arm is the most densely stained region of the karyotype, analogous to the fluorescence. It is also variable in length and polymorphic. Two bands may be visible, the proximal band generally being darker. The short arm and the proximal long arm are usually likewise positively stained.

4.2. C Bands

C bands occur as constitutive heterochromatin at four typical locations: (i) at centromeres, (ii) at acrocentric regions, (iii) at the secondary constriction sites of chromosomes 1, 9, and 16, (iv) as Y-chromatin at the distal long arm of the Y chromosome corresponding to the brilliant fluorescent part. They are described separately because they are less suitable for individual characterization of a given chromosome, but in combination with other techniques and owing to their polymorphic nature they may be useful. Their localization in the centric region of human chromosomes is different from the mouse where it flanks the centromere of bi-armed chromosomes. This is important for the recognition of human chromosomes in interspecific somatic cell hybrids. The typical C banding pattern has been described by Arrighi and Hsu (1971), Chen and Ruddle (1971),

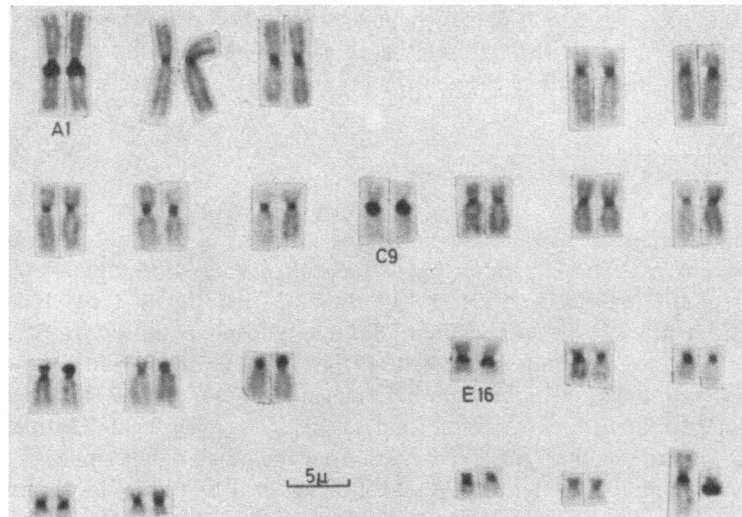

Fig. 10. C banding pattern (courtesy of Drs. T. R. Chen and F. H. Ruddle, and *Chromosoma*). The prominent constitutive heterochromatin at the secondary constrictions of chromosome 1, 9 and 16 are designated

Gagné *et al.* (1971), Yunis *et al.* (1971), and is demonstrated in Fig. 10. Chromosomes 1, 9, and 16 exhibit large amounts of heterochromatin in their proximal long arm at the site of the secondary constriction; this is polymorphic like the distal heterochromatin of the Y chromosome. They are highly polymorphic and described in Section 5 (see Figs. 10 and 15).

The regular C pattern, according to Arrighi and Hsu (1971), and Chen and Ruddle (1971), shows heterochromatin on the proximal end of the short arm in chromosomes 6, 8, 11, and 12, whereas in chromosomes 7, 10 and X the heterochromatin is centrally located over the centromere at the convergence of the short and the long arm.

The D group shows three patterns: chromosome 13 with two distinct regions of heterochromatin over the proximal long and the proximal short arm and distal over the satellite region, in chromosome 14 continuously throughout the short arm, and in chromosome 15 the heterochromatin region tends to be limited to the region directly over the centromere.

In the E group, where chromosome 16 is most distinct as described in section 5, chromosome 17 tends to show its heterochromatin distribution towards the short arm whereas in chromosome 18 it appears to be more evenly distributed. The F group is not distinct by this technique.

In the G group one can distinguish two pairs, one where hetero-chromatin is spread throughout the short arm (chromosome 21) and one where it is limited to the centromeric region (chromosome 22).

4.3. T Bands

Specific staining of the telomeric (terminal) regions of chromosomes can produce a pattern (T bands) that is helpful in analysing anomalies in this region (Dutrillaux, 1973; Dutrillaux et al., 1973). The T bands are produced either by heat exposure to 87° C in phosphate buffer at pH 6.7 in the presence of commercial Giemsa stain followed by acridine orange stain (5 mg in 100 ml phosphate buffer at pH 6.7), or by the same heat treatment followed by Giemsa stain at pH 5.1. After heat treatment for 30 min and longer the T bands disappear and R bands appear.

According to Dutrillaux (1973), following heat treatment for 15 min and staining with acridine orange T bands are typically present as green bands at the end of the *short* arm of chromosome 1, 4, 11, and 19, whereas they are most prominent at the end of the *long* arm of chromo-some 8, 9, 10, 17 and to some extent also of number 5, 12, 14, 16, 20, 21, and 22. No T bands have been recognized in chromosome 3, 6, 18, and the X and the Y chromosome. Furthermore, typical intercalary green bands can be produced in chromosome 11 (proximal long arm), 19 (proximal short arm), and 22 (proximal long arm), if heat treatment persists for about 15–25 min. Dutrillaux (1973) emphasized that the long arm of chromosome 22 can be particularly well identified by this method.

4.4. Autoradiography

Autoradiography of ³H-thymidine-labeled chromosomes permits the identification of chromosomes 4 and 5 (B group), 13, 14, 15 (D group), 16, 17, 18 (E group), the Y chromosome, and one X chromosome in cells of female origin. In addition, the chromosomes of the A group display a characteristic replication pattern which can be used for identifi-cation or as reference points. Previously autoradiographic analysis was the only special method for identification of individual chromosomes or chro-mosome pairs. The introduction of the new staining techniques has reduced the importance of the time-consuming and usually less unequivocal autoradiographic identification, which has, however, retained definite usefulness as an additional and independent identification procedure, and is indispensable for the identification of X-chromosomal anomalies in female cells, especially X-autosomal translocations (Fig. 11). Auto-

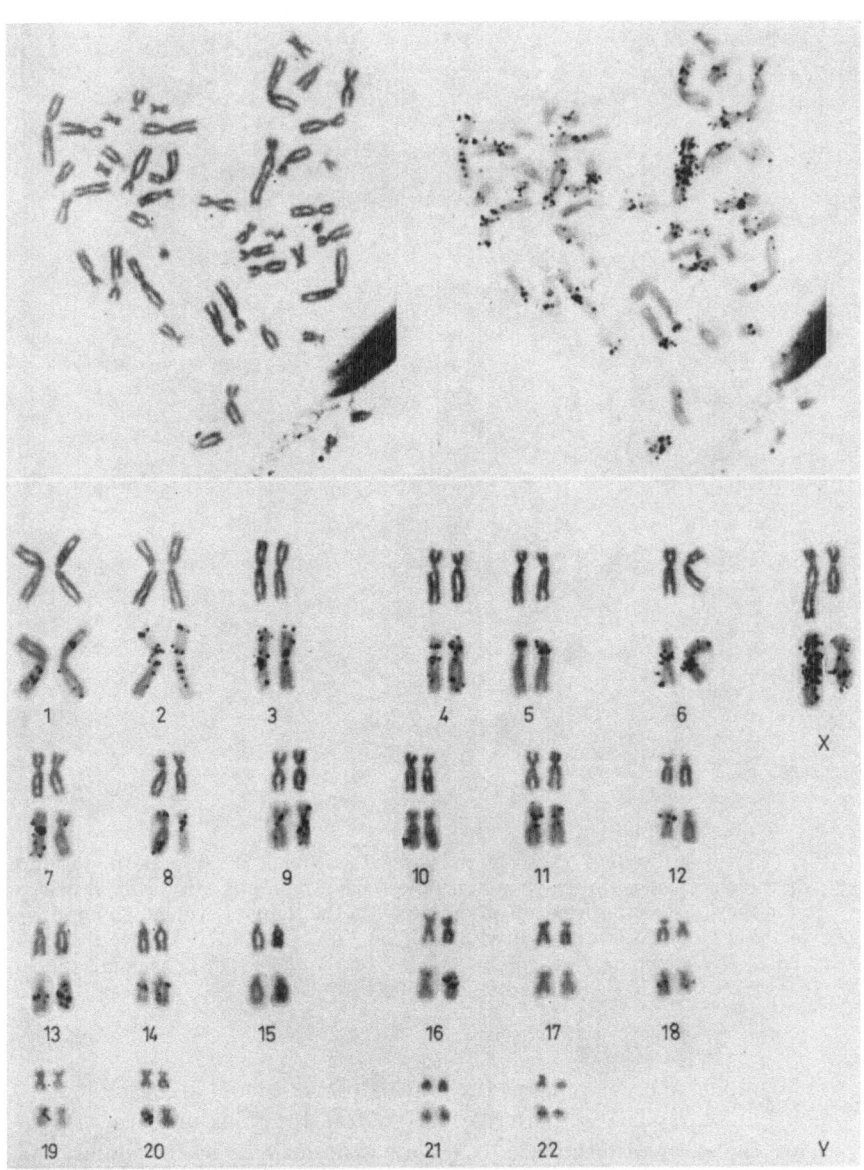

Fig. 11. Autoradiogram of an X-autosomal translocation in a female cell from a fibroblast culture. The long arm of a late replicating X chromosome has been extended by the translocation of an unidentified, presumably autosomal segment which replicates relatively early. Other characteristic patterns of terminal labeling are apparent in chromosomes 3, 4, 5, 13, 14, 15, and the E group (16–18). Chromosomes 21 and 22 should now be reversed, because the smaller, later replicating pair has been identified as number 21 (see text). (Passarge and German, unpublished data; case identical with patient HG 230 of German, 1967)

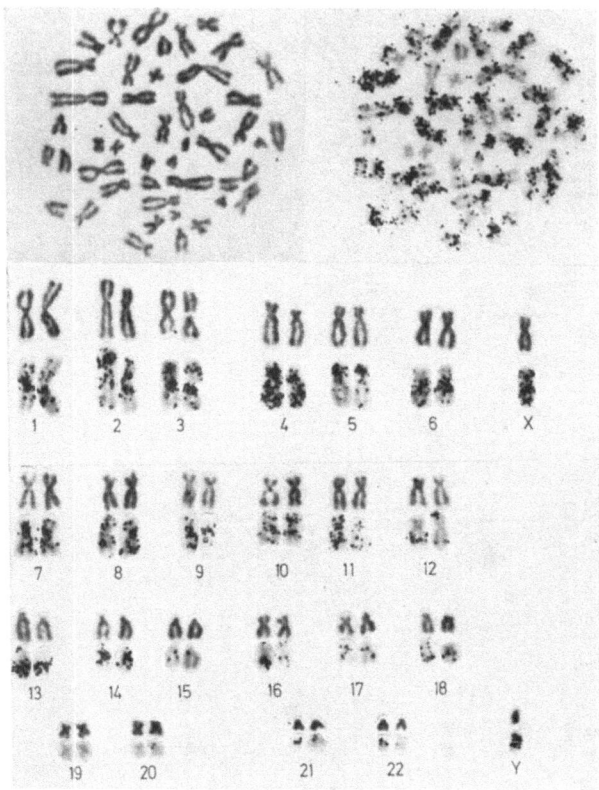

Fig. 12. Autoradiogram showing the typical terminal DNA pattern during "intermediate" stage in nearly all chromosomes. In addition, a terminal deletion in the short arm of one chromosome 4 (on the right) is evident. (Lymphocyte culture, ^{3}H-TdR for the final 6 hours, 0.3 μCi per ml, 1.9 Ci/mMol spec. activity, AR stripping film for 10 days, Aceto orcein stain.) Magnification about × 1,000

radiographic analysis remains the only cytogenetic method for the analysis of DNA replication patterns, e.g. in translocations and insertions. Thus, the study of terminal DNA synthesis remains a useful adjunct to other methods.

Excellent recent descriptions of replication patterns in normal and some abnormal human metaphase chromosomes have been provided by Giannelli (1970) and Miller (1970). Comparative studies using autoradiographic techniques and the Giemsa and Quinacrine techniques on the same cells have unambiguously identified the replication patterns of *all* chromosomes (Ganner and Evans, 1971; D. A. Miller *et al.*, 1971; Breg *et al.*, 1972; Calderon and Schnedl, 1973). Useful comments on the application of autoradiography are recorded in the Chicago Report (1966). The following Section describes the autoradiographic analysis of chromosome pairs No. 4 and 5, C group, pairs 13, 14, and 15, and one X chromosome in cells of female origin in mitoses derived from cultures where ^{3}H-thymidine was continually available for 6–7 hours before harvest. Prior to the G_2-phase (cf. Chapter VII), these cells could have incorporated the isotope for approximately the last $1^1/_2$–3 hours of the S-phase, when the relative asynchrony of replication in different chromosome segments is most prominent, as in Fig. 12.

It is convenient to divide the end of the S-phase into an "early", "intermediate", and "late" stage (Schmid, 1963; Gey, 1966), as illustrated in Figs. 11 and 12. The replication pattern of a morphologically distinct chromosome can serve as point of reference, for example *chromosome 1* whose pattern of replication can be divided into the following three stages: 1. early stage: replication in the entire chromosome, except for the distal quarter of the long arm; 2. intermediate stage: replication in the proximal long arm and the distal short arm (Fig. 12); 3. late stage: replication in the proximal long arm near the centromere (Fig. 11). The long arm is defined by radioactivity near the centromere during the late stage and the secondary constriction there (cf. Section 5). The morphologically identifiable chromosomes Nos. 2, 3, 16 can also be used as indicators of the replication stage (Schmid, 1963; Gey, 1966). In our experience many cells defy a clear classification into stages. Different, and more detailed stages of the late S-phase were defined by German (1964a).

Replication of the B Group
These morphologically similar chromosomes can be easily distinguished autoradiographically in the intermediate and late stages of the S-phase. The pair that is late replicating in its entire length is regarded as No. 4, while pair No. 5 completes synthesis in the long arm relatively early so that only the short arm is labeled (Fig. 11 and 12). Although this characteristic pattern is not present in all cells, up to 90% of metaphases in the intermediate stage are suitable for analysis (Gey, 1966).

Replication of the C Group
The fluorochrome identification has now permitted determination of the DNA labeling patterns of specific C group chromosomes (Breg *et al.*, 1972) which had not previously been possible with certainty.

Based on a total of 33 cells studied by double karyotypy using fluorescent patterns and terminal DNA labeling from corresponding cells, Breg *et al.* (1972) described the C group as follows: chromosome 6 labeled latest in the centromeric region; chromosome 7 also labeled late in the centromeric region, but additionally labeled late in the long and the short arm; chromosome 8 showed late replication in both the long and the short arm, but not in the centromeric region; chromosome 9 and 11 were latest labeling in the short arm; chromosome 10 labeled latest in the long arm, sometimes close to the centromere; chromosome 12 labeled very early and was nearly unlabeled in late S-phase (see also Fig. 12).

Replication of the D Group

The acrocentric chromosomes of this group can be fairly reliably distinguished in the late stages of the S-period: pair 13 replicates in most of its long arm, while pair 15 exhibits only very little or no activity, and pair 14 replicates in intermediate fashion near the centromere and the proximal third of the long arm (Fig. 11 and 12). However, in many cells chromosomes 14 and 15 are difficult to distinguish and it is likely that misclassifications will occur. Gey (1966) found characteristic terminal DNA replication in only 60% of cells.

The detailed investigations of the D group by Giannelli and Howlett (1966), which were based on grain counts and measurements, have recently been expanded by comparative studies combining quinacrine fluorescence and autoradiography (D. A. Miller *et al.*, 1971; Ganner and Evans, 1971). The results confirmed the earlier autoradiographic classification of this group and demonstrated that brightly fluorescent chromosome segments were also late replicating.

Replication of the G Group

The differences in the replication pattern of this group are so minimal that differentiation has not been possible with certainty, although the two pairs of this group complete replication at different times at the end of the S-phase. The fluorescent patterns have confirmed these earlier reports that the smaller pair, which is trisomic in Down's syndrome, and designated as number 21, completes terminal DNA replication later than pair 22 (Ganner and Evans, 1971; Breg *et al.*, 1972). However, in view of the unequivocal differentiation of both pairs by their fluorochrome patterns, identification of G group chromosomes by means of autoradiography alone should no longer be attempted.

Late Replicating X Chromosome

In cells of female origin, one chromosome of the C (6–X–12) group becomes strongly radioactive in the late stages of the S-phase, when replication in most parts of other chromosomes has been completed (Fig. 11). Various observations indicate that this is one of the two X chromosomes (for review see German, 1967). This late replication permits reliable identification of this chromosome in the large majority of metaphases during the late stages. Pulse-labeling (see Chapter VII) at the beginning of the S-phase has shown that the late replicating X chromosome begins DNA synthesis later than the other chromosomes.

Although the late replication of one X chromosome in XX cells involves both arms, the most intense activity tends to occur in the long arm, and in very late stages the short arm may take up relatively little label. Supernumerary and structurally abnormal X chromosomes generally replicate late, which allows proper differentiation from anomalies of the autosomes of the C group (Fig. 11).

The direct comparison of the same cells by autoradiographic and fluorescent techniques revealed no differences in the distribution or intensity of fluorescence in the late-replicating X chromosome and its homologous partner (Ganner and Evans, 1971; Breg *et al.*, 1972). The terminal DNA replication pattern thus remains the most reliable method to identify X-chromosomal anomalies in female cells. Abnormal X chromosomes, however, are not always late replicating, as illustrated by an X-autosomal translocation [t(14q; Xq)] with late replication of the normal X chromosome (Grzeschik *et al.*, 1972). In addition, Calderon and Schnedl (1973) have shown that the pattern of the late replicating, heterochromatic X chromosome is identical to that of the euchromatic X chromosome.

Replication of other Chromosomes

Chromosomes 1, 2, 3, 16, 17, 18, and the Y chromosome display a characteristic replication pattern which will be only briefly described, because they can be identified morphologically. Detailed descriptions can be found in the works mentioned earlier, in particular by Giannelli (1970) and Miller (1970). The replication pattern of these chromosomes may, however, be informative in some structural anomalies and may lead to proper identification of an abnormal chromosome.

Chromosome 1 has already been described above in connection with the division at the end of the S-phase (Fig. 12).

Chromosome 2 remains radioactive in its entire length when the distal ends of chromosome 1 have already completed synthesis. Relatively late activity is usually present in the proximal part of both arms, near the centromere.

Chromosome 3 is active near the centromere and the distal parts of both arms in the middle and late stages, so a symmetrical replication pattern may result.

Chromosome 16 replicates very actively in the long arm, mostly near the centromere during the early and late stages.

Chromosome 17 completes replication very early.

Chromosome 18 is still active in its entire length even during the late stages and can be easily distinguished from chromosome 17.

The *Y chromosome* replicates later than the chromosomes of the G group, but Craig and Shaw (1971) have shown that it is actually not a particularly late-replicating chromosome. The fluorescent part of the long arm can be easily identified, but the identification of the faint parts, i.e. the proximal long arm and the short, requires autoradiography as an additional identification procedure (see Fig. 12).

In general, the foregoing sections clearly show that late DNA replication and intense fluorescence may or may not coincide in one and the same chromosome segment. Late replication and intense fluorescence are correlated in chromosome 3 (centromere), the D group, chromosomes 17 and 18, chromosomes 21 and 22, and the Y chromosome. Differences are apparent in the B group (long arm of chromosome 4 and 5), the X chromosomes in female cells, the F group, where segments of bright fluorescence are not correlated with late replication, and the secondary constrictions of chromosomes 1, 9, and 16 where late replication is not correlated with bright fluorescence. The reasons for these differences (unknown at present) are discussed by Ganner and Evans (1971), Comings (1973), de la Chapelle *et al.* (1973), Miller *et al.* (1973).

Giemsa staining patterns and fluorescence have been compared by Evans *et al.* (1971), Dutrillaux *et al.* (1972), Aula and Saksela (1972), Comings (1973) and others. The most apparent homology concerns the secondary constriction of chromosome 9 which stains negatively with both techniques Miller *et al.* (1973), Schnedl (1973), and Ferguson-Smith (1974) reviewed in detail the differences in staining and their implications.

4.5. Chromosome Measurement

A considerable amount of data on measurements of chromosomes have accumulated (see Chicago Conference, 1966; Paris Conference, 1971 and Table 3), but in general its reliability has remained disputed. Patau (1965) and Hughes (1966) have described the inherent difficulties in some detail. Variation in length of single chromosomes may be considerable, even in good preparations. The minimal coefficient of length variation is 5.3% with a lower standard deviation of 4,4% (at $p=0.05$) according to Patau (1965). Differences in the degree of contraction in both arms of the chromosome and other factors related to preparation prevent measurements of absolute chromosome length for identification of homologous chromosomes in single cells. In air-dried preparations, chromosomes situated at the periphery tend to be somewhat longer than their homologous partners (Patau, 1965).

However, chromosome measurements may be a useful parameter within a given cell and in combination with other studies. Adequate standardization and controls are most important. Percentage measurements of relative length of the haploid autosomal complement (see Table 3) are usually obtained by measuring enlarged photographic prints of exactly corresponding magnification, or by projecting the negative on to a screen or other surface at a fixed distance. The use of a map-measuring instrument is recommended. Some systems employ an automated pencil reader which can digitalize the coordinates of a censor (Gilbert and Muldal, 1971). The centromere and/or the achromatic regions of a chromosome may cause problems; they often cannot be-determined with sufficient accuracy. Standardization and uniformity of procedures as well as careful controls must compensate for many inevitable inaccuracies. For detailed information on the methodology of chromosome measurements the reader is referred to Ferguson-Smith *et al.* (1962), Penrose (1964), Patau (1965), Giannelli and Howlett (1966), Hughes (1966), Bender and Kastenbaum (1969), Neurath and Enslein (1969), Perry (1969), and Giannelli (1970).

Basically, determining the DNA content of a chromosome is the only means of exact measurement, provided the packing of DNA is known and can be taken into account. In any event, no entirely satisfactory method of measurement has yet been worked out although determination of the so-called chromatin areal (Hughes, 1966) or microspectrophotometric methods (Rudkin, 1965) represent useful compromises which can be applied to certain problems. Actual measurements of relative length and centromere index from few different sources are given in Table 3 of the Appendix.

5. Variability of the Karyotype

In its early stages, a decade ago, studies in human cytogenetics were mostly concerned with the criteria that have defined the standard karyotype described in section 3, and with the major deviations in number and structure which are associated with various developmental anomalies and reproductive wastage. In recent years it has become apparent that studies of randomly selected healthy individuals are necessary in order to obtain unbiased knowledge of the actual frequency and distribution within the karyotype of chromosomal anomalies as well as certain variants, now commonly called *minor variants*, or *polymorphisms*. The

Edinburgh study (Court Brown *et al.*, 1966, 1967, 1969; Jacobs *et al.*, 1970) and the New Haven study (Lubs and Ruddle, 1970 and 1971), respectively, based on several thousand unselected adults and newborn babies, were the first systematic studies providing quantitative data about the variability of the human karyotype in normal adult and newborn populations, and much of what follows is derived from these data. More recently, another large newborn series has been published (Hamerton *et al.*, 1972) in addition to others (Leisti, 1971; Zankl and Zang, 1971; Mikelsaar *et al.*, 1973).

The emerging picture is that the human karyotype is a highly polymorphic system. Initially, using standard stains at metaphase, minor autosomal structural variants of the standard karyotype, as described in Section 3 were noted in about 2–3% of an unselected population of newborn babies (Lubs and Ruddle, 1970). More recent evidence, based on the special Giemsa staining procedures, more than doubles this figure (Craig-Holmes and Shaw, 1971 and 1973; Bobrow *et al.*, 1971; Ferguson-Smith, 1974), while the fluorescent variants occur in as much as 50% of the population (Mikelsaar *et al.*, 1973; Schnedl, 1974).

Knowledge of the type and distribution of variants is important in differentiating them from pathological situations and determining the precise location of break points in chromosomal rearrangements. Type, frequency, and site of these variants are characteristic for each chromosome and serve as distinct cytological marker when present. They are valuable for genetic studies, in particular mapping genes on chromosomes, because they are inherited in Mendelian fashion as demonstrated for example in the study of Wikramanyake, Renwick and Ferguson-Smith (1971). Furthermore, the fluorescent variants of the acrocentric chromosomes may permit to identify the origin of the duplicated chromosome in trisomic individuals (Breg *et al.*, 1971; Licznerski and Lindsten, 1972; Robinson, 1973; Ferguson-Smith, 1974).

In the following section some important general properties of the different human chromosomal polymorphisms mentioned above are considered, followed by a detailed description of specific variants and their frequency and an outline of their role in the identification of individual chromosomes.

5.1. General Properties of Chromosomal Variants and Their Basis for Recognition

The following types of chromosomal variants have been recognized:

1. Variations at the satellited regions of the acrocentric chromosomes of the D and the G group, consisting of enlargement of the short arm

proximal to the satellite, enlargement of the satellites, duplication of the satellites (tandem satellites), absence of satellites, diminished short arm (see Fig. 13a and b);

2. polymorphisms at secondary constriction sites, consisting of increased length, ability to stain rather selectively with certain modifications of the denaturation-reassociation Giemsa technique, constriction fragility, pericentric inversions (see Fig. 13c and d, and Fig. 14);

3. intense or brilliant fluorescence of the heterochromatin located at the centromere and satellites of some autosomes (e.g. 3 and 4, and the acrocentrics) and the long arm of the Y chromosome. These are particularly common variants, but detectable only by the fluorescence technique.

Arguments for the homology of these types have been advanced by Ferguson-Smith (1974). It should be noted that the data of Lubs and Ruddle (1970) in the New Haven study were derived from the analysis of two perfect metaphases (no overlap, no bent chromosomes) from each of 2,444 of a total of 4,400 unselected newborn babies. Individual chromosomes had not been identified by special means other than measurements. Racial differences were evident in the New Haven study where a number of variants occurred more frequently in negroes (Lubs and Ruddle, 1971).

Using the staining reactions producing Q and C bands in particular, but also the G bands as well as standard stains, the following important chromosomal variants have been recognized. These variants, when present, serve as a distinct feature of the particular chromosome involved although their polymorphic character does not always allow reliance upon them for diagnosis. The following section describes the site and specific properties of chromosomal variants recognized to date. The figures on frequencies are summarized in Table 4 of the Appendix.

5.2. Sites and Frequencies of Chromosomal Variants

Chromosome 1. In about 25% of metaphases, chromosome 1 displays a clearly visible secondary constriction in the paracentric region of the long arm (lqh+)[3] which may lead to a slight lengthening of the long arm. This constriction stains deeply with the C banding technique (Fig. 15) which confirmed the polymorphic character of this region and clearly demonstrated that there can be additional heterochromatin (Craig-Holmes and Shaw, 1971; Ferguson-Smith, 1974), although most variants show decreased heterochromatin (Craig-Holmes *et al.*, 1973). This region also stains darkly with the G banding and remains dark or weakly fluorescent with the Q banding method.

---

[3] For nomenclature see Section 7 of this chapter.

Previously this inherited variant had been thought to be due to differential contraction or uncoiling. Close linkage with the Duffy blood group locus led to the assignment of the *Fy* gene to chromosome 1 (for review see Renwick, 1971), which in turn led to the assignment of several other genes to this chromosome (Ruddle *et al.*, 1972).

An increased amount of heterochromatin at the secondary constriction of chromosome 1, leading to an increase in arm length, was observed by Ferguson-Smith (1974) in 8.8% (19 of 216) individuals thus constituting the site of the most frequent polymorphism in man that is detectable on Giemsa stains (fluorescent markers of chromosomes 3, 4, and the acrocentrics are much more common). Previous studies by Lubs and Ruddle (1970) and by Ferguson-Smith, using standard techniques, had shown an incidence of 0.62% and 0.5%, respectively. This region, as do most common polymorphic regions, shows late labeling with autoradiographic techniques using tritiated thymidine, during the terminal S-phase (see Section 4.3 and Chapter VII).

Chromosome 2. The distal third of the long arm is very rarely demarcated by a secondary constriction of which the frequency has been given as 0.17% by Lubs and Ruddle (1970) and 0.004% by Ferguson-Smith (1974) who observed increased fragility at this site. A paracentric secondary constriction with increased fragility has been observed as a familial trait (Lejeune *et al.*, 1968; Leisti, 1971; Ferguson-Smith, 1974). The C-band pattern of chromosome 2 is rather constant (Craig-Holmes *et al.*, 1973).

Chromosome 3. A narrow band of brilliant fluorescence occurs on the long arm immediately adjacent to and extending into the centromere (Fig. 5) in about 50% of individuals in one homolog, and in about 25% in both homologs (Schnedl, 1971 and 1974). It segregates as a Mendelian inherited trait. Leisti (1971) reported a secondary constriction at this site. Lubs and Ruddle (1970) reported a somewhat shortened chromosome 3 in 0.29% of 2,444 normal newborns which were studied in detail for minor variants. Ferguson-Smith (1974) has seen increased fragility in a medially located constriction of the long arm.

Chromosomes 4 and 5. Morphologically distinct variants on routine stain are extremely rare in the B group (Lubs and Ruddle, 1970) and their occurrence would appear more likely to be the result of a translocation or insertion unless an alteration of the otherwise rather constant C band could be demonstrated. The centromeric region of chromosome 4 shows bright fluorescence in approximately 35–50% of cells and appears to be consistent within an individual.

Chromosome 9. Chromosome 9 is the only chromosome of the C group which exhibits a frequent and characteristic polymorphism. Its paracentric secondary constriction at the proximal long arm has long been known as the site of considerable variation. The C banding technique revealed this region as a dense heterochromatic block whose size may vary by about 100% between individuals (Fig. 15). It is remarkable that this region usually remains rather pale and poorly stained with the G banding methods. However, it stains metachromatically in a very dramatic and rather selective fashion with a modification using Giemsa at pH 11 (Bobrow, Madan, and Pearson, 1972; Gagné and Laberge, 1972). Bobrow *et al.* (1972) observed obvious length variation between homologs in 7 out of 32 random blood samples. It remains negative as a Q band, but also displays variable length.

The secondary constriction can be relatively long and the centromere so indistinct that using standard stains a "translocation chromosome" may be erroneously diagnosed (Fig. 13c–d). This region, which replicates DNA late during the terminal phase of DNA synthesis, is of interest, because it may occur as a site of increased spontaneous chromosomal fragility (Schmid and Vischer, 1969; Schwanitz *et al.*, 1972). The increased susceptibility of this and other secondary constrictions (e.g. chromosomes 1 and 16) to breaks by mitomycin C had been known for many years (for review see Brøgger and Johansen, 1972; Brown *et al.*, (1972).

An interesting polymorphism of chromosome 9 consists of a pericentric inversion of its secondary constriction, resulting in an almost metacentric chromosome (Wahrman *et al.*, 1972). Ferguson-Smith (1974) observed this variant in 1.8% (4 of 216) individuals studied by the special Giemsa techniques. This is probably the same variant as that described by Lubs and Ruddle (1970, 1971) as a metacentric C group chromosome, then considered to be a number 11. Ferguson-Smith (1974) reported an incidence of an elongated secondary constriction of chromosome 9 (9qh+)[4] in 1.2% (27 of 2,270) of individuals when using standard stains and in 3.7% (8 of 216) when studying the G banding pattern.

Chromosome 13, 14, and 15. The D group has the highest incidence of minor variants in the karyotype. They occur almost exclusively in the satellite region and are detectable by standard techniques (Fig. 13a) although they are revealed in greater detail and clarity by virtually all banding techniques, especially by fluorescence. As mentioned earlier, the variants observed consist of elongated short arms (ph+), large satellites (ps+), double satellites (pss), shortened short arms with or without satellites (ph−). The Q banding pattern shows intense or brilliant fluorescence of either the satellites or the short arm, or both, in addition to a variable intense band of the proximal long arm of chromosome 13. Intensity and distribution of fluorescence tends to be remarkably con-

---

[4] See footnote on p. 169.

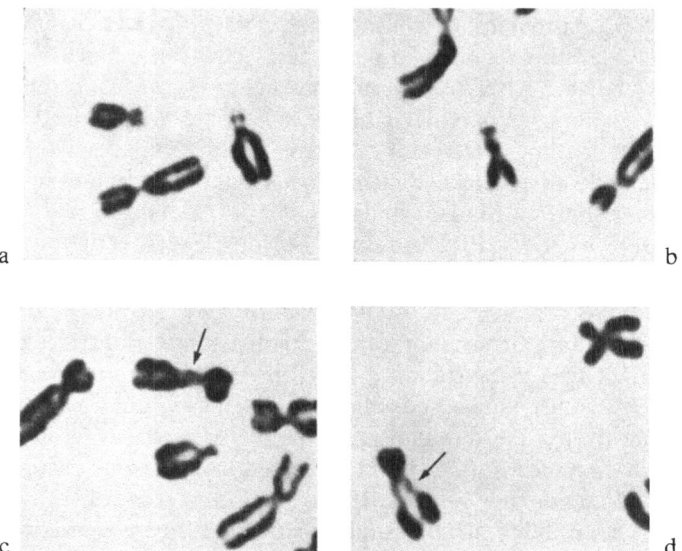

Fig. 13 a–d. Variants of chromosomes of the D group (a), G group (b), and prominent secondary constriction of a chromosome 9 (c and d). Standard Giemsa stain

sistent within an individual and serves as an excellent cytological marker. The G and the C banding pattern likewise demonstrates a high degree of variability of the satellite region, consisting mainly of differences in size of the centromeric heterochromatin block and differences in the size of the stained satellites.

In standard preparations, the following criteria were used by Lubs and Ruddle (1970) for defining variants in the D and the G group: The length of the short arm of an acrocentric chromosome is compared to the short arm of chromosome 18; normally the short arm of the acrocentric chromosome will be shorter than the short arm of chromosome 18. The short arm of an acrocentric chromosome is called "long" if it is as long as the short arm of a chromosome 18, and "very long" if it exceeds this length. The term "giant satellite" refers to satellites whose size exceeds that of the corresponding short arms.

The frequency of polymorphism in the D group was given as 3.7% (8 of 216) using C and G banding methods (Ferguson-Smith, 1974) and 2.3% (411 of 2,444) by standard methods (Lubs and Ruddle, 1970). Fluorescent techniques, however, reveal variants considerably more often, namely in about 50% of the population (Schnedl, 1974) which

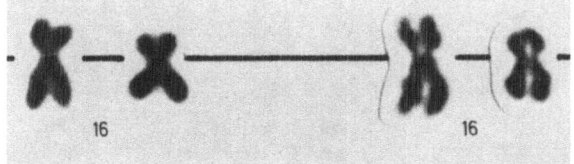

Fig. 14. Variation in length of chromosome 16 as a result of an extended
secondary constriction (left number of the pair on the right)

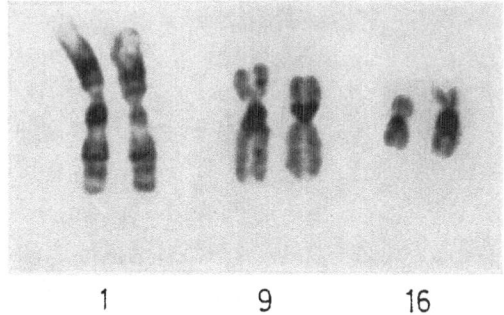

1   9   16

Fig. 15. Polymorphism of constitutive heterochromatin at secondary constriction sites of chromosomes 1, 9 and 16 as revealed by the C banding technique
(arranged from karyotypes provided by Dr. M. A. Ferguson-Smith)

renders them very important for tracing the origin of a particular
chromosome or its cell. Ferguson-Smith (1974) reported the following
distribution of the different types of polymorphisms for each of the
D group chromosomes: 7 ph−, 0 ph+, 5 ps+, 0 pss for chromosome
13; 0 ph−, 0p h+, 3 ps+, 2 pss for chromosome 14; 0 ph−, 7 ph+,
2 ps+, 1 pss for chromosome 15 (for nomenclature see Section 7) among
a total of 62 variant D group chromosomes in 2,486 individuals using
both standard (2,270) and banding (216) techniques.

Chromosome 16. The variable paracentric secondary constriction at
the proximal long arm of chromosome 16 (16qh) is a rather common
site of a human chromosomal polymorphism. Morphologically, it usually
appears as an extended, lighter staining region. It stains intensely with
the G and C banding methods (Figs. 6, 10, 14, and 15) and remains a
negative region with the Q banding technique (Pawlowitzki, 1972). The
C banding pattern may reveal a decreased heterochromatic region
at this site (Craig-Holmes and Shaw, 1971).

A different site (16 qh') located in the distal third of the long arm,
has displayed increased fragility as an inherited dominant trait (Magenis,

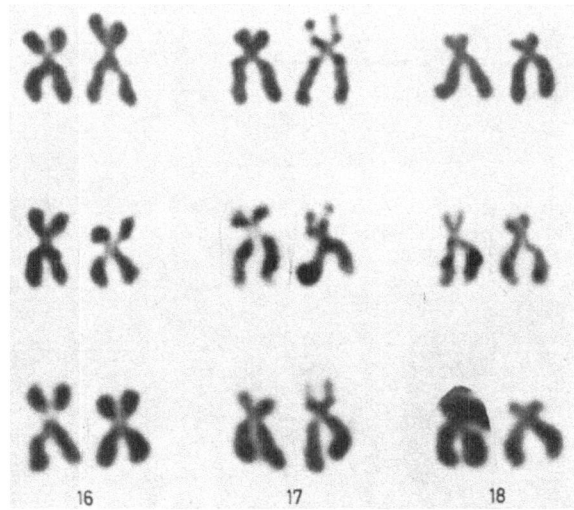

Fig. 16. Prominent secondary constriction in the short arm of chromosome 17
(from Schmid, 1969, courtesy of Dr. W. Schmid and *Cytogenetics*)

Hecht, and Lovrien, 1970), and is most susceptible to breakage by
mitomycin C (see Brøgger and Johansen, 1972). Both sites, 16 qh and
16 qh' have been used advantageously to determine the map position of
the haptoglobin α-locus (*Hp*α) on the long arm of chromosome 16, with
a probability of 9:1 that *Hp*α is located between the two sites rather
than distal to 16qh' (Renwick, 1971).

About 10% of cells usually show the secondary constriction in the
proximal long arm which leads to the frequently observed variation in
the length of this chromosome. It becomes visible in up to 80% of cells
after 7 hours in a calcium-free medium (Saksela and Moorhead, 1962;
Sasaki and Makino, 1963). Using banding techniques, Ferguson-Smith
(1974) observed a prominent secondary constriction on the proximal
long arm near the centromere in 5.5% (12 of 216) individuals whereas,
with standard techniques, a frequency of 0.88% (Ferguson-Smith, 1974)
and about 4% (Lubs and Ruddle, 1970), respectively, has been given.

Chromosome 17. Variants of this chromosome are rare. A pronounced
achromatic secondary constriction can lead to the formation of pseudosatellites in the short arm (Fig. 16) which may occur as a familial trait
and serve as a useful marker. It does not participate in association as
do the acrocentrics (Schmid, 1969; Leisti, 1971). Ferguson-Smith (1974)
found this variant in 0.46% (1 of 216) and 0.75% (17 of 2,270) individuals
using G banding and standard techniques, respectively. Lubs and Ruddle

(1970) found no such variant in their study and we have not seen it in preparations from over 2,000 individuals during the past five years.

A secondary constriction may also occur medially in the long arm of chromosome 17 (Ferguson-Smith, 1964 and 1974). This site may be identical to the area where an increased rate of chromatid and some isochromatid breakage by adenovirus type 12 has been demonstrated in human embryonic kidney cells (Zur Hausen, 1967), and near the *Tk* locus (thymidine kinase).

Chromosome 18. No variants have been recognized to date.

Chromosome 19 and 20. Craig-Holmes and Shaw (1971) described an F group chromosome with an increased amount of centromeric heterochromatin with the C banding procedure of Arrighi and Hsu (1971). Lubs and Ruddle (1970) reported a secondary constriction in the distal short arm resulting in pseudosatellites, and one chromosome with an elongated long arm, both corresponding to a frequency of 0.04%.

Chromosome 21, and 22. The satellite region of the acrocentric chromosomes of the G group is subject to polymorphic variation about as frequently as that of the D group chromosomes. Satellites as well as the short arms may display pale, medium, intense or brilliant fluorescence, the pattern being constant and heritable within an individual. In the G banding pattern, there may be either positive or negative heteropyknosis of prominent satellites consistent for each variant, whereas the satellite stalks and other secondary constrictions remain pale and negatively heteropyknotic areas in both the G and the C banding techniques. The centromeres stain darkly with C banding methods and reveal variations in width of the heterochromatic block (Craig-Holmes and Shaw, 1971; Ferguson-Smith, 1974). Standard staining techniques reveal a lower incidence of recognizable variants, but even these methods have given rather high incidence figures. Court Brown et al. (1966) observed a prolonged short arm in 1.2% of 756 persons in the Edinburgh study while Lubs and Ruddle (1970) observed 3.5% of the 2,444 newborns studied for minor variants in the New Haven study. Prominent satellites (Fig. 13b) were observed in 2.5% in the New Haven study.

Other variants are considerably less frequent and consist mostly of giant satellites or extraordinarily prolonged or shortened short arms (0.04–0.08 each, according to Lubs and Ruddle, 1970). For a definition of these variants see above under D group. In a number of cases, giant satellites were often unexpectedly associated with congenital malformations and they may not always represent normal variants. A secondary constriction occurs in the long arm of one chromosome of this group in about 5%. The familial occurrence of double satellites has recently been reviewed by Leisti (1971). Ferguson-Smith (1974) gave an overall inci-

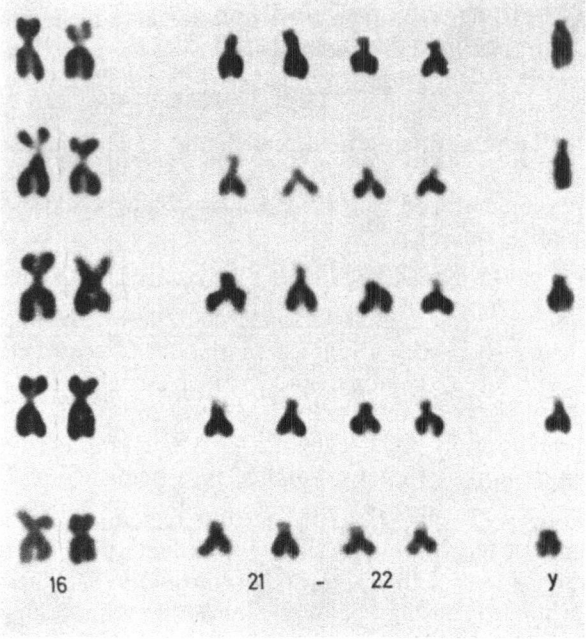

Fig. 17. Normal variation in length of the Y chromosome

dence of variant G group chromosomes of 1.6% and 1.8%, respectively, using standard and banding techniques. Among 2,486 individuals examined, he obtained the following distribution according to chromosome and type of variant: chromosome 21: 1 ph−, 2 ph+, 3 ps+, 1 pss; chromosome 22: 0 ph−, 1 ph+, 2 ps+, 0 pss.[5]

X Chromosome. No definite variants have been recognized so far. Lubs (1969) described the familial occurrence of an X chromosome with satellites on the distal long arm, but this was probably not a normal variant since there was some evidence for loss of chromosomal material.

Y Chromosome. The considerable variability in length of the long arm of the Y chromosome (Fig. 17) has been known for over a decade, and its remarkable consistency within one individual and strictly heritable nature has been well recognized and documented in numerous reviews (Cohen *et al.*, 1966; Unnérus *et al.*, 1967). This polymorphism has been utilized to detect foreign cells following an exchange transfusion (Naiman *et al.*, 1966), to exclude paternity (Jonasson *et al.*, 1972) or to trace male descendants in a large pedigree (Borgaonkar *et al.*, 1969).

---

[5] See footnote on p. 169.

In general the length of the Y chromosome in the population follows a normal distribution curve. Lubs and Ruddle (1970) found a variant Y chromosome in 5.6% of 2,444 newborns, most of which were elongated (longer than an F group chromosome in 5%, longer than a chromosome 18 in 0.33%) whereas it was very small (smaller than a G group chromosome) in 0.25%. Similar figures were obtained by Court Brown *et al.* (1966) who reported variants in 2–3% of a normal adult population. A short Y chromosome was found in less than 1% of males. Considerable advances in understanding the polymorphic nature of the Y chromosome resulted from the introduction of the Q, C and G banding techniques and the demonstration that the Y chromosome can be recognized as a brilliant fluorescent spot, the "F-body" or Y-chromatin, in interphase nuclei (Caspersson *et al.*, 1970c; Pearson, Bobrow and Vosa, 1970) (see Chapter IX).

The variable distal part of the long arm of the Y chromosome displays brilliant fluorescence and stains very darkly with both the C and the G banding techniques. It is now generally accepted that this is a genetically inactive region of heterochromatin (Arrighi and Hsu, 1971; Chen and Ruddle, 1971) and that the polymorphism of the Y chromosome results from variation of the length of this part (Schnedl, 1971; Bobrow *et al.*, 1971; Borgaonkar and Hollander, 1971; Robinson and Buckton, 1971). A short Y chromosome may lack the fluorescent part of the long arm, but this observation seems to be rare.

6. Evaluation of Cytogenetic Data

The analysis and evaluation of cytogenetic data must take into account differences between laboratories and individuals. No definitive approach exists, but certain rules should be observed to render results reliable and allow meaningful comparisons. Any communication should contain an adequate answer to the following basic questions: How are the data obtained? How extensive are they? How certain and reproducible are they? The following section is a brief outline of specific points deemed necessary in this context.

6.1. Microscopic and Photographic Analysis

The analysis of the cytogenetic results consists of microscopic and photographic analysis. Microscopic analysis means direct analysis through the microscope, usually with the help of drawing, by counting and checking each chromosome for proper classification and structural integrity. The importance of direct microscopic observations cannot be overestimated. It is recommended that at least one drawing of a good metaphase be made from every patient, because many small variants are not visible in photographs. Moreover, the observer is forced to examine a cell more carefully when he prepares a drawing. This drawing is schematic and designed to reproduce the relative position of individual chromosomes to one another while their identity according to pair number or group is diagnosed and recorded. This is done at maximal magnification. It can be recorded in many ways; the method employed in our laboratory is reproduced in Fig. 18. Obviously one has to record the position of the cell on the slide, the observer, and the microscope, for example, if the coordination of several microscopes differ. Even simple counting through the microscope without drawing should always be accompanied by looking for variants or structural anomalies. A partial or limited analysis may often also be necessary to screen many cells for certain chromosomes or chromosome groups in detail in order to confirm or discard a questionable observation.

The number of metaphases required to be examined can fluctuate between a few and far over 100 cells. About five clearly differentiated intact metaphases, or even two, may allow the recognition of a structural chromosomal anomaly if it does not occur as a mosaic (see Section 6.2). Analysis of about 10–15 good metaphases is recommended for routine analysis of patients, unless a mosaic is suspected on clinical grounds. If one or more cells with a conspicuous finding are found, more cells must be analysed. For example, in a particularly well-designed cytogenetic study of abnormal children (K. Patau), 11 cells were routinely examined directly through the microscope; if no anomaly was found the analysis was terminated; if a conspicuous cell occurred, the analysis was extended to 17 cells; if two identical anomalies occurred, a total of 23 cells were analysed (Summitt, 1969).

Photographic analysis forms the basis for documentation, further analysis, and the preparation of karyotypes. The number of karyotypes required for a single analysis depends on the problem investigated. It is again stressed that any doubt about a photographic finding calls for direct microscopic observation, because photographs reveal less details of chromosome structures than can be observed directly. Q bands can only be evaluated in photographs because of fading and G banding patterns are generally easier to recognize in photographs.

We consider it to be important that every karyotype should contain a photograph of the metaphase from which it has been derived, in order to reveal the quality of the cell, the degree of spreading, the position of individual chromosomes, homogeneity of the stain, background, possible artifacts, and other factors. The intact metaphase can be reproduced at

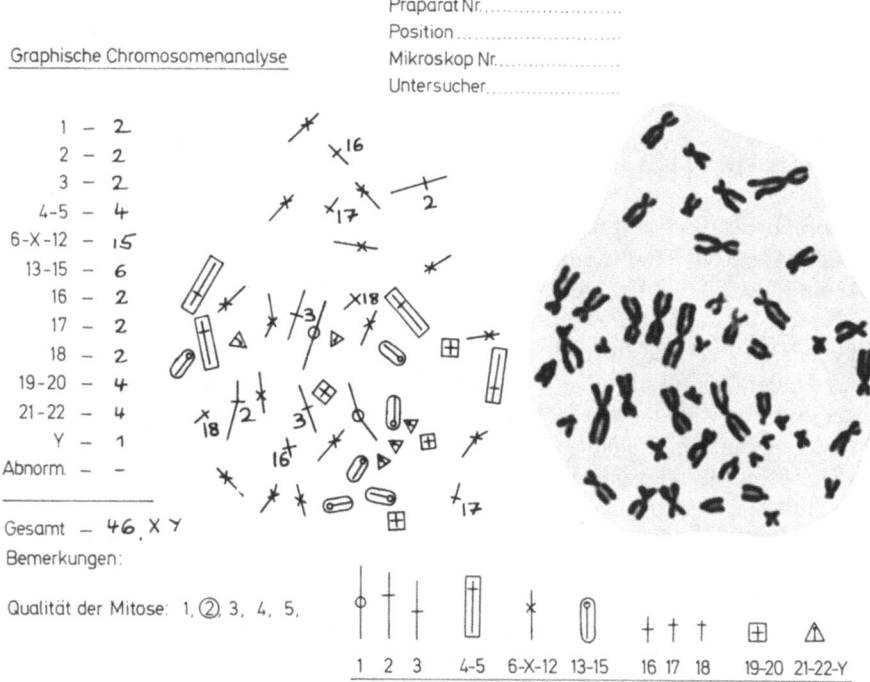

Fig. 18. Schematic drawing of a metaphase (left half of figure) which is, also shown as a photograph for comparison and projected on an analysis sheet, as routinely used in this laboratory (introduced originally from the laboratory of Dr. James German, New York)

smaller magnification than the karyotype for this purpose, as in Figs. 3 and 4. Photographs intended for preparations of karyotypes are generally enlarged 2,500–3,000 times. Instructions for microphotography usually come with the microscope. For a concise review see Christenson (1965). The proper evaluation of cytogenetic data requires the following information:

1. Number, type, medium, and duration of cultures. In lymphocyte cultures, method of mitotic stimulation and arrest, possibly the degree of transformation and mitotic index; in fibroblasts, the origin and age of culture, the number of subcultures, and other pertinent information;

2. the total number of metaphases examined;

3. the proportion of cells analysed by microscopy, including drawing and counting, and by photography and karyotypy;

4. in special studies, pertinent details of the technique involved, the number of informative and non-informative cells, data about the cell cycle, e.g. the proportion of unlabeled mitoses;

5. criteria for the selection of cells analysed.

A simplified method for recording a cytogenetic diagnosis is provided by the Chicago and Paris Nomenclature (see Section 7).

6.2. Diagnosis of Chromosomal Mosaics

Chromosomal mosaicism is the presence of more than one cell line with different chromosome complements of post-zygotic origin in one and the same individual. In a chimera the cells originate from more than one zygote (for review see Ford, 1969). The evidence for mosaicism or chimerism becomes difficult to prove if one cell line is in the minority, e.g. under 10% of all cells in the tissue examined. Definitive exclusion of a mosaic is, therefore, theoretically impossible. In practice one would apply an appropriate statistical procedure and state the confidence limits of exclusion in relation to the number of cells examined. Tables 5 and 6 in the Appendix correlate the probability of exclusion with the percentage of a different cell line and the number of cells examined. One Table is based on the Poisson distribution and the other on the binomial distribution.

An age-dependent XO/XX mosaic was discussed in Section 2 and a relevant table is listed in the Appendix (Table 1).

6.3. Analysis of Chromosomal Breaks

The study of chromosomal breakage and arrangements is relevant to its spontaneous occurrence in certain hereditary disorders (German, 1972; Passarge, 1972) and its induction by certain physical or chemical agents in vivo and in vitro (Bloom, 1972; Evans, 1972). A review of the different types of breaks and their implications is beyond the scope of this chapter and in the following section some comments restricted to the methodology of analysis will be made.

The exact procedure of analysis depends largely on the problem and the purpose of the investigation. Nevertheless, there are some guide lines which should be observed, such as those set forth by Cohen and Shaw (1965). In view of their eminent importance for a meaningful analysis of chromosomal breaks in mitosis and meiosis, they are summarized below.

1. Adequate controls are essential. The choice of controls depends on the experimental design. They may come from the same individual in certain experiments in vitro, or from control persons of the same age, sex, race etc. when evaluating in vivo effects or spontaneous chromosomal breaks. The control cultures should, of course, run concurrently

and be handled in the same way and at the same time. Experimental and control preparations should be coded, using random numbers.

2. In vitro experiments should be carried out on parallel cultures and their results compared. Two observers should analyse each cell. The cells examined should come from a large number of slides.

3. Selection and analysis of cells to be examined should aim at reducing undue bias. The observer should not know whether he is examining a test preparation or a control. It is often necessary, therefore, to let another person select the cells for analysis. This approach would reduce the possibility that the observer recognizes the test preparations from certain characteristics, such as mitotic rate, overall cytological impression, etc., prior to analysis. The criteria for the selection of cells cannot be given too much attention.

4. The procedure of analysis must be standardized. The observer should not be asked to interpret his observations during analysis. Criteria and definitions used throughout the study must be established prior to the initiation of the experiments and should not be changed without good reason.

5. For the subsequent statistical analysis of the data it would be prudent to design the experiment in accordance with established principles or advice from a statistician. Concerning point 4, we would consider the direct analysis of at least 100–200 metaphases per experiment to be the absolute minimum, although this number would usually be sufficient for simple screening. The metaphases analysed should be intact. The localization of breaks or other defects can be recorded on a simple diagram (Fig. 19), or e.g. as shown by Brøgger and Johansen (1972). As the duration of the lymphocyte culture appears to influence the type and frequency of breaks (Court Brown, 1967), an adequate analysis will include cultures of different duration, say 48, 72, and 120 hours.

6.4. Chromosomal Analysis by Computer

Certain stages of chromosomal analysis can be done advantageously by a computer to improve speed, efficiency, and accuracy. However, karyotyping represents a serious bottleneck because of the time required. Even a skilled technician working with excellent material will rarely need less than 8–10 min to prepare one karyotype.

The analysis of metaphase chromosomes essentially consists of the following stages, each of which could be subject to computer analysis: 1. culture and processing of the preparation; 2. finding and selection of suitable cells; 3. analysis by measurements and calculations; 4. karyotyping by pattern recognition and alignment; 5. presentation and com-

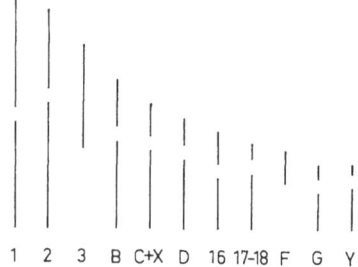

1   2   3   B C+X   D   16 17-18 F   G   Y

Fig. 19. Schematic representation of chromosomes or chromosome arms that can be positively identified on routine stains for marking the localization of breaks and rearrangements. However, with the identification methods now available, a more sophisticated diagram would be needed

parison of data, including print-out of tables, histograms, graphs, statistical tests etc.

As pointed out by Shaw (1972) and others, pattern recognition remains a major problem because the computer has difficulty discerning overlapping chromosomes and structural anomalies or artifacts. For this reason, a number of semi-automated procedures have been devised (Bender and Kastenbaum, 1969; Neurath and Enslein, 1969; Lubs and Ruddle, 1970; Wald *et al.*, 1970; Gilbert and Muldal, 1971; Lubs and Ledley, 1973; Castleman and Wall, 1973).

Of particular interest to the cytologist are recent developments of a so-called facsimile karyotype and a light microscope system, developed at the Department of Biomathematics, University of Texas, M. D. Anderson Hospital and at the Jet Propulsion Laboratory, California Institute of Technology, respectively (for detailed description see p. 311ff. of Wright, Crandall, and Boyer, 1972). In these systems, chromosomes are scanned either from a photograph or directly from the slide and digitized, isolated into individual images, and measured according to length, centromere index, area, integrated optical density, and other suitable parameters, normalized, and finally reproduced as pictorial karyotype.

All systems currently in use require the supervision by a human operator at one point, although this need not be a highly skilled cytologist.

The error rate of computer karyotyping has been estimated by Shaw (1972) as about 1% (two errors per 184 chromosomes or one misqualification per every four cells compared to 0.2% human inaccuracy). However, this involved excellent cells and presumably the computer will be highly inaccurate in mediocre preparations. The inferior qualitative accuracy of the computer is compensated by a higher degree of consistency and the provision of quantitative measurements.

The time required for complete computer analysis lies at present at about 8–12 min per cell, whereby the actual computing time is less than 1 min. This compares favorably with conventional analysis by microscopy and photographic karyotyping which takes about 15 min per cell and does not yield comparable quantitative data.

Costs of computer analysis have been estimated at approximately $ 1–10 per cell, but this probably does not include the initial investment.

Future progress and more widespread use can be expected from the development of fully automated analysis and reductions in speed and relative cost.

6.5. Storage and Retrieval of Cytogenetic Data

In medical cytogenetics, it would be desirable to maintain pertinent family and clinical data together with the results of cytogenetic studies. Although few laboratories use computer cards and tapes, some form of card file or log book in simple numerical or alphabetical order will sooner or later become necessary. It will usually be advantageous to maintain family rather than individual records.

The following information would appear pertinent for inclusion in any record system:
1. Basic individual data, i.e. family and given name, marital status, date of examination, birth date.
2. Basic genetic data, i.e. sibs, parents, grandparents, with their basic individual data, other familial relationships.
3. Clinical data and diagnosis.
4. Photographs of patient.
5. Cytogenetic data.
6. Other genetic data.
7. Documentation of other and previous professional examinations.

Any retrieval system must be able to yield the name of the propositus and the diagnosis. Detailed descriptions of traditional or computer type of records are available (Yesley, 1966; Newcombe, 1969; Kimberling, 1972).

7. Nomenclature

Four international conferences, held at Denver in 1960, London in 1963, Chicago in 1966, and Paris in 1971, have attempted to achieve certain standardizations in cytogenetics and to facilitate communication. The conferences in 1960 and 1963 defined the karyotype as described in Section 3; the conference in 1966 introduced a convenient shorthand notation for cytogenetic findings which is currently in use and adaptable to computer techniques; the conference in 1971 expanded and partially altered the shorthand notation of the Chicago Conference and established criteria for the definition and designation of chromosome bands as revealed by fluorochromes and the various Giemsa techniques. Notes on autoradiography were provided by the conferences in 1966 and 1971; all conferences on nomenclature dealt with problems of chromosome measurements; the 1966 conference commented on cytogenetic data in general. Identification of human male meiotic chromosomes is included in the Paris Conference (1971). The main features of the Chicago (1966) and Paris (1971) nomenclature are given below, certain alterations to the Chicago nomenclature by the Paris Conference included.

7.1. Designation of the Normal Karyotype and of Numerical Alterations

Autosomes are numbered in pairs 1 through 22, and groups A through G. The karyotype is designated by stating (i) the total number of chromosomes, (ii) the sex chromosomes identified, separated from (i) by a comma, (iii) listing any missing or additional whole chromosome after another comma by placing a minus ($-$) or plus ($+$) sign before the symbol indicating the alteration, e.g.

46,XX or 46,XY = normal female or male karyotype.

47,XY,$+21$ = a male with 47 chromosomes and an additional chromosome 21.

45,XY,$-G$ = a male with 45 chromosomes and a missing G group chromosome which has not been positively identified.

(Placing the $+$ or $-$ sign *before* the symbol was introduced in 1971, after it had been previously placed behind it; a $+$ or $-$ sign placed *after* a symbol now means increase or decrease in length, see below). There is no empty space behind any comma or other punctuation mark.

Mosaics are designated by separating the karyotype of cell lines recognized by an oblique (/), e.g.

47,XY,$+G$/45,X = a mosaic consisting of two cell lines, a male with 47 chromosomes and an additional G group chromosome, and a cell line with 45 chromosomes and one missing sex chromosome (either X or Y).

7.2. Designation of Structural Alterations (Chicago Conference, 1966)

All structural changes are designated by lower case letters. The following single or three letter symbols were adopted by the Chicago Conference (1966) and are now in use:

p   = short arm of a chromosome (from petit)
q   = long arm of a chromosome (next letter from p)
h   = secondary constriction
i   = isochromosome
r   = ring chromosome
s   = satellite
t   = translocation
inv = inversion
mar = marker chromosome
mat = maternal origin
pat = paternal origin
ace = acentric chromosome fragment (no centromere)
cen = centromere
dic = dicentric chromosome (two cen)
tri = tricentric chromosome (three cen)
end = endoreduplication
?   = question mark placed after the symbol indicates that definite identification of the symbol is lacking (originally the question mark was placed before the symbol).

These symbols are placed either before or after the chromosome or chromosome group to which they are referring. The Paris Conference (1971) recommends that all symbols for rearrangements be placed *before* the designation of the chromosomes involved in the rearrangement, and the rearranged chromosome placed in parenthesis.

The plus (+) and minus (−) signs placed *after* a symbol now indicate an *increase (+) or decrease (−) in length of the symbol* they follow, e.g.

46,XY,16q+ = a male with 46 chromosomes and an increase in length of the long arm of one chromosome 16.

47,XX,+16q+ = a female with 47 chromosomes and an additional chromosome 16 which has an increase in length of its long arm.

46,XX,16p− = a female with 46 chromosomes and a decrease in length of the short arm of a chromosome 16.

An increase or decrease in length of secondary constrictions or negatively staining regions are designated by placing an h after the symbol for the arm in order to distinguish this situation from increase or decrease in length due to other structural alterations, e.g.

46,XY,16qh+ = a male with 46 chromosomes and an increase in length of the secondary constriction in the long arm of chromosome 16.

46,XY,13ph− = a male with 46 chromosomes and a decrease in length of the negatively staining region of the short arm of chromosome 13.

Translocations are designated by placing a t before the affected chromosomes, which are placed in parenthesis. The designation of translocations and certain other structural alterations (see below) have been refined at the Paris Conference 1971 and now permits inclusion of information about the type of translocation and the site of the break points (see Section 7.5 below).

Balanced reciprocal translocations are designated by listing the normal chromosomes missing as a result of their participation in the translocation, e.g.

46,XX,t(2q− ;5p+) = a female with 46 chromosomes and a balanced reciprocal translocation between the long arm of a chromosome 2 which has been shortened, and the short arm of a chromosome 5, which has been lengthened. If desired the type of translocation may be specified by writing "rcp" (reciprocal) instead of the t, i.e. 46,XX,rcp(2q− ;5p+) (see Section 7.4 below).

An *unbalanced reciprocal translocation* of this type would be written 46,XX,− 5,+ t(5p+ ;2q), indicating that one chromosome 5 was replaced by the translocation chromosome which consists of a chromosome 5 with short arm elongated as a result of a translocation onto it of the whole or part of the long arm of a chromosome 2.

Robertsonian translocations (fusion at the centromere of acrocentric chromosomes) are written without semicolon, e.g. 45,XX,t(13q14q) in the balanced state, whereas they are written 46,XX,− 14,+ t(13q14q) in the unbalanced state, indicating the presence of an additional chromosome 13 as a result of translocation. The symbol "rob" may replace the t to indicate the type of translocation (see Section 7.4). If a small centric fragment is also present, one would write 46,XX,t(13p14p;13q14q) to indicate the presence of a reciprocal translocation.

Other symbols indicating structural alterations are

46,XY,r(16) = a male with 46 chromosomes and a ring chromosome 16.

46,X,i(Xq) = a female with 46 chromosomes with only one normal X chromosome and one X chromosome represented by an isochromosome (i) for the long arm (q).

46,X,dic(Y) = a male with 46 chromosomes and a dicentric Y chromosome.

46,XY,inv(2p+q−) = a pericentric inversion in chromosome 2, break points unidentified (see below), causing an increase in length of the short arm and decrease in length of the long arm (or 2p−q+ for the opposite).

7.3. Alterations of the Chicago Nomenclature and New Symbols of the Paris Conference (1971)

Some alterations proposed in the Paris Report (1972) were mentioned in Section 7.2. However, the improved ability to identify and localize anomalies by the banding techniques called for additional symbols which are listed below (from Table 2, Paris Conference, 1971).

del	deletion
der	derivative chromosome
dup	duplication
ins	insertion
inv ins	inverted insertion
rcp	reciprocal translocation (optional instead of the "t")
rec	recombinant chromosome
rob	Robertsonian translocation (optional instead of the "t")
tan	tandem translocation (optional instead of the "t")
ter	terminal or end, thus e.g. "qter" for end of long arm
:	break (no reunion, as in terminal deletion)
: :	break and join
→	from − to.

The basic rules adopted at the Chicago Conference (1966) have been retained, that is, additional designations should be taken from the first three letters of the word required, used in lower case and clearly defined, and placed immediately before the chromosome symbol to which they refer. If simple letters are used, they should not duplicate the capital letters A to G, X and Y or other lower case letters already in use. Lower case letters such as "1" or "o" are too easily confused with numerals to be useful.

7.4. Chromosome Band Nomenclature

The Paris Conference (1971) developed a system of nomenclature which allows precise designation of break points in any chromosome rearrangement by introducing certain new symbols and utilizing the banding pattern (Q,G,R bands) as a means of designating whole or rearranged chromosomes. The proposed system, nevertheless, still rests on the symbols and rules of the Chicago Conference (1966) which were outlined above.

7.4.1. Identification and Diagrammatic Representation of Landmarks and Bands

The chromosome map shown in Fig. 9 is based on the Q, G, and R banding patterns, but without regard to the intensity of fluorescence or staining. However, intensity has influenced the decision which band should be used as landmark on a chromosome.

The size and position of the chromosome bands in Fig. 9 reflects qualitative impressions rather well, although it was not possible to base them on accurate and reproducible measurements. The agreement of the three banding techniques appeared good enough to allow the construction of a single map, except for the centromere which is mapped in relation to the Q technique only. The centromere is represented by a line which is considered to bisect the centromere at its mid-point and to divide the adjacent regions. The C-bands were not included in preparing this chromosome map (see p. 152).

The definitions of a *chromosome landmark*, a *chromosome band*, and a *chromosome region* were given above in Section 4 (p. 148). The chromosome arms are designated by the symbols p and q as previously described (Chicago Conference, 1966, see Section 7.2 above).

7.4.2. Designation of Band and Region Numbering

Regions and bands are numbered from the centromere outwards along each chromosome arm. Consequently, there are regions and bands on the short and on the long arm, respectively, which have the same number. This necessitates the use of the arm designation in every case.

Bands and regions are designated as follows:
1) chromosome number, e.g. 2
2) arm designation, e.g. 2p
3) region number, e.g. 2p2
4) band number, e.g. 2p21 (= chromosome 2, short arm, region 2, band 1).

The chromosome map of Fig. 9 and Table 2 follow this system of regional and band numbering.

The Paris Conference (1971) makes certain provisions in the event that an established landmark requires to be subdivided: (i) each landmark should always retain its original region number, even if they come to lie in an adjacent region as a result of the subdivision; (ii) each band should always retain its original band number, which can be subdivided by adding a decimal point as designation of sub-bands, e.g. 2p31.1, 2p31.2,

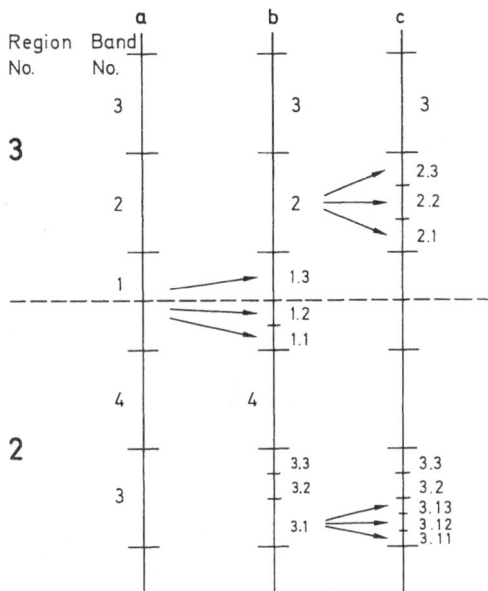

Fig. 20. Example illustrating the convention for numbering the subdivisions of a landmark, as described in the text. (Modified from Fig. 6a and b from the Paris Conference, 1971)

2p31.3 would indicate that band 1 in region 3 of the short arm of chromosome 2 had been subdivided into 3 further bands (Fig. 20).

It is important to realize that each band is designated by its midline rather than its margins and that all areas of a chromosome belong to a band region; there are no interband regions. As a consequence, subdivisions of bands may lie in an adjacent region of the original band (Fig. 20), because in each region, band 1 intersects two adjacent regions as illustrated in Fig. 9 and Fig. 20. A definition of the actual regions and bands of the human complement is given in the Appendix (Table 2) and in Section 4.

7.5. Designation of Structural Changes According to Break Point

7.5.1. Simple Breaks

The symbol "del" designates a deletion, e.g.
64,XX,del(5)(p14) = a *terminal deletion* in chromosome 5 with a break point in region 1 band 4 of the short arm.

Isochromosomes are designated as shown in Section 7.2, because their break points in or close to the centromere are not specified.

7.5.2. Two-break Rearrangements

When this type of rearrangement involves both arms of a single chromosome, the break point in the short arm should be written first, e.g. (2)(p21q31). Break points in the same chromosome should not be separated by a semicolon (;). When two breaks occur in the same chromosome arm, the proximal one should be written first, e.g. (3)(p13p22).

46,XX,del(2)(q22q31) = An *interstitial deletion* in chromosome 2 with break points in region 2 band 2 and region 3 band 1, with a deletion of the intervening segment.

46,XY,inv(2)(q13q23) = A *paracentric inversion* in the long arm of chromosome 2 with break points in region 1 band 3 and region 2 band 3.

46,XY,inv(2)(p13q23) = A *pericentric inversion* in chromosome 2, with break points in region 1 band 3 of the short arm and region 2 band 3 of the long arm.

46,XY,r(2)(p21q31) = A *centric ring chromosome* 2 with break points in region 2 band 1 of the short arm and region 3 band 1 of the long arm.

Translocations involving two break points are written accordingly, the letter "t" being used as proposed by the Chicago Conference (1966) or using the abbreviation of the type of translocation ("rcp", "rob", "tan", see above, Section 7.3).

Reciprocal translocations: 46,XY,t(2;5)(q21;p14) would indicate a reciprocal translocation between chromosome 2 and chromosome 5, with break points at region 2 band 1 of the long arm of chromosome 2 and region 1 band 4 of the short arm of chromosome 5. The "t" may be substituted for "rcp", thus 46,XY,rcp(2;5)(q21;p14).

Robertsonian translocations: 45,XY,t(13;21)(p11;q11) would indicate a balanced centric fusion type (Robertsonian) translocation between chromosomes 13 and 21, break points being in region 1 band 1 of the short arm of chromosome 13 and region 1 band 1 of the long arm of chromosome 21.

An unbalanced karyotype as a result of such a translocation would be written 46,XY,$-21,+$t(13;21)(p11q11) which would indicate a duplication of chromosome 21.

If break points cannot be identified, the translocation would be written as outlined in Section 7.2, i.e. 45,XY,t(13q21q) in the balanced state, or 46,XY,−21,+t(13q21q) in the unbalanced state. If a translocation results in a *dicentric chromosome* one would write e.g. 45,XY, tdic(13;21)(p11;p11), which would indicate a Robertsonian translocation between chromosome 13 and 21 with break points in the short arms in regions 1 bands 1 of both chromosomes.

Other *translocations of whole arms*, where information about the centromere that is involved is lacking, would be written as follows: 46,XY,t(2q5p;2p5q), indicating a derivative translocation chromosome consisting either of the long arm of a chromosome 2 and the short arm of a chromosome 5 or vice versa.

Dicentric chromosomes resulting from an interchange are designated by e.g. 46,X,dic(Y)(q12), which would indicate a dicentric Y chromosome with a break in its long arm in region 1 band 2. In this case the symbol t is omitted, because no other chromosome is involved.

7.5.3. Three-break Rearrangements

In this type of rearrangement, 1, 2 or 3 chromosomes may be involved.

In the Paris Conference (1971) they are referred to as "complex translocations" when 3 chromosomes, and as "insertions" when 1 or 2 chromosomes are involved. The order of the bands in the inserted segment in relation to the centromere at a new site may be the same as in the original site (*direct insertion*) or may be reversed (*inverted insertion*). The symbol "ins" was adopted in the Paris Report (1972) for a direct insertion and "inv ins" for an inverted insertion.

The following explanations are taken verbatim from the Paris Conference (1971):

Direct insertion within a chromosome

46,XY,ins(2)(p13q21q31) = an insertion in chromosome 2 with the segment between break points q21 and q31.

Note: The break point into which the chromosome segment is inserted is always written first in the second set of parenthesis. This departs from the convention established above that the lowest break point number in the short arm will always be written first.

Inverted insertion within a chromosome

46,XY,inv ins(2)(p13q21q31) = indicating an inversion of the segment 21 to 31 of the long arm of chromosome 2 inserted into break point p13 with band q31 proximal and band q21 distal to the centromere.

Direct insertion between chromosomes

46,XX,ins(5;2)(p14;q13q31) = an insertion into chromosome 5 at break point p14 of the region lying between band q13 and band q31 of chromosome 2.

Note: that the receptor chromosome for the inserted region is written first in the first set of parenthesis and the receptor break is written first in the second set of parenthesis. The semicolon is used to separate the break points in different chromosomes but not to break points in the same chromosome.

Inverted insertion between chromosomes

46,XX,inv ins(5;2)(p14;q13q31) = this is written in the same way as for an inverted insertion within a chromosome except that a semicolon separates the chromosomes and the break points in the different chromosomes.

Complex translocation

46,XX,t(2;5;7)(p21;q23;q22) = a translocation between chromosomes 2,5, and 7 at break points p21 on chromosome 2, q23 on chromosome 5, q22 on chromosome 7. The segment of chromosome 2 distal to 2p21 is translocated onto chromosome 5 at q23; the segment of chromosome 5 distal to 5q23 is translocated to chromosome 7 at break point q22; the segment of chromosome 7 distal to q22 is translocated to chromosome 2 at break point q21.

Note: the chromosome with the lowest number will always be written tirst in the parenthesis and the chromosome whose segment is translocated fo this first chromosome will be written last.

7.5.4. Four-break Rearrangements

In view of the large number of possibilities of this type which could all be written according to the principles outlined above, the Paris Conference (1971) gives just one example of a human case, i.e. a *double reciprocal translocation involving 3 chromosomes:* 46,XX,t(1;3)(3;9) (p12;p13q25;q22) indicating a translocation between chromosome 1 and chromosome 3, and a second translocation between the same chromosome 3 and chromosome 9. The breaks occurred in p12 of chromosome 1, p13 and q25 of chromosome 3, and q22 of chromosome 9. The exchanges involved the segments distal to these bands and regions. Note that the two break points of chromosome 3 which is involved in two translocations are *not* separated by a semicolon or by parenthesis. In contrast, two rearrangements involving two chromosomes each, would be written 46,XX,t(1;3)(3;9)(p12;p13)(q25;q22).

7.5.5. Marker Chromosomes

A marker chromosome of completely unknown origin is designated by the symbol "mar" according to the Chicago Conference (1966), as mentioned in the Table in Section 7.2, but if a part of it can be identified, the designation would be as follows, e.g.

46,XX,t(12;?)(q15;?), indicating a rearranged chromosome 12 in which a segment of the long arm distal to q15 cannot be identified. If the marker chromosome is longer than the original chromosome, it would be written e.g.

46,XX,t(12q+;?)(q15;?) if the rearranged chromosome has a longer arm (q) than normal, due to an unkown segment attached to it distal of q15.

7.5.6. Derivative and Recombinant Chromosome

The Paris Report (1972) defines a *derivative chromosome* ("der") as "a structurally altered chromosome originating from one or more chromosomes in the complement as the direct result of a structural rearrangement without any intervening process of meiotic crossing-over". A *recombinant chromosome* ("rec") is defined as "a structurally altered chromosome resulting from the process of meiotic crossing-over within a rearranged chromosome segment". A recombinant chromosome may be indistinguishable from the derivation chromosome of a rearrangement, because the crossing-over in a reciprocal translocation may be interstitial. The most usual form of a recombinant chromosome, according to the Paris Conference (1971) will originate from crossing-over in an inverted or inserted chromosome segment.

The designation of an unbalanced karyotype containing a derivative chromosome is written as follows: (i) chromosome number, (ii) sex chromosomes, (iii) the missing chromosome indicated in the usual way, (iv) the derivative chromosome, (v) the arrangement derived from iv. Finally, the source "mat" or "pat" is indicated, if known, according to the Chicago nomenclature (cf. Section 7.2).

A balanced reciprocal translocation between e.g. chromosome 2 and chromosome 5 would be written t(2;5)(q21;q31) and the derivative chromosome designated der(2) and der(5). The Paris Conference (1971) gives in Table 5 full details of the possible 16 unbalanced karyotypes resulting from disjunction of the gametes. The four examples from adjacent I and adjacent II disjunction are sufficient to illustrate the principle of designation here:

adjacent I: 46,XX,−5,+der(5),t(2;5)(q21;q31)mat
          46,XY,−2,+der(2),t(2;5)(q21;q31)mat

adjacent II: 46,XY, − 5, + der(2),t(2;5)(q21;q31)
46,XY, − 2, + der(5),t(2;5)(q21;q31)mat

The Paris Conference (1971) further points out that the full karyotype designation need only be written once in any report. It suggests an abbreviated version for subsequent use, e.g. for the first line of adjacent I: 46,XX, − 5 + der(5)mat. It might even be possible to leave out the designation of the sex chromosomes if the sex of the individual is clear from the pedigree and preceding descriptions.

Time will tell how these somewhat cumbersome designations work out in practice. For the full details concerning the designation of derivative and recombinant chromosomes as well as designations of whole rearranged chromosomes, the reader is referred to the original publication of the Paris Conference (1971).

Acknowledgement. The author's own research is supported by grants from the Deutsche Forschungsgemeinschaft. I thank Mrs Lis Berenbrok for preparing the photographs and Mrs Mechthild Roloff and Miss Nicole Vangijsel for secretarial help. Drs. M. A. Ferguson-Smith (Glasgow) and W. Schnedl (Wien) kindly made manuscripts available prior to publication, and I thank Drs. Daniel Bergsma (New York), T. R. Chen (Houston), H. J. Evans (Edinburgh), M. A. Ferguson-Smith (Glasgow), P. H. LaMarche (Providence), F. H. Ruddle (New Haven), W. Schmid (Zürich), and L. Zech (Stockholm) for figures and tables as also acknowledged in the text.

Appendix

Table 1. The diagnosis of XO/XX mosaicism, when the presumptive XO line is the minor line. The recorded numbers of presumptive XO cells to establish mosaicism are the numbers which would be found by chance in 0.1 per cent of the general population (see Text on p. 140)

Age group	Total cells	No. of presumptive XO cells	
		Early cultures	Late cultures
15–54	30	4	5
	50	5	5
	100	6	7
55–64	30	7	8
	50	9	10
	100	13	14
65+	30	8	9
	50	10	11
	100	15	17

(Table from Court Brown *et al.*, 1966 and 1967.)

Table 2. Bands serving as landmarks which divide the chromosomes into cytologically defined regions. The omission of an entire chromosome or a chromosome arm indicates that either both arms or the arm in question consists of only one region, delimited by the centromere and the end of the chromosome arm

Chromosome No.	Arm	Number of regions	Landmarks (the numbers in parentheses are the region and band numbers as shown in Fig. 9 on p. 152)
1	p	3	Proximal band of medium intensity (21), median band of medium intensity (31)
	q	4	Proximal negative band (21) distal to variable region, median intense band (31), distal medium band (41)
2	p	2	Median negative band (21)
	q	3	Proximal negative band (21), distal negative band (31)
3	p	2	Median negative band (21)
	q	2	Median negative band (21)
4	q	3	Proximal negative band (21), distal negative band (31)
5	q	3	Median band of medium intensity (21), distal negative band (31)
6	p	2	Median negative band (21)
	q	2	Median negative band (21)
7	p	2	Distal medium band (21)
	q	3	Proximal medium band (21), median band of medium intensity (31)
8	p	2	Median negative band (21)
	q	2	Median band of medium intensity (21)
9	p	2	Median intense band (21)
	q	3	Median band of medium intensity (21), distal band of medium intensity (31)
10	q	2	Proximal intense band (21)
11	q	2	Median negative band (21)
12	q	2	Median band of medium intensity (21)
13	q	3	Median intense band (21), distal intense band (31)
14	q	3	Proximal intense band (21), distal medium band (31)
15	q	2	Median intense band (21)
16	q	2	Median band of medium intensity (21)
17	q	2	Proximal negative band (21)
18	q	2	Median negative band (21)
21	q	2	Median intense band (21)
X	p	2	Proximal medium band (21)
	q	2	Proximal medium band (21)

Table 3. Measurements of relative length (in percentage of the total haploid autosome length) and centromere index (length of short arm divided by total chromosome length × 100). Chromosomes stained with orcein or the Giemsa 9 method and pre-identified by Q band patterns

Chromo-some No.	Relative length				Centromere index			
	A	B	C	D	A	B	C	D
1	9.08	9.08 ± 0.611	9.11 ± 0.53	8.44 ± 0.433	48.0	49.4 ± 3.04	48.6 ± 2.6	48.36 ± 1.
2	8.45	8.17 ± 0.250	8.61 ± 0.41	8.02 ± 0.397	38.1	39.4 ± 2.05	38.9 ± 2.6	39.23 ± 1.
3	7.06	6.96 ± 0.352	6.97 ± 0.36	6.83 ± 0.315	45.9	47.6 ± 2.10	47.3 ± 2.1	46.95 ± 1.
4	6.55	6.62 ± 0.403	6.49 ± 0.32	6.30 ± 0.284	27.6	29.2 ± 2.97	27.8 ± 3.3	29.07 ± 1.
5	6.13	6.34 ± 0.366	6.21 ± 0.50	6.08 ± 0.305	27.4	29.2 ± 3.03	26.8 ± 2.6	29.25 ± 1.
6	5.84	6.19 ± 0.516	6.07 ± 0.44	5.90 ± 0.264	37.7	39.1 ± 2.63	37.9 ± 2.5	39.05 ± 1.
7	5.28	5.60 ± 0.435	5.43 ± 0.47	5.36 ± 0.271	37.3	35.3 ± 2.90	37.0 ± 4.2	39.05 ± 1.
X	5.80	5.45 ± 0.377	5.16 ± 0.24	5.12 ± 0.261	36.9	41.4 ± 6.16	37.5 ± 2.7	40.12 ± 2.
8	4.96	5.13 ± 0.307	4.94 ± 0.28	4.93 ± 0.261	35.9	32.7 ± 2.80	32.8 ± 2.8	34.08 ± 1
9	4.83	4.81 ± 0.194	4.78 ± 0.39	4.80 ± 0.244	33.3	37.0 ± 3.04	32.7 ± 4.1	35.43 ± 2.
10	4.68	4.66 ± 0.512	4.80 ± 0.58	4.59 ± 0.221	31.2	35.4 ± 3.81	32.3 ± 2.9	33.95 ± 2.
11	4.63	4.70 ± 0.289	4.82 ± 0.30	4.61 ± 0.227	35.6	40.7 ± 3.07	40.5 ± 3.3	40.14 ± 2.
12	4.46	4.66 ± 0.410	4.50 ± 0.26	4.66 ± 0.212	30.9	30.5 ± 3.64	27.4 ± 4.0	30.16 ± 2.
13	3.64	3.22 ± 0.310	3.87 ± 0.26	3.74 ± 0.236	14.8	–	16.6 ± 3.6	17.08 ± 3
14	3.55	3.09 ± 0.212	3.74 ± 0.23	3.56 ± 0.229	15.5	–	18.4 ± 3.9	18.74 ± 3
15	3.36	2.83 ± 0.262	3.30 ± 0.25	3.46 ± 0.214	14.9	–	17.6 ± 4.6	20.30 ± 3
16	3.23	3.46 ± 0.353	3.14 ± 0.55	3.36 ± 0.183	40.6	42.2 ± 3.57	42.5 ± 5.6	41.33 ± 2
17	3.15	3.06 ± 0.377	2.97 ± 0.30	3.25 ± 0.189	31.4	36.6 ± 5.86	31.9 ± 3.3	33.86 ± 2
18	2.76	2.98 ± 0.316	2.78 ± 0.18	2.93 ± 0.164	26.1	31.5 ± 4.15	26.6 ± 4.2	30.93 ± 3
19	2.52	2.55 ± 0.269	2.46 ± 0.31	2.67 ± 0.174	42.9	48.1 ± 2.48	44.9 ± 4.0	46.54 ± 2.
20	2.33	2.61 ± 0.144	2.25 ± 0.24	2.56 ± 0.165	44.6	46.5 ± 3.59	45.6 ± 2.5	45.45 ± 2.
21	1.83	1.34 ± 0.189	1.70 ± 0.32	1.90 ± 0.170	25.7	–	28.6 ± 5.0	30.89 ± 5
22	1.68	1.53 ± 0.178	1.80 ± 0.26	2.04 ± 0.182	25.0	–	28.2 ± 6.5	30.48 ± 4
Y	1.96	1.82 ± 0.353	2.21 ± 0.30	2.15 ± 0.137	16.3	–	23.1 ± 5.1	27.17 ± 3

A Previous Denver-London data (not pre-identified by Q-staining method).

B Data from 20 cells provided by Dr. P. Pearson. Cells stained with orcein. The short arms in groups D and G and in the Y were excluded.

C Data from 10 cells provided by Drs. T. Caspersson, M. Hultén, J. Lindsten, and L. Zech. Cells stained with orcein.

D Data from 95 cells provided by Drs. H. Lubs, T. Hostetter, and L. Ewing from 11 normal subjects (6–10 cells per person). Average total length of chromosomes per cell: 176 microns. Cells stained with orcein or Giemsa 9 technique.

Cells in B, C, and D were measured from projected negatives of metaphase cells. Standard deviations in samples B and C are based on the total sample of measurements. Standard deviations in sample D are an average of the standard deviations found in each of 11 subjects (6–10 cells per subject.)

(From: Paris Conference, 1971, reproduced with permission from The National Foundation, New York.)

Table 4. Frequency of minor chromosomal variants based on the different techniques[a]

Chromosome	Standard stains (1)	Q bands (2)	C bands (3)
1	0.005	0.06–0.1	0.25
2	0.002		
3	0.003	0.5	0.1
4		0.12	
5	0.0004		
6	0.0004		
7			
8			
9	0.1	0.04–0.2	0.32
10			
11			
12			
13	⎫	0.3	⎫
14	⎬ 0.003–0.17	0.2	⎬ 0.15
15	⎭	0.2	⎭
16	0.04	0.06 0.07	0.35
17	0.007	0.005	
18			
19	0.0004		⎫
20			⎬ 0.1
21	⎫	0.06–0.3	⎫
22	⎬ 0.004–0.06	0.2	⎬ 0.25
Y	0.06	0.08	

[a] Based on data of Lubs and Ruddle (1970) and Hamerton *et al.* (1972) for (1), Schnedl (1974), Bobrow and Pearson (personal communication 1972 to H. G. Schwarzacher), Ferguson-Smith (1974) for (2), and Craig-Holmes, Moore, and Shaw (1973) for (3).

Table 5. Diagnosis of a chromosomal mosaic. Correlation of distribution of two cell lines and total number of cells examined

% of cells in minority	Number of cells to be examined to find at least one different cell			Expected number of different cells for $p = 0.80$ if the following number of cells are examined		
	$p = 0.80$	$p = 0.95$	$p = 0.99$	25	30	50
1	160	300	460	0	0	0
2	80	150	230	0	0	0
3	53	100	154	0	0	1
4	40	75	115	0	0	1
5	32	60	92	0	1	1
6	27	50	77	1	1	2
7	23	43	66	1	1	2
8	20	38	58	1	1	2
9	18	33	52	1	1	3
10	16	30	46	1	1	3
11	15	27	42	1	2	3
12	13	25	38	2	2	3
13	12	23	35	2	2	4
14	11	21	33	2	2	4
15	11	20	31	2	2	4
16	10	19	29	2	2	4
17	9	18	27	2	2	5
18	8	17	25	2	2	5
19	8	16	24	2	2	5
20	8	15	23	2	3	6
25	6	12	18	3	3	7
30	5	10	15	4	3	9
40	4	8	11	5	3	12
50	3	6	9	6	4	15

Table prepared by Paul H. LaMarche, M.D. and Horace Martin, Ph.D. (Providence, Rhode Island) according to the Poisson distribution. (Published with kind permission of Dr. Paul H. LaMarche.)

Table 6. Probability of missing any cell from a minor component in a random sample of mosaic tissue

Frequency of minor component	Number of cells examined		
	10	20	50
0.05	0.598	0.358	0.077
0.1	0.349	0.122	0.005
0.2	0.107	0.012	0.001
0.3	0.028	0.001	0.001
0.4	0.006	0.001	0.001
0.5	0.001	0.001	0.001

(Table adapted from Ford, 1969.)

References

Arrighi, F. E., Hsu, T. C.: Localization of heterochromatin in human chromosomes. Cytogenetics **10**, 81–86 (1971).

Aula, P., Nichols, W. W.: The cytogenetic effects of mycoplasma in human leukocyte cultures. J. cell. Physiol. **70**, 281–290 (1968).

Aula, P., Saksela, E.: Comparison of areas of quinacrine mustard fluorescence and modified Giemsa staining in human metaphase chromosomes. Exp. Cell Res. **71**, 161–167 (1972).

Barton, D. E., David, F. N., Merrington, M.: The position of the sex chromosomes in the human cell in mitosis. Ann. hum. Genet. **28**, 123–128 (1964).

Bender, M. A., Kastenbaum, M. A.: Statistical analysis of the normal human karyotype. Amer. J. hum. Genet. **21**, 322–351 (1969).

Bloom, A. D.: Induced chromosomal aberrations in man. In: Advances in human genetics, vol. 3, p. 99–172, H. Harris and K. Hirschhorn, eds. New York-London: Plenum Press 1972.

Bobrow, M., Madan, K., Pearson, P. L.: Staining of some specific regions of human chromosomes, particularly the secondary constriction of No. 9. Nature New Biol. (Lond.) **238**, 122–124 (1972).

Bobrow, M., Pearson, P. L., Pike, M. C., El-Alfi, O. S.: Length variation in the quinacrine-binding segment of human Y chromosomes of different sizes. Cytogenetics **10**, 190–198 (1971).

Bobrow, M., Collacott, H. E. A., Madan, K.: Chromosome banding with acridine orange. Lancet **1972 II**, 1311.

Borgoankar, D. S., Hollander, D. H.: Quinacrine fluorescence of the human Y chromosome. Nature (Lond.) **230**, 52 (1971).

Borgoankar, D. S., McKusick, V. A., Herr, H. M., Cobos, de los, L., Yoder, O. C.: Constancy of the length of human Y chromosome. Ann. Génét. **12**, 262–264 (1969).

Breg, W. R., Allderdice, P. W., Miller, D. A., Miller, O. J.: Quinacrine fluorescence patterns and terminal DNA labelling of human C group chromosomes. Nature New Biol. (Lond.) **236**, 76–78 (1972).

Breg, W. R., Miller, O. J., Miller, D. A., Allerdice, P. W.: Distinctive fluorescence of quinacrine-labelled human G group chromosomes. Nature New Biol. (Lond.) **231**, 276–277 (1971).

Brøgger, A., Johansen, J.: A model for the production of chromosome damage by mitomycin C. Chromosoma (Berl.) **38**, 95–104 (1972).

Brown, J. A., Palmer, C. G., Yu, P. L.: The distribution and localization of drug-induced secondary constrictions in human chromosomes. Canad. J. Genet. Cytol. **14**, 81–93 (1972).

Calderon, D., Schnedl, W.: A comparison between quinacrine fluorescence banding and ^{3}H-thymidine incorporation patterns in human chromosomes. Humangenetik **18**, 63–70 (1973).

Caspersson, T., Lomakka, G., Zech, L.: The 24 fluorescence patterns of the human metaphase chromosomes—distinguishing characters and variability. Hereditas (Lund) **67**, 89–102 (1971).

Caspersson, T., Zech, L.: Fluorescent labeling and identification of human chromosomes. In: Perspectives in cytogenetics, p. 163–185, S. W. Wright, B. F. Crandall, L. Boyer, eds. Springfield: Charles C. Thomas 1972.

Caspersson, T., Zech, L. (eds.): Chromosome identification—technique and applications in biology and medicine. Nobel Symposium **23**. New York and London: Academic Press 1973.

Caspersson, T., Zech, L., Johansson, C.: Differential binding of alkylating fluorochromes in human chromosomes. Exp. Cell Res. **60**, 315–319 (1970a).

Caspersson, T., Zech, L., Johansson, C., Lindsten, J., Hultén, M.: Fluorescent staining of heteropycnotic chromosome regions in human interphase nuclei. Exp. Cell Res. **61**, 472–474 (1970c).

Caspersson, T., Zech, L., Johansson, C., Modest, E. J.: Identification of human chromosomes by DNA-binding fluorescent agents. Chromosoma (Berl.) **30**, 215–277 (1970b).

Castleman, K. R., Wall, R. J.: Automatic systems for chromosome identification. In: Chromosome identification, p. 77–84, T. Caspersson and L. Zech (eds.), Nobel Symposium **23**. New York and London: Academic Press 1973.

Chapelle, A. de la, Schröder, J., Selander, R.-K., Stenstrand, K.: Differences in DNA composition along mammalian metaphase chromosomes. Chromosoma (Berl.) **42**, 365–382 (1973).

Chen, T. R., Ruddle, F. A.: Karyotype analysis utilizing differentially stained constitutive heterochromatin of human and murine chromosomes. Chromosoma (Berl.) **34**, 51–72 (1971).

Chicago Conference (1966): Standardization in human cytogenetics. Birth defects: Original article series, vol. II, No. 2. New York: The National Foundation 1966.

Christenson, L. P.: Applied photography in chromosome studies. In: Human chromosome methodology, p. 129–153, J. J. Yunis, ed. New York: Academic Press 1965.

Cohen, M. M., Shaw, M. W.: Specific effects of viruses and antimetabolites on mammalian chromosomes. In Vitro **1**, 50–66 (1965).

Cohen, M. M., Shaw, M. W., MacCluer, J. W.: Racial differences in the length of the human Y chromosome. Cytogenetics **5**, 34–52 (1966).

Comings, D. E.: The rationale for an ordered arrangement of chromatin in the interphase nucleus. Amer. J. hum. Genet. **20**, 440–460 (1968).

Comings, D. E., Avelino, E., Okada, T. A., Wyandt, H. E.: The mechanism of C- and G-banding of chromosomes. Exp. Cell Res. **77**, 469–493 (1973).

Cooke, P.: Patterns of secondary association between the acrocentric autosomes of man. Chromosoma (Berl.) **36**, 221–240 (1972).

Court Brown, W. M.: Human population cytogenetics. Amsterdam: North-Holland Publ. Co. 1967.

Court Brown, W. M., Buckton, K. E., Jacobs, P. A., Tough, I. M., Kuensberg, E. V., Knox, J. E. D.: Chromosome studies on adults. Eugenics laboratory memoir series XLII, The Galton laboratory. Cambridge: University Press 1966.

Court Brown, W. M., Smith, P. G.: Human population cytogenetics. Brit. med. Bull. **25**, 74–80 (1969).

Craig, A. P., Shaw, M. W.: Autoradiographic studies of the human Y chromosome. Chromosoma (Berl.) **32**, 364–377 (1971).

Craig-Holmes, A. P., Moore, F. B., Shaw, M. W.: Polymorphism of human C-band heterochromatin. I. Frequency of variants. Amer. J. hum. Genet. **25**, 181–192 (1973).

Craig-Holmes, A. P., Shaw, M. W.: Polymorphism of human constitutive heterochromatin. Science **174**, 702–704 (1971).

Denver Report 1960: A proposed standard system of nomenclature of human mitotic chromosomes. Lancet **1960 I**, 1063–1065.

Drets, M. E., Shaw, M. W.: Specific banding patterns of human chromosomes. Proc. nat. Acad. Sci. (Wash.) **68**, 2073–2077 (1971).

Dutrillaux, B.: Nouveau système de marquage chromosomique: les bandes T. Chromosoma (Berl.) **41**, 395–402 (1973).

Dutrillaux, B., Grouchy, de, J., Finaz, C., Lejeune, J.: Mise en évidence de la structure fine des chromosomes humains par digestion enzymatique (pronase en particulier). C. R. Acad. Sci. (Paris) **273**, 587–588 (1971).

Dutrillaux, B., Laurent, C., Robert, J. M., Lejeune, J.: Inversion péricentrique, inv(10), chez la mère et aneusomie de recombinaison, inv(10), rec(10), chez son fils. Cytogenet. Cell Genet. (Basel) **12**, 245–253 (1973).

Dutrillaux, B., Lejeune, J.: Sur une nouvelle technique d'analyse du caryotype humain. C. R. Acad. Sci. (Paris) **272**, 2638–2640 (1971).

Evans, H. J.: Actions of radiations on human chromosomes. Phys. in Med. Biol. **17**, 1–13 (1972).

Evans, H. J., Buckton, K. E., Sumner, A. T.: Cytological mapping of human chromosomes: results obtained with quinacrine fluorescence and the acetic-saline Giemsa technique. Chromosoma (Berl.) **35**, 310–325 (1971).

Ferguson-Smith, M. A.: The techniques of human cytogenetics. Amer. J. Obstet. Gynec. **90**, 1035–1054 (1964).

Ferguson-Smith, M. A.: Chromosomal abnormalities. II. Sex chromosome defects. In: Medical genetics, p. 16–26, V. A. McKusick and R. Clairborne (eds.). New York: HP Publishing Co., Inc. 1973.

Ferguson-Smith, M. A.: Autosomal polymorphism. In: Medical genetics today, D. L. Rimoin, R. N. Schimke, eds. Birth defects: Original article series. New York: The National Foundation (in press).

Ferguson-Smith, M. A., Ferguson-Smith, M. E., Ellis, P. M., Dickson, M.: The sites and relative frequencies of secondary constrictions in human somatic chromosomes. Cytogenetics **1**, 325–343 (1962).

Ferguson-Smith, M. A., Handmaker, S. D.: The association of satellited chromosomes with specific chromosomal regions in cultured human somatic cells. Ann. hum. Genet. **27**, 143–156 (1963).

Ford, C. E.: Mosaics and chimaeras. Brit. med. Bull. **25**, 104–109 (1969).

Ford, E. H. R.: Human chromosomes. London and New York: Academic Press 1973.

Gagné, R., Laberge, C.: Specific cytological recognition of the heterochromatic segment of number 9 chromosome in man. Exp. Cell Res. **73**, 239–242 (1972).

Gagné, R., Tanguay, R., Laberge, C.: Differential staining patterns of hetero-chromatin in man. Nature New Biol. (Lond.) **232**, 29–30 (1971).

Galperin, H.: Relative positions of homologous chromosomes or groups in male and female metaphase figures. Humangenetik **7**, 265–274 (1969).

Ganner, E., Evans, H. J.: The relationship between patterns of DNA replication and of quinacrine fluorescence in the human chromosome complement. Chromosoma (Berl.) **35**, 326–341 (1971).

German, J.: The pattern of DNA synthesis in the chromosomes of human blood cells. J. Cell Biol. **20**, 37–55 (1964a).

German, J.: Identification and characterization of human chromosomes by DNA replication sequence. In: Cytogenetics of cells in culture, p. 191–207, R. J. C. Harris, ed. New York-London: Academic Press 1964b.

German, J.: The chromosomal structural load in man. Tex. Rep. Biol. Med. **24**, 347–364 (1966).

German, J.: Autoradiographic studies of human chromosomes. I. A review. Proceedings of the Third Int. Congr. Human Genetics, p. 123–136, J. F. Crow, J. V. Neel, eds. Baltimore: Johns Hopkins Press 1967.

German, J.: Genes which increase chromosomal instability in somatic cells and predispose to cancer. Progr. med. Genet. **8**, 61–101 (1972).

Gey, W.: Untersuchungen über die DNS-Replikationsmuster der Chromosomengruppen 4–5, 13–15 und 21–22 an in vitro gezüchteten menschlichen Lymphocyten. Humangenetik **2**, 246–261 (1966).

Giannelli, F., Howlett, R. M.: The identification of the chromosomes of the D group (13–15) Denver: An autoradiographic and measurement study. Cytogenetics **5**, 186–205 (1966).

Giannelli, F.: Human chromosomes DNA synthesis, Monographs in human genetics, vol. 5, L. Beckman, M. Hauge, eds. Basel: Karger 1970.

Gilbert, C. W., Muldal, S.: Measurement and computer system for karyotyping human and other cells. Nature (Lond.) **230**, 203–207 (1971).

Grzeschik, K. H., Allderdice, P. W., Grzeschik, A., Opitz, J. M., Miller, O. J., Siniscalco, M.: Cytological mapping of human X-linked genes by use of somatic cell hybrids involving an X-autosome translocation. Proc. nat. Acad. Sci. (Wash.) **69**, 69–73 (1972).

Hamerton, J. L.: Human cytogenetics, vol. I, General cytogenetics, and vol. II, Clinical cytogenetics. New York-London: Academic Press 1971.

Hamerton, J. L., Ray, M., Abbott, J., Williamson, C., Ducass, G. C.: Chromosome studies in a neonatal population. Canad. med. Ass. J. **106**, 776–779 (1972).

Hausen, H., zur: Induction of specific chromosomal aberrations by adenovirus type 12 in human embryonic kidney cells. J. Virol. **1**, 1174–1185 (1967).

Hirschhorn, K.: Chromosomal abnormalities. I. Autosomal defects. In: Medical genetics, p. 3–14, V. A. McKusick and R. Clairborne (eds.). New York: HP Publishing Co., Inc. 1973.

Hughes, D. T.: Quantitative studies in karyotype analysis. In: Chromosomes today, vol. 1, p. 188–210, C. D. Darlington, K. R. Lewis, eds. Edinburgh-London: Oliver & Boyd 1966.

Jacobs, P. A.: Structural abnormalities of the sex chromosomes. Brit. med. Bull. **25**, 94–103 (1969).

Jacobs, P. A., Frackiewicz, A., Law, P.: Incidence and mutation rates of structural rearrangements of the autosomes in man. Ann. hum. Genet. **35**, 301–319 (1972).

Jacobs, P. A., Price, W. H., Law, P., eds.: Human population cytogenetics, Pfizer medical monographs 5. Edinburgh: Edinburgh Univ. Press 1970.

Kimberling, W. J.: Computers and gene localization. In: Perspectives in cytogenetics, p. 131–147, S. W. Wright, B. F. Crandall, L. Boyer, eds. Springfield, Illinois: Charles C. Thomas 1972.

Leisti, J.: Structural variation in human mitotic chromosomes. Ann. Acad. Sci. fenn. A, IV Biologica **179**, 1–69 (1971).

Lejeune, J., Dutrillaux, B., Lafourcade, J., Berger, R., Abonyi, D., Rethoré, M. O.: Endoréduplication sélective du bras long du chromosome 2 chez une femme et sa fille. C. R. Acad. Sci. (Paris) **266**, 24–26 (1968).

Licznerski, G., Lindsten, J.: Trisomy 21 in man due to maternal non-disjunction during the first meiotic division. Hereditas (Lund) **70**, 153–154 (1972).

Littlefield, L. G., Goh, K.-O.: Cytogenetic studies in control men and women. I. Variations in aberration frequencies in 29,709 metaphases from 305 cultures obtained over a three-year period. Cytogenet. Cell Genet. (Basel) **12**, 17–34 (1973).

London Report 1963: The London Conference on the normal human karyotype. Cytogenetics **2**, 264–268 (1963).

Lubs, H. A.: A marker X chromosome. Amer. J. hum. Genet. **21**, 231–244 (1969).

Lubs, H. A., Ledley, R. S.: Automated analysis of differentially stained human chromosomes. In: Chromosome identification, p. 61–76, T. Caspersson and L. Zech (eds.). Nobel Symposium **23**. New York and London: Academic Press 1973.

Lubs, H. A., Ruddle, F. H.: Applications of quantitative karyotypy to chromosome variation in 4,400 consecutive newborns. In: Human population cytogenetics, p. 120–142, P. A. Jacobs, W. H. Price, P. Law, eds. Pfizer medical monographs 5. Edinburgh: Edinburgh Univ. Press 1970.

Lubs, H. A., Ruddle, F. H.: Chromosome polymorphism in American negro and white populations. Nature (Lond.) **233**, 134–136 (1971).

Magenis, R. E., Hecht, F., Lovrien, E. W.: Heritable fragile site on chromosome 16: probable localisation of haptoglobin locus in man. Science **170**, 85–87 (1970).

Manolov, G., Manolova, Y., Levan, A.: The fluorescence pattern of the human karyotype. Hereditas (Lund) **69**, 273–286 (1971).

Mikelsaar, A.-V. N., Tüür, S. J., Käosaar, M. E.: Human karyotype polymorphism. I. Routine and fluorescence microscopic investigation of chromosomes in a normal adult population. Humangenetik **20**, 89–101 (1973).

Miller, D. A., Allderdice, P. W., Miller, O. J.: Quinacrine fluorescence patterns of human D group chromosomes. Nature (Lond.) **232**, 24–27 (1971).

Miller, O. J.: Autoradiography in human cytogenetics. In: Advances in human genetics, vol. 1, p. 35–130, H. Harris and K. Hirschhorn, eds. New York-London: Plenum Press 1970.

Miller, O. J., Breg, W. R., Mukherjee, B. B., Gamble, A. V. N., Christakos, A. C.: Non-random distribution of chromosomes in metaphase figures from cultured leucocytes. II. Peripheral location of chromosomes 13, 17–18, and 21. Cytogenetics **2**, 152–167 (1963).

Miller, O. J., Miller, D. A., Warburton, D.: Application of new staining techniques to the study of human chromosomes. Progr. med. Genet. **9**, 1–47 (1973).

Milunsky, A., Littlefield, J. W., Atkins, L.: Tetraploidy in amniotic-fluid cells. Lancet **1970** II, 979.

Naiman, J. L., Punnett, H. H., Destiné, M. L., Lischner, H. W.: Yy chromosomal chimaerism. Lancet **1966** II, 590.

Neurath, P. W., Enslein, K.: Human chromosome analysis as computed from arm lengths measurements. Cytogenetics **8**, 337–354 (1969).

Newcombe, H. B.: Record linking; the design of efficient systems for linking records into individual and family histories. Amer. J. hum. Genet. **19**, 335–359 (1967).

Ockey, C. H.: The positions of chromosomes at metaphase in human fibroblasts and their DNA synthesis behavior. Chromosoma (Berl.) **27**, 308–320 (1969).

Pardue, M. L., Gall, J. G.: Chromosomal localization of mouse satellite DNA. Science **168**, 1356–1358 (1970).

Paris Conference (1971): Standardization in human cytogenetics. Birth defects: Original article series, vol. VIII, No 7. New York: The National Foundation 1972.

Passarge, E.: Advances in human cytogenetics. I. Basic considerations. In: Human genetics, p. 26–37, Birth defects: Original article series, vol. IV, No 6. New York: The National Foundation 1968.

Passarge, E.: Spontaneous chromosomal instability. Humangenetik 16, 151–157 (1972).

Patau, K.: Identification of chromosomes. In: Human chromosome methodology, p. 155–186, J. J. Yunis, ed. New York-London: Academic Press 1965.

Patil, S. R., Lubs, H. A.: Non-random association of human acrocentric chromosomes. Humangenetik 13, 157–159 (1971).

Patil, S. R., Merrick, S., Lubs, H. A.: Identification of each human chromosome with a modified Giemsa stain. Science 173, 821–823 (1971).

Pawlowitzki, I. H.: The correspondence between quinacrine banding patterns and sites of secondary constrictions in human chromosomes. Humangenetik 15, 236–244 (1972).

Pawlowitzki, I. H., Cenani, A.: Sporadic triploid cells in human blood and fibroblast cultures. Humangenetik 5, 65–69 (1967).

Pearson, P. L., Bobrow, M., Vosa, C. G.: Technique for identifying Y chromosomes in human interphase nuclei. Nature (Lond.) 226, 78–80 (1970).

Penrose, L. S.: A note on the mean measurement of human chromosomes. Ann. hum. Genet. 28, 195–196 (1964).

Perry, J.: System for semi-automatic chromosome analysis. Nature (Lond.) 224, 800–803 (1969).

Pierre, R. V., Hoagland, H. C.: 45,X cell lines in adult men: Loss of Y chromosome, a normal aging phenomenon? Mayo Clin. Proc. 46, 52–55 (1971).

Powsner, E. R.: Frequency of endoreduplication in short-term cultures of human blood cells. J. Lab. clin. Med. 67, 610–614 (1966).

Renwick, J. H.: The mapping of human chromosomes. Ann. Rev. Genet. 5, 81–120 (1971).

Robinson, J. A.: Origin of extra chromosome in trisomy 21. Lancet 1973 I, 131–133.

Robinson, J. A., Buckton, K. E.: Quinacrine fluorescence of variant and abnormal human Y chromosomes. Chromosoma (Berl.) 35, 342–352 (1971).

Ruddle, F. H., Riccuti, F., McMorris, F. A., Tischfeld, J., Creagan, R., Darlington, G., Chen, T.: Somatic cell genetic assignment of peptidase C and the Rh linkage group to chromosome A-1 in man. Science 176, 1429–1431 (1972).

Rudkin, G. T.: Photometric measurements of individual metaphase chromosomes. In Vitro 1, 12–20 (1965).

Saksela, E., Moorhead, P. S.: Enhancement of secondary constrictions and the heterochromatic X in human cells. Cytogenetics 1, 225–244 (1962).

Sasaki, M. S., Makino, S.: The demonstration of secondary constrictions in human chromosomes by means of a new technique. Amer. J. hum. Genet. 15, 24–33 (1963).

Schmid, E., Bauchinger, M.: Structural polymorphism in chromosome 17. Nature (Lond.) 221, 387–388 (1969).

Schmid, W.: DNA replication patterns of human chromosomes. Cytogenetics 2, 175–193 (1963).

Schmid, W.: Satellites on the long Y chromosome arm: a familial Y autosome translocation in man. Cytogenetics 8, 415–426 (1969).

Schmid, W., Vischer, D.: Spontaneous fragility of an abnormally wide secondary constriction region in a human chromosome No. 9. Humangenetik 7, 22–27 (1969).

Schnedl, W.: Fluorescenzuntersuchungen über die Längenvariabilität des Y-Chromosoms beim Menschen. Humangenetik 12, 188–194 (1971).

Schnedl, W.: Banding pattern of human chromosomes. Nature New Biol. (Lond.) 233, 93–94 (1971).

Schnedl, W.: Analysis of the human karyotype by the recent banding technique. Arch. Klaus-Stiftg Vererb.-Forsch. (in press).

Schwanitz, G., Rott, H.-D., Köllermann, M.: Vergrößerte sekundäre Einschnürung des Chromosoms C_9 bei Mutter und Kind. Humangenetik 11, 258–263 (1971).

Schwarzacher, H. G., Schnedl, W.: Endoreduplication in human fibroblast cultures. Cytogenetics 4, 1–18 (1965).

Seabright, M.: A rapid banding technique for human chromosomes. Lancet 1971 II, 971–972.

Seabright, M.: The use of proteolytic enzymes for the mapping of structural rearrangements in the chromosomes of man. Chromosoma (Berl.) 36, 204–210 (1972).

Shaw, M. W.: Computers and chromosomes: The state of the art. In: Perspectives in cytogenetics, p. 309–315, S. W. Wright, B. F. Crandall, L. Boyer, eds. Springfield: Charles C. Thomas 1972.

Spence, M. A., Francke, U., Forsythe, A. B.: Evidence against the peripheral location of the Y chromosome in human metaphase cells. Cytogenet. Cell Genet. (Basel) 12, 49–52 (1973).

Summitt, R. L.: Cytogenetics in mentally defective children with anomalies: a controlled study. J. Pediat. 74, 58–66 (1969).

Sumner, A. T., Evans, H. J., Buckland, R. A.: New technique for distinguishing between human chromosomes. Nature New Biol. (Lond.) 232, 31–32 (1971).

Turpin, R., Lejeune, J.: Les chromosomes humains. Caryotype normal et variations pathologiques. Paris: Gauthier-Villars 1965.

Unnérus, V., Fellman, J., de la Chapelle, A.: The length of the human Y chromosome. Cytogenetics 6, 213–227 (1967).

Wahrman, J., Atidia, J., Goitein, R., Cohen, T.: Pericentric inversions of chromosome 9 in two families. Cytogenetics 11, 132–144 (1972).

Wald, N., Ranshaw, R. W., Herron, J. M., Castle, J. G.: Progress on an automatic system for cytogenetic analysis. In: Human population cytogenetics, p. 263–280, P. A. Jacobs, W. H. Price, P. Law, eds., Pfizer medical monographs 5. Edinburgh: Edinburgh Univ. Press 1970.

Wang, H. C., Fedoroff, S.: Banding in human chromosome treated with trypsin. Nature New Biol. (Lond.) 235, 52–53 (1972).

Wikramanyake, E., Renwick, J. H., Ferguson-Smith, M. A.: Chromosomal heteromorphisms in the assignment of loci to particular chromosomes: a study of four pedigrees. Ann. Génét. 14, 245–256 (1971).

Wright, S. W., Crandall, B. F., Boyer, L., eds.: Perspectives in cytogenetics. The next decade. Springfield, Illinois: Charles C. Thomas 1972.

Yesley, G. J.: Methods of coding and filing family records. In: The metabolic basis of inherited disease (J. B. Stanbury, J. B. Wyngaarden, D. S. Fredrickson, Eds.), 2nd edit., p. 1373–1379. New York: McGraw-Hill Book Co. 1966.

CHAPTER IX

Analysis of Interphase Nuclei

HANS GEORG SCHWARZACHER

With 17 Figures

1. Introduction

The presence of certain chromosomes in a cell, particularly the sex chromosomes, can be observed not only during mitosis but also in interphase. Previous chapters (V–VIII), have mentioned that the sex chromosomes show some peculiarities: One of the two X chromosomes in female cells is facultatively heterochromatic; the Y chromosome in male cells is also heterochromatic and is, moreover, distinguished by a bright fluorescence after quinacrine staining, and dark staining after applying the Giemsa banding techniques.

Heterochromatic chromosome segments may be heteropyknotic under certain conditions, which means they are strongly condensed and, if sufficiently so, are visible in interphase as tight bodies among the remaining more loosely packed chromatin of a cell nucleus. This is the case in one of the two X chromosomes in female cells, forming the "X-chromatin body". Chromosome segments which strongly fluoresce after quinacrine staining can be seen as bright spots in interphase nuclei. This is the case in the Y chromosome, forming the "Y-chromatin body". Investigation of interphase nuclei allows provisional designation of the sex chromosome status. Since this can be done on easily accessible tissues, such as buccal mucosa, hair root or blood cells, it is indicated whenever an anomaly of the sex chromosomes is suspected. It is also a useful screening procedure for a large number of individuals. Analysis of cell nuclei from amniotic fluid enables a prenatal sex diagnosis. Finally, it should be emphasized that examination of the interphase nuclei may sometimes assist in the interpretation of complicated sex chromosome anomalies.

Besides the sex chromosomes, certain parts of some autosomes show peculiar staining properties (see Chapters V, VI). Staining of the centromeric heterochromatin (Arrighi and Hsu, 1971; Yunis *et al.*, 1971) will produce corresponding small dark spots in interphase nuclei. The special Giemsa methods at a high pH (Bobrow *et al.*, 1972; Gagné and Laberge, 1972) preferentially stain the heterochromatic segment of chromosome 9, and produce 2 tiny spots in diploid interphase nuclei. Although these methods are of considerable value for basic theoretical problems, they are of limited use in medical practice at present.

2. Y-Chromatin

As has already been described in Chapters V and VIII, part of the Y chromosome (the distal part of the long arm) fluoresces very strongly after quinacrine staining and retains its fluorescence during interphase, when it can be perceived as a small luminous spot of about 0.3–1 µ diameter (Zech, 1969; Caspersson *et al.*, 1969; Pearson *et al.*, 1970; George, 1970). In most cases, no other chromosome exhibits a fluorescent part of similar intensity and size when stained with quinacrine or similar dyes. The presence of such a small fluorescing body in the cell nucleus can, therefore, be taken to be a fairly certain sign of the presence of a Y chromosome in that particular cell. Variations in the size of the Y chromosome are reflected in the size of the Y-chromatin. In cases of an extremely small Y chromosome, a distinctly fluorescing Y-chromatin body may be absent in interphase nuclei. On the other hand, in cases of an unusual strong quinacrine fluorescence of other chromosome regions (as e.g. the centromeric region of chromosome 3, or the satellites of acrocentric chromosomes, see Chapter VIII) the corresponding fluorescing small spots may confuse the accurate diagnosis of interphase nuclei. In unclear cases, a further examination (X-chromatin, chromosome analysis) is required.

Where more than one Y chromosome is present, the number of Y-chromatin bodies in interphase nuclei usually corresponds to the number of Y chromosomes. It should also be kept in mind that not every cell nucleus containing a normal-sized Y chromosome will show Y-chromatin for technical reasons, such as position of the Y within the nucleus, pyknotic nuclei, insufficient staining, etc.

3. X-Chromatin

The X-chromatin (Barr body, sex chromatin) of man is a small heterochromatic body that is present in somatic cell nuclei of normal females, but absent in male tissues. Its normal size is about 1 μ and it is preferentially located at the periphery of the cell nucleus. In segmented nuclei of granulocytes, it may form a characteristic appendage, the so-called drumstick. This special form will be described in detail in Chapter X.

The X-chromatin can be regarded as a heterochromatic X chromosome which is mostly or entirely heteropyknotic and condensed during interphase. When only one X chromosome is present in a cell, as in the normal male (with 44 autosomes and XY sex chromosomes), this single X chromosome is isopyknotic or euchromatic, i.e. it is *not* condensed, so the cell nucleus contains no X-chromatin. If more than one X chromosome is present in a cell, then one per set of diploid chromosomes is isopyknotic and thus not visible as a distinct chromatin body, while all remaining X chromosomes be can heterochromatic and able to form an X-chromatin body. Normal diploid female cells (with 44 autosomes and XX sex chromosomes) thus contain at most one X-chromatin body. The formation of an X-chromatin body (by one of the two X chromosomes) is, however, not an invariable phenomenon; the frequency of its occurrence in the cells of normal female organisms varies from tissue to tissue.

The rule that at least one X chromosome per diploid chromosome set is euchromatic, and all remaining X chromosomes may be heteropyknotic, is also valid for cells with anomalous numbers of X chromosomes. The number of X-chromatin bodies in a cell with a diploid autosome set is at most one less than the number of X chromosomes present. It may thus be possible to use the analysis of X-chromatin for the diagnosis of abnormalities in the number of X chromosomes (for reviews see Barr and Carr, 1962; Grumbach *et al.*, 1963; Barr, 1966; Hamerton, 1969 and 1971). Structural abnormalities of X chromosomes in female cells result in altered forms or sizes of X-chromatin bodies (see, for example, Klinger *et al.*, 1965).

4. Techniques for Preparing Interphase Nuclei

4.1. Introductory Remarks

The *X-chromatin* can be observed in living cells with the phase contrast microscope under favorable conditions (De Mars, 1963; Schwarzacher, 1963). It can also be visualized by fluorescent dyes (e.g. acridine orange or quinacrine, see e.g. Mukherjee *et al.*, 1972; Klinger and Moser, 1972). For an exact demonstration of the X-chromatin, however, the cells are better stained with one of the specific methods.

The *Y-chromatin* can be visualized after staining with quinacrine or similar acting fluorescent dyes, which should be applied after a short fixation period (Pearson *et al.*, 1970). A special X-chromatin staining can usually readily be done on the same cells after quinacrine staining. *The following procedure is recommended:*

1. Obtaining the material and preparing slides.
2. Fixation (96% alcohol, or methyl alcohol, or acetic acid alcohol).
3. Staining with quinacrine (see Chapter V) and evaluation of the Y-chromatin in the aqueous preparation under the fluorescence microscope.
4. Dissolving the superfluous quinacrine dye in 96% ethyl alcohol, which also serves as additional fixative.
5. Staining for X-chromatin (thionine, Feulgen or other).

Of course, quinacrine or X-chromatin staining, can be applied separately, but some tissues are more suitable for demonstrating the Y- or the X-chromatin body.

4.2. Obtaining the Material

Interphase cell nuclei from many tissues are suitable for examining the Y- and the X-chromatin. Preparations of whole cells are always superior to sections. It is important to obtain fresh and undamaged cells. Total preparations are made from cell smears (easily accessible are mucosa smears), hair roots, embryonic membranes, teased tissues, and cell cultures. Peripheral blood cells can be used for both Y- and X-chromatin (the latter in the special shape of the drumstick, see Chapter X). In spermatozoa, only the Y body can be visualized.

Buccal smears are recommended for clinical use because they are easily obtained. In gynecological or urological practice, vaginal or urethral smears might be preferred.

Very suitable material are cells of *hair roots*. From newborn babies and embryos the *amnion* gives excellent whole cell preparations. Amniotic fluid cells obtained by *amniocentesis* are also suitable. In examining *post mortem* material, connective tissue cells and smooth muscle fibers have proved favorable, because the nuclei of these cells may remain well-preserved for more than 24 hours after death. *Histological sections* from material embedded in paraffin or similar, usually give less satisfactory results, especially for the Y-chromatin.

4.3. Fixation

The demands made on the fixative in preserving the nuclear structure are considerable, and only very few of the customary mixtures are suitable for a satisfactory preparation of the Y- and the X-chromatin body. In any event, it is important to standardize the fixation in order to obtain comparable results. Since chromatin particles can be recognized in only a relatively small percentage of living cells, fixation with agents that cause the nuclear structures to become moderately coarse through precipitation and coacervation of the proteins will yield more distinct preparations. However, too coarse nuclear structures and clumping produced by strong acid fixative or unbuffered formol, is just as undesirable. Strong acid solutions also cause uncontrolled acid hydrolysis of the nucleoproteins, especially during longer fixations.

In general, a mixture of 1 part acetic acid and 3 parts alcohol (methanol or 96% ethanol) is a very good fixative for both the quinacrine staining of the Y body and the staining of the X-chromatin body. 96% ethanol or absolute methanol (without ether which may alter the nuclear structure) is also highly recommended. Since these fixatives are widely used, the results can be standardized and compared with those from different laboratories. For quinacrine staining, 50% acetic acid is also a suitable fixative. However, it should not be applied for longer than 20 min. For subsequent X-chromatin staining, the preparations should be refixed in 96% ethanol for at least 30 min.

4.4. Preparing Different Tissues

4.4.1. Buccal Smears (Marberger et al., 1955; Moore and Barr, 1955)

The smears are taken from the inner surface of the cheek. They can be taken from both sides in order to detect possible discrepancies between left and right, e.g. in mosaics. For these smears, cells from the deep epithelial layers with well delineated nuclei should be used. They are spread in a thick layer on a slide and fixed immediately in 96% ethyl alcohol or acetic acid alcohol 1:3 (see Appendix). Fixation should always be carried out directly *without prior air drying* which will alter the nuclear structure of most cells. Even in thin smears, sufficient cells will adhere to the slide. For fluorescent staining of the Y body the time required for fixation is 10 min, for X-chromatin staining, 30 min. Preparations for X-chromatin may be kept in alcohol for several days. After alcohol fixation the preparations should be air-dried, which makes the cells adhere firmly to the slide.

4.4.2. Smears from Other Mucosae

Smears from other mucosae are basically handled in the same manner. Depending on the type of mucosa, careful scraping is necessary in order to avoid damage. Particularly good cells can be obtained from deeper layers of vaginal mucosa by the usual Papanicolaou technique. Smears from portions of malformed urethra must naturally be obtained with great care (see Carpentier, 1966).

4.4.3. Teasing Preparations

All tissues amenable to isolating single cells (e.g. connective tissue, liver, muscle tissue) can yield good preparations. The cells are isolated by teasing the tissue in balanced salt solution, not too thinly spread on slides, and further handled like smears from mucosae.

4.4.4. Hair Roots

Very good material can be readily obtained from cells of hair roots (Schmid, 1967). The external root sheath of a plucked hair is removed with forceps under a preparation microscope. The root sheath is fixed in acetic acid alcohol (1:3) and stained. This method was modified by McKee-Katz et al. (1970) and by Engel et al. (1972). A simplified procedure was developed by Schwinger et al. (1971) which is particularly useful for fluorescence staining of the Y-chromatin (see Appendix).

4.4.5. Amniocentesis Material and Other Cell Suspensions

Cell suspensions such as amniotic fluid obtained by amniocentesis or from suspension cultures, are centrifuged, (about 1,000 rev/min for 10 min) and resuspended in fixative (acetic acid ethanol 1:3, at least 20 times the volume of the cells). This is repeated, then the cells are suspended in a small amount of fixative (approximately 5 times the volume of the pellet), and smears made from this suspension.

4.4.6. Preparations from Membranes

Amnion is a particularly suitable source and preparations of this type are the method of choice for examination of aborted or removed embryos and newborn babies.

A strip of amnion, about $1^1/_2$ times the width and $^3/_4$ the length of a slide, is spread out on a slide with the epithelial side upwards. The edges are tucked underneath the slide and smoothed if necessary. Fixation (96% ethanol is to be preferred) is carried out in grooved cuvettes (e.g. Coplin jars) without prior drying. After at least 1 hour of fixation, small pieces can be cut from the amnion strip, attached to new slides with celloidin, and processed for different stains, etc.

Flat pieces of chorion can be prepared in the same way. The thin layer of tissues of the fetal side must be used. It can be stripped off either before or after fixation.

4.4.7. Tissue Cultures

Cultured cells from skin, fascia or organ explants (see Chapter II) growing in monolayer produce ideal preparations for the examination of X-chromatin. The Y-chromatin, however, is less clearly seen, since the quinacrine dyes sometimes have a tendency to stain other cell components as well in this material. The X-chromatin may also stain more prominently with quinacrine in cultured fibroblasts (Mukherjee et al., 1972b). Nevertheless, it is a good rule to make interphase preparations from all cultures used for chromosomal analysis, since these preparations can later also be used for other cytological studies.

Primary cultures on cover slips or cells which have settled on coverslips from the suspension of a subculture can be used. For this type of preparation, small culture vessels with a groove for a coverslip are particularly convenient (Leighton tubes, see Chapter II). When the cultures are grown in larger bottles, one or more coverslips are placed on the bottom of the bottle before adding the cell suspension. The coverslip can be secured to the bottom with a drop of chick plasma. Enough cells have usually accumulated on the coverslips 24–48 hours after transfer; these can then be removed from the bottle without disturbing the culture. The size of the coverslips must be accomodated to the size of the bottle neck, ensuring easy passage under sterile conditions. The coverslips should be fixed (96% ethyl alcohol, or methyl alcohol, or a 1:3 mixture of acetic acid and alcohol) without drying after brief rinsing (1 min) in an isotonic salt solution (e.g. Hanks' solution).

Further handling of the coverslips for fixation and staining is facilitated by using a coverslip holder or a rack to protect them from breakage. We recommend short test tubes with a flat bottom and a hole (enabling staining and fixing fluids to enter), and a groove (to hold the coverslip in place). Such tubes permit easy transfer either singly, or on a larger carrying rack. After successful fixation and staining, the coverslips are mounted on slides, the cell layer facing downwards.

4.4.8. Peripheral Blood Cells

Blood films or smears, as routinely prepared for hematological investigation, are fixed in absolute methanol or 96% ethanol. After quinacrine staining, the Y-chromatin can be demonstrated in all types of white blood cells. The X-chromatin can be found in Giemsa, Pappenheim or similarly stained preparations as a small nuclear appendage, the so-called drumstick in polynucleated granulocytes (see Chapter X).

4.4.9. Spermatozoa

A small drop of ejaculate is either directly spread on a clean slide or diluted 1:1 in balanced salt solution before preparing a smear. The smears should not be dried but fixed directly in a 1:3 mixture of acetic acid and alcohol for 10 min, then air dried, and stained with quinacrine. In Y carrying sperms, the Y body can be demonstrated. An X-chromatin body is not visible in sperms.

4.4.10. Section Preparations

X-chromatin can be studied in sections prepared according to the usual histological techniques from many tissues. For fixation, a mixture of acetic acid, ethanol, and formalin is recommended (Davidson's mixture, see Appendix). This should not be applied for longer than 24 hours and must then be replaced by 70% ethanol. Special care should be taken with dehydration and embedding, because the nuclear structure is particularly sensitive.

Fluorescent quinacrine staining of the Y-chromatin is usually much less successful than X-chromatin staining in sections. Paraffin or celloidin material is not suitable. Sufficient staining is achieved in frozen sections or cryostat sections after a fixation in acetic acid/alcohol for at least 4 hours (personal communication of Dr. H. Cramer, Marburg). The sections should be postfixed in 96% ethanol for 30 min, washed in buffer, and then stained with quinacrine. To improve the quality of simple frozen sections the material can be embedded in a saturated gelatine solution for 3–6 days at 37°.

5. Staining

5.1. Fluorescence Staining with Quinacrines

There is no difference in fluorescence staining and microscopy techniques between chromosome preparations and interphase cell nuclei preparations. Therefore, the procedures for staining with quinacrine dyes (quinacrine dihydrochloride and quinacrine mustard) should be applied as described in detail in Chapter V. Brief directions for laboratory use are given in the Appendix. Quinacrine staining presents the Y-chromatin in particular. The X-chromatin is usually not sufficiently delineated to allow a certain diagnosis in such preparations. An improvement may be effected by a short hydrolysis in HCl (3 N HCl at 21–22° C for 2 min) before the staining and a dehydration in alcohol and mounting in a neutral non-fluorescent mounting medium (e.g. DPX, Gurr Ltd.) after staining (Klinger and Moser, 1972). In preparations so treated it may be possible to diagnose either the Y or the X-chromatin.

Any non-fluorescent staining of X-chromatin can easily be carried out on the same preparation following standard quinacrine fluorescence staining of the Y body. In our experience, non-fluorescent staining of the X-chromatin is superior to quinacrine.

5.2. Specific Staining Methods for X-Chromatin

In principle, all methods which stain the chromatin of the cell nucleus and differentiate it from the nucleolus are suitable, though methods which just stain the chromatin and neither the cytoplasm nor the nucleoli are superior.

Specific staining of deoxyribonucleic acid according to *Feulgen* is considered the most reliable method, but is also somewhat laborious. Histochemical textbooks provide detailed information on the theory and practice of Feulgen staining. In the Appendix a Feulgen method is given which has proved to be excellent for the analysis of X-chromatin.

Particularly well differentiated preparations are obtained by the *thionine* method of Klinger and Ludwig (1957, see Appendix) which is somewhat easier to perform than Feulgen and almost as reliable. Hydrolysis in HCl precedes staining. The acid treatment requires sufficient adhesion of the cells to the slides in both of these methods. If smears are air-dried following fixation, enough cells will usually remain on the slide. Other material which floats off easily (e.g. sections) must be fixed to the slide with celloidin.

Correct timing and intensity of hydrolysis are most important for the thionine method. Too little hydrolysis will also stain cytoplasm, nucleoli and whatever bacteria may be present (e.g. in oral smears), too much hydrolysis may result in weakly stained cell nuclei. Unsatisfactory staining can often be improved by varying the duration of hydrolysis. Another reason for poor quality may be the differentiation after staining, which should be done under microscopic monitoring until the method has been standardized.

Owing to their specificity for the chromatin of the cell nucleus, the Feulgen and thionine methods have the advantage that X-chromatin cannot be confused with other structures, as in buccal smears contaminated with bacteria. In practice, however, there are other stains which have proved suitable, inasmuch as they produce stronger contrast and are much simpler; two of these are particularly worth mentioning: *cresyl violet* (Moore and Barr, 1955) and *carbol fuchsin* (Eskelund, 1956; Barr, 1965; see Appendix).

Carbol fuchsin is more reliable, but making up the solution is somewhat laborious. Both methods stain several cell components, but chromatin is stained more intensely than nucleoli and cytoplasm. Microorganisms in buccal smears are usually intensely stained and may obscure diagnosis, but after some practice, they will not be confused with the X-chromatin body. These stains are particularly suitable for routine examinations of buccal smears.

Two other stains, *pinacyanol* (Klinger and Hammond, 1971) and *diamond-fuchsin* may also produce suitable results, but HCl pretreatment is necessary (5 N HCl for 5 min at room temperature) to demonstrate the X-chromatin clearly. In contrast to the Feulgen and thionine stains, the HCl treatment in these methods produces a much coarser appearence of the chromatin, and they can serve for a quick orientation only.

Staining with *aceto-orcein* (Sanderson, 1960), performed in the same way as in chromosomal preparations (see Chapter IV) usually yields rather low-contrast pictures of X-chromatin in comparison to the stains mentioned above. Better results can be achieved by simultaneously fixing and staining fresh cell smears with orcein. This procedure is similar to the orcein squash technique described for isolated cells (see Chapter IV).

Other stains especially developed for the study of X-chromatin (e.g. Guard, 1969; Lennox, 1956; Hienz, 1959; Cuadrillero, 1959) are too laborious for routine use and yield preparations not better than an accurately performed Feulgen reaction or the other methods mentioned.

Among the more common histological stains, the *hematoxylin-eosin* stain is particularly suitable if the preparation is well fixed (Fig. 4), but requires some practice to avoid confusion of small nucleoli with the X-chromatin; experience should be gained by comparing the hematoxylin-eosin preparations with Feulgen preparations from the same material.

If it is necessary to examine material lacking a distinct X-chromatin pattern following routine use of hematoxylin-eosin or some other stain (e.g., sections of biopsies), it may be possible to bring out the X-chromatin more distinctly by hydrolysis in HCl followed by re-staining with thionine.

6. Evaluation of the Preparations

6.1. Y-Chromatin

The quinacrine fluorescing portion of the Y chromosome is readily seen in most interphase nuclei. The number of the Y-chromatin bodies corresponds directly to the number of Y chromosomes present in the cell nucleus in question. In some tissues, a Y-chromatin body cannot be observed in every cell nucleus because of its position, or because it is obscured by other quinacrine-stained particles. In most cases, evaluation of about 50 cell nuclei will be sufficient if no abnormality is found.

In a normal male cell nucleus, the Y-chromatin appears as a bright dot about $0.3–1,0\,\mu$ in diameter (Fig. 1). In some cell nuclei, a double structure can be found (Fig. 2a). Such twin bodies are observed in G1 nuclei as well as in G2 nuclei (Schwinger and Pera, 1972). A slight dispersion of the Y body can sometimes be seen, particularly in large cell nuclei or in actively multiplying cells such as stimulated lymphocytes.

The Y-chromatin is often situated at the periphery of the cell nucleus, although not so frequently as the X-chromatin.

Measurements of the Y chromosomes of different lengths have shown that the fluorescing segment also differs in length (Schnedl, 1971; Bobrow et al., 1971). Hence, the size of a Y-chromatin body is an indicator of the size of the Y chromosome. Some variations in the appearance of the Y-chromatin may also be caused by different forms of cell nuclei and by a variable position of the Y chromosome within the nucleus.

As has been already mentioned, other strongly fluorescing chromosome segments can appear also in the form of small dots in interphase. Usually they do not attain the size of the Y-chromatin, but, in some cases, other chromosomes may have such strongly fluorescent regions that a Y-chromatin body may be simulated (see e.g. Caspersson et al., 1971). On the other hand, in cases of extremely small Y chromosomes with a very short fluorescing segment, the Y-chromatin body may be difficult or impossible to diagnose. Although these cases seem to be rare, one should keep in mind that a small fluorescent body does not always imply the presence of a Y chromosome, and convertely the absence of a typical Y-chromatin body may not necessarily imply the absence of the Y chromosome. In doubtful cases, X-chromatin staining should follow

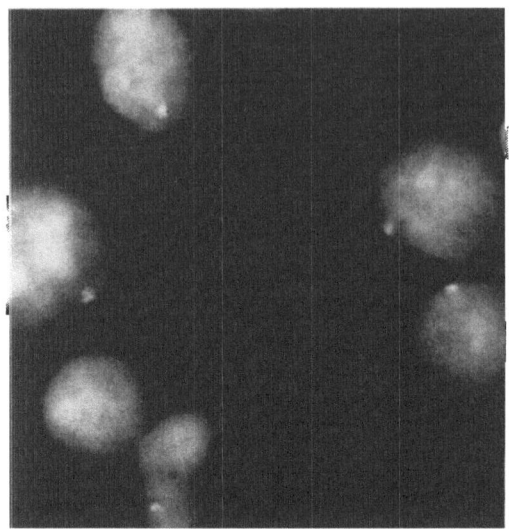

Fig. 1. Smear of bone marrow from a chromosomally normal man. Staining with quinacrine dihydrochloride. Distinct Y-chromatin in almost every cell nucleus. Microscope: Zeiss "Universal", epi-illumination. × 2.400. (Courtesy of Dr. D. Mutton and Prof. P. Polani)

as next step. A-chromosomal analysis will, of course, decide any remaining questions. This problem should be particularly kept in mind with respect to prenatal sex diagnosis. Some authors even assert that by simply looking for Y-chromatin bodies in amniotic fluid cells, a normal male may be missed (Rook *et al.*, 1971).

6.1.1. Oral Mucosa

In oral mucosal smears from normal 46, XY men, a Y-chromatin is present in 25–50% of cell nuclei (Pearson *et al.*, 1970). Müller *et al.* (1971) reported 20–60%, and Majewski *et al.*, (1971) found Y-chromatin bodies in up to 80% with careful staining. The frequency of observed Y-chromatin bodies depends on the quality of the preparation, the stain, and the fluorescence microscope. If only lightly stained larger nuclei are analysed, this frequency goes up to 100% in well-prepared smears. Buccal smears from females may show fluorescent bodies resembling a Y-chromatin body in a maximum 5% of cells (Majewski *et al.*, 1971). These discrepancies between different groups of observers indicate the necessity of establishing one's own standard for the frequency of Y-chromatin bodies in normal individuals. Sometimes the evaluation of oral mucosa smears can be made difficult by bacteria and other contamination.

6.1.2. Hair Root Cells

Hair root cells are readily accessible, and easy to prepare (Schwinger *et al.*, 1971; see Appendix). The evaluation is usually not hampered by contamination. A Y-chromatin body is found in 70–90% of normal males except for occasional cases with quite low frequencies which occur with a very small Y chromosome.

6.1.3. Blood Cells

The Y-chromatin can be observed in all types of white blood cells of normal males (Fig. 2b and c). In lymphocytes, it is usually somewhat easier to detect than in granulocytes. Polani and Mutton (1971) give ranges from 61–87% positive lymphocytes and from 27–60% positive polymorphs in normal males. In granulocytes, the Y-chromatin body is sometimes situated in a small protrusion of the nucleus (Fig. 2b), the so-called "small club" (see Chapter X). In true drumsticks of female granulocytes, which contain the heterochromatic X chromosome, no prominent fluorescence can be observed.

A fluorescent body of the size of a small Y-chromatin has been reported in 0.6% of female white blood cells (Hellriegel *et al.*, 1973). This can be due to cases with larger quinacrine positive autosomal spots. Such fluorescing spots are usually situated within the nucleus, in granulocytes they do not form extra protrusions like the small clubs. The finding of fluorescing small clubs is, therefore, a strong indication of the presence of a Y chromosome.

Feto-maternal transfer of leukocytes may increase the frequency of Y-chromatin bodies in pregnant women carrying a male fetus. If more than 4% Y-chromatin positive cells are found in the mother, the diagnosis of a male fetus seems to be fairly certain. However, in 13 out of 101 pregnant women carrying a male fetus, no Y-chromatin was found. A prenatal sex diagnosis is therefore possible only in Y-positive cases. The Y-chromatin positive granulocytes disappear within a few days after delivery. Y-chromatin-positive lymphocytes have been reported to persist for much longer in the circulating blood of the mother (Hellriegel *et al.*, 1973).

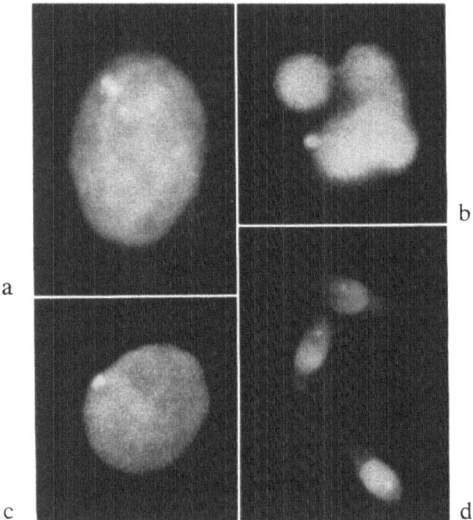

Fig. 2a–d. Quinacrine mustard stainings of different cell nuclei of chromosomally normal men. a From a buccal smear. The Y-chromatin appears as a double structure. Besides the Y-chromatin there are a few other tiny fluorescent spots. b Granulocyte from a blood smear. The Y-chromatin protruding in a small appendage. c Large lymphocyte from a blood smear. d Spermatocytes. The Y-chromatin situated near the borderline of the strongly fluorescent part of the sperm head. Staining with quinacrine mustard. Microscope: Reichert "Zetopan", dark field illumination. × 2.400

The Y-chromatin can also be diagnosed in dried blood spots and marks under certain conditions (Müller *et al.*, 1971), which may be important in forensic medical problems.

6.1.4. Spermatozoa

In normal healthy men, a Y-chromatin body (Fig. 2d) has been reported in 40–47.5% of the spermatozoa (Barlow and Vosa, 1970; Pearson and Bobrow, 1970; Sumner *et al.*, 1971). Assuming a 1:1 sex ratio 50% of sperms would be expected to contain a Y-chromosome. The observed deficiency is presumably due to technical factors, but it should be kept in mind that nothing is known about the true ratio of X- to Y-bearing spermatozoa. The Y-chromatin body is preferentially seen at the boundary between the dense and less dense regions of the sperm head. If it lies within the dense part, it may easily be missed.

It is noteworthy that two Y bodies have been reported in 0.9–1.4% of sperms of chromosomally normal males (Pearson and Bobrow, 1970;

Sumner *et al.*, 1971). This could reflect the rate of non-disjunction of the sex chromosomes in meiosis. However, a certain overestimation of this rate because of technical errors should be taken in consideration. More than one Y-chromatin body is also present in diploid or higher polyploid sperms which can be found occasionally (Pearson and Bobrow, 1970).

6.1.5. Other Tissues

Most of tissues suitable for X-chromatin analysis (see following paragraphs) can also be used to study the Y-chromatin. In some, however, the diagnosis of the Y-chromatin may be difficult, because also the cytoplasm may have a tendency to stain with quinacrine.

For the evaluation of cells from amniocentesis see 6.3.

6.2. X-Chromatin

6.2.1. General Considerations

The percentage of cell nuclei containing an X-chromatin body ("X-chromatin positive" cells) can be determined in most cases with sufficient confidence by analysing about 200–400 nuclei. It should also be determined whether more than one X-chromatin body is present per nucleus and whether its size and shape is normal.

In diploid female cell nuclei, the X-chromatin has a maximum diameter of about 1 µ. Its shape may resemble a short rod or it can appear triangular, as one edge is frequently attached to the nuclear membrane (Figs. 8 and 9). A bipartite structure (Fig. 3) can occasionally be seen, as if the X-chromatin were formed of two rods, or of one which has been bent and folded back on itself (see Klinger, 1966). In most tissues, the X-chromatin occupies a peripheral position within the cell nucleus and lies next to the nuclear membrane in a high percentage of cells. Nerve cells are a notable exception to the rule of peripheral location of X-chromatin (Barr and Bertram, 1949). In order to avoid confusion of the X-chromatin with other chromatin bodies or artefacts, it is recommended to regard cell nuclei as "positive" only if the X-chromatin body is of normal size and shape and lies at the periphery of the nuclear membrane.

Typical X-chromatin is absent from all cell nuclei of male tissues. Chromatin bodies unrelated to sex may simulate an X-chromatin. Occasionally, even in males, a very small percentage of cells may be falsely interpreted as "positive".

The frequency of X-chromatin is not the same in all tissues of female origin and, usually, a certain percentage of the cell nuclei is always X-chromatin negative. For unknown reasons, it does not appear to be obligatory for the inactive X chromosome to be visibly heteropyknotic.

It should also be noted that the cell nuclei of many tissues from both sexes can contain, in addition to the X-chromatin, other heteropyknotic areas which can render the diagnosis rather difficult. It may thus be necessary to examine preparations of the same tissue from both sexes to correctly evaluate a pathological situation. This should be repeated and extended to other tissues if abnormal findings seem to be present, and supplemented by chromosomal analysis. It should always be remembered that all investigations of X-chromatin are prey to subjective errors of judgment. Every worker should establish his own standards by studying normal tissues.

6.2.2. Mucosa Smears

Only well-preserved cells with an evenly granulated nuclear texture and without signs of pyknosis, wrinkles or other damage should be selected for study (Figs. 8–10). All suitable cell nuclei within a given microscopic field should be examined in order to avoid biased selection.

In *buccal smears* fixed in 96% alcohol and stained with carbol fuchsin, we find an X-chromatin body in 25–60% of all cell nuclei in normal female subjects, whereas a chromatin body resembling an X-chromatin may be present in maximally 1% of the cell nuclei of normal males. We do not count the tiny heteropycnotic granules which are sometimes seen in male cell nuclei since they do not lie at the periphery. There are contrasting statements in the literature, because the evaluation of X-chromatin is greatly influenced by subjective factors and the technique used, as emphasized above. In any event, in good preparations the difference between normal female and normal male cells can always be unequivocally established. The percentage of X-chromatin positive cells appears to be reduced in female newborns during the first two days after birth (Smith *et al.*, 1962; Taylor, 1963), although Hsu *et al.* (1967) have shown these differences to be minor. Most probably, they are due to technical reasons, because the nuclear membrane of buccal mucosa cells of newborn babies may be poorly delineated and difficult to evaluate.

Figs. 8–16 may serve to illustrate several important technical points which should be considered in the evaluation of buccal smears stained with carbol fuchsin. The X-chromatin can be recognized as a very distinct heterochromatic body (Figs. 8 and 9) and confusion with chro-

Fig. 3   Fig. 4   Fig. 5

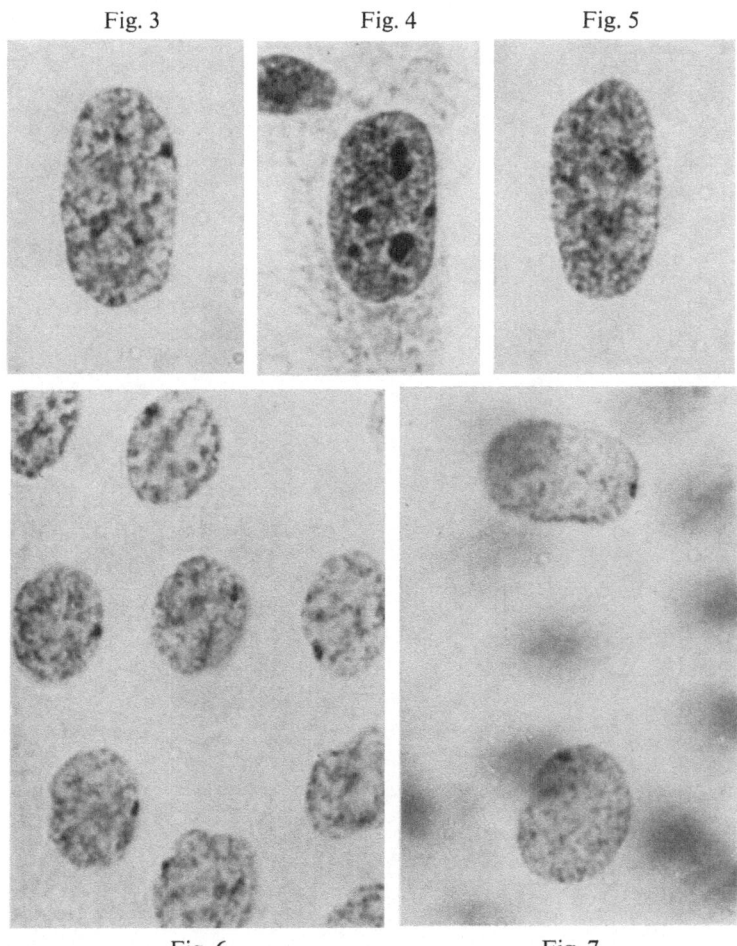

Fig. 6   Fig. 7

Fig. 3. Cell nucleus of a fibroblast-like cell in culture from a skin biopsy of a 22 year old healthy female individual. Fixation in 95% ethyl alcohol, Feulgen stain. X-chromatin located at the periphery of the nucleus. × 2.000

Fig. 4. As in Fig. 3, but stained with hematoxylin-eosin. In addition to the peripherally located X-chromatin, 4 nucleoli of different sizes are present

Fig. 5. As in Fig. 3. X-chromatin located inside the nucleus. Feulgen stain

Fig. 6. Total preparation from the amnion of a newborn girl. Focused at the epithelial cells. Fixation in 95% ethanol, Feulgen stain. Each nucleus contains an X-chromatin body. × 2.000

Fig. 7. Same amnion preparation as in Fig. 6, but focused at connective tissue cell nuclei

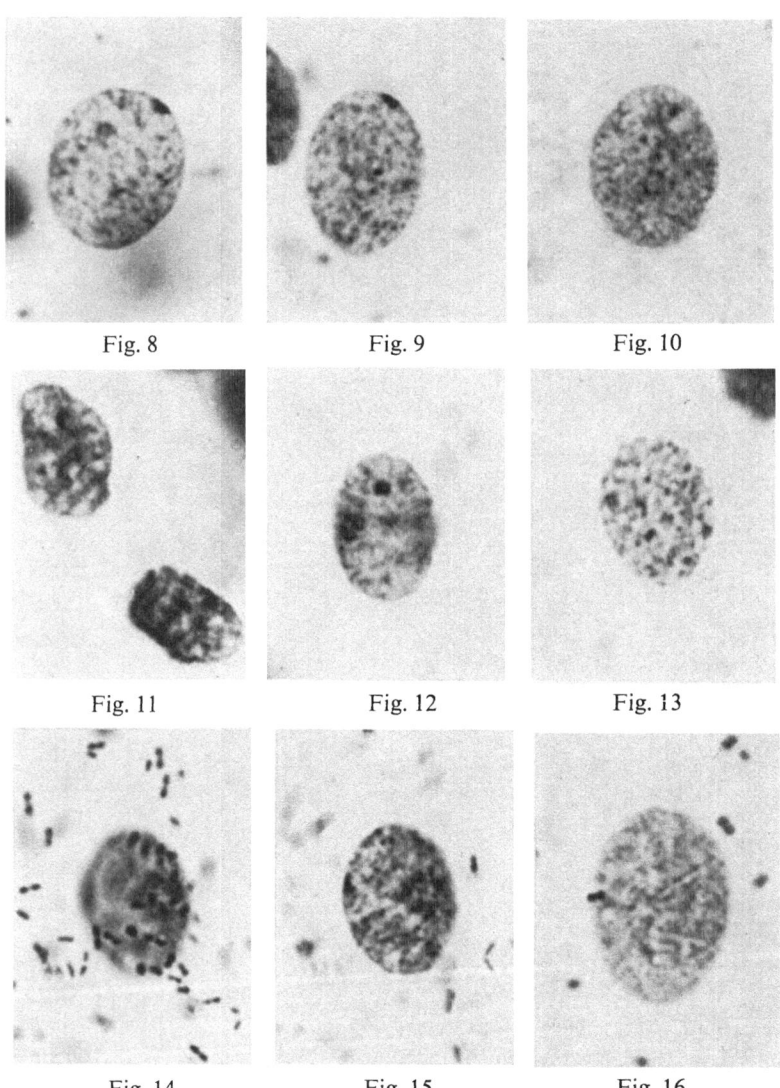

Fig. 8   Fig. 9   Fig. 10

Fig. 11   Fig. 12   Fig. 13

Fig. 14   Fig. 15   Fig. 16

Figs. 8–16. Cell nuclei from buccal smears. Fixation in 95% ethyl alcohol, stained with carbol fuchsin. × 2.000. Figs. 8 and 9: Normal adult female (22 years old). Peripheral location of the X-chromatin. Fig. 10: Normal adult male (24 years old). The small heteropycnotic granule in the upper right quadrant should not be confused with an X-chromatin. Figs. 11–13: Examples of cell nuclei of the smear from a normal female which should not be scored as X-chromatin-positive, because of pycnosis. (Fig. 11), absent peripheral location of the X-chromatin (Fig. 12), and coarsly granulated nuclear structure (Fig. 13). Figs. 14 and 15: Cell nucleus (X-chromatin positive) in different planes of focus. The X-chromatin can hardly be confused with bacteria. Fig. 16: A bacterium at the periphery of the nucleus which could be mistaken for a X-chromatin by an inexperienced observer matin particles unrelated to sex in both sexes is hardly possible (Fig. 10). Bacteria and other contamination can usually be readily distinguished from X-chromatin by focussing on a different plane (Figs. 14 and 15). Fig. 16 shows an X-chromatin negative cell from a normal male and bacteria simulating an X-chromatin body at the peripery of the nucleus. The experienced worker would encounter no difficulty in differentiating this from X-chromatin. Sometimes smears contain so many bacteria that they seriously hinder the analysis, which will require a repeated examination after a hydrolysis of the preparation by removing the coverslip, passing it through decreasing concentrations of alcohol into distilled water, treating with HCl and restaining with Thionine.

6.2.3. Tissue Cultures

X-chromatin is easy to determine in cultures of fibroblastlike cells (Fraccaro and Lindsten, 1959; Miles, 1960; Schnedl, 1964). Normal female tissue will exhibit 40–80% X-chromatin-positive cells (Figs. 3–5), whereas male cell nuclei never display typical X-chromatin. The frequency of X-chromatin positive nuclei in female tissue cultures is dependent on the type of culture, growth activity, and cell density (Miles, 1960; Therkelsen, 1963; Schnedl, 1964; Klinger *et al.*, 1968), but independent of the cell cycle (Klinger *et al.*, 1966).

6.2.4. Membrane Preparations from Amnion

Fetal and newborn amnion consists of a single layer of cubic to flat epithelial cells and a layer of connective tissue. In young embryos (smaller than 50 mm crown-rump length), the connective tissue is relatively rich in cells and forms an epithelium-like layer separating it from the magma reticulare. In many instances, especially when the amnion has been freshly fixed, the round nuclei of the densely packed epithelial cells are extremely suitable for evaluation (Fig. 6). In preparations fixed in alcohol and stained by the Feulgen method, 90–100% of these cells may be X-chromatin positive in female amnia; amnion epithelial cells of male origin are X-chromatin negative. The cell nuclei of the connective layer can usually be properly evaluated (Fig. 7) even if the epithelial cells are pyknotic as a result of mechanical stress, cell death, or drying.

6.2.5. Section Preparations

The frequency of X-chromatin positive cells in section preparations depends to some extent on their thickness, because sections do not reveal the entire nucleus (Hienz, 1959; James, 1960). In addition, the usual methods of embedding may alter the nuclear structure, thus making the analysis of X-chromatin somewhat difficult. In order to compensate for the numerous factors influencing the frequency of X-chromatin in sections, it is wise to use normal control material from the same tissue of both sexes and treat in the same manner.

6.3. Prenatal Sex Diagnosis

During pregnancy, it is possible to examine the desquamated cells suspended in the amniotic fluid. This material can be obtained by *amniocentesis* (see Chapter III). As described earlier, the preparations are made from cell centrifugates.

After quinacrine staining, the Y-chromatin body is usually visible in at least some cell nuclei in the case of a normal male embryo. Double Y cases can also be diagnosed. Y-chromatin is, of course, also visible in all abnormal cases where a Y chromosome is present. Previous comments on small Y chromosomes or other fluorescing chromosome components also apply to amniocentesis cells.

The X-chromatin body is more difficult to demonstrate. Only well-preserved nuclei without folds and without signs of pyknosis should be examined. The frequency of X-chromatin positive nuclei may be quite low (5%) in normal females (Fuchs and Riis, 1956; Sachs *et al.*, 1955; Abbo and Zellweger, 1970).

Despite these technical handicaps, it is possible to obtain a clearly positive Y or X diagnosis in many cases. The karyotypic analysis will, of course, give an exact diagnosis, but since the evaluation of interphase nuclei only requires a few hours, this should be always tried when a quick diagnosis of sex is desired.

6.4. Polyploid Cells

Polyploid cell nuclei are rare in buccal mucosae and hair roots, but in amniotic epithelium or in tissue cultures they may account for up to 5% of cells.

The number of Y-chromatin bodies is directly related to the number of Y chromosomes, so a tetraploid cell (with 88 autosomes and XXYY sex chromosomes) will show 2 Y-chromatin bodies. The degree of ploidy is therefore directly visible from the number of Y-chromatin bodies in a normal male. If more than one Y-chromatin body is found in a cell nucleus, its size and staining intensity should be checked to recognize a

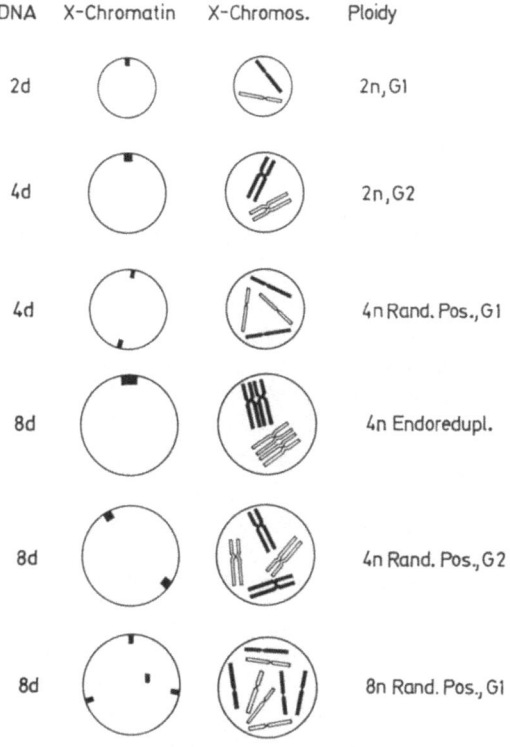

DNA   X-Chromatin   X-Chromos.   Ploidy

2d   2n, G1

4d   2n, G2

4d   4n Rand. Pos., G1

8d   4n Endoredupl.

8d   4n Rand. Pos., G2

8d   8n Rand. Pos., G1

Fig. 17. Diagrammatic representation of the forms in which X-chromatin bodies appear in cell nuclei of various classes of DNA (2d, 4d, 8d) and their relative chromosome status. *DNA:* DNA content of the cell nucleus. *X chromat:* X-chromatin. Size is diagrammatically reproduced. *X chromos:* Status of X chromosomes. Heterocyclic X chromosomes black. *Ploidy:* Ploidy (2n, 4n, 8n) and phase of the cell cycle (G1, G2). *Random pos.:* Random arrangement of the chromosomes. *Endoredupl.:* State according to endoreduplication. (From Schwarzacher, 1966)

polyploid normal nucleus before concluding that an additional Y chromosome is present.

In respect of the X-chromatin body, the situation is as follows: Since one X chromosome per diploid chromosome set is euchromatic and all other X chromosomes are heterochromatic and may display heteropyknosis, the number of X-chromatin bodies in female cells is maximally doubled with every additional diploid set of chromosomes, while even highly polyploid cell nuclei of male origin do not show an "X-chromatin". The type of chromosomal arrangement determines whether an X-chromatin of double or multiple size or whether two or more X-chromatin bodies of normal size occur (Fig. 17). If several X-chromatin bodies or a rather large one are present in one cell nucleus, one should attempt to determine its ploidy by estimating the size of the nucleus and the intensity of staining.

6.5. Malignant Tissues

The examination of tumors is beset with numerous obstacles. Cell smears are often produced only with difficulty and result in altered topography, and one has to rely on section preparations with limited reliability owing to lack of control material for comparison.

Moreover, it must be emphasized that the process of heterochromatization and X-chromatin formation in tumor cells can deviate completely from normal cells. A chromatin body in the cell nucleus of a malignant cell need in no way be related to the X chromosomes and should not be referred to as "X-chromatin". The problem of X-chromatin in tumors is discussed by Atkin (1964), Kallenberger (1964), Gropp *et al.* (1965 and 1967), for example.

Nothing is known so far on the occurrence of Y-chromatin bodies in tumor cells of male origin.

Appendix

Laboratory Procedures

1. Preparation of Hair Follicle Cells for Y Chromatin

The shaft of a plucked hair with a conspicuous root is gripped with a pincette and the root dipped for one to two min in 25 % acetic acid. Immediately afterwards, the fixed root is dragged lightly over the surface of a well cleaned slide, so that enough cells are scraped off and adhere to the slide. The cells are air-dried and stained. The same hair-root can be utilized for several preparations. It is advisable to mark the underside of the slide and deposit the cells near this mark. Smears may be made in parallel lines across the slide in order to locate their position on the slide more easily when examining under the microscope.

2. Preparation and Fixation of Buccal Smears

1. Preparing the smear: Scrape the buccal mucosa firmly several times with a metal spatula or the end of a slide. Use a properly cleaned slide. Spread the smear over the slide, leaving it rather thick. Mark the side of the slide bearing the smear with a glass pen. Obtain a second smear from the same side of the cheek (thus

reaching the deep epithelial layer with the intact cells). Also take smears from the other side.

2. Fixation: The smear is not permitted to dry, but is immediately fixed in 96% alcohol for at least 30 min (where it can be kept for several weeks).

3. After fixing, the smear is air dried so that the cells adhere firmly to the slide. Avoid exposure to dust. For long-term (months) storage, smears should be kept refrigerated in dust-proof boxes. Transport of unstained smears is best carried out after fixation and air drying.

3. Staining with Quinacrine (see also Appendix to Chapter V)

1. Immerse slides for 5 min in buffer solution at pH 6.0.

2. Place slides for 5 min in solution of quinacrine dihydrochloride or quinacrine mustard 0.2% in phosphate buffer at pH 6.0.

3. Place slides for 5 min in buffer solution at pH 6.0.

4. Cover the cells on the wet slides with a cover glass. Avoid air bubbles.

5. Check under the fluorescence microscope: if staining is too weak, wait for 10 min; if still too weak, remove cover glass and restain in the dye solution for 5 min, shorten the time of differentiation in buffer (point 3), put on a new cover slip; if stained too strongly (no banding pattern of chromosomes, no distinct Y-chromatin body), remove cover slip and place slides again for a few minutes in fresh buffer, and mount.

6. If preparation is satisfactory, seal with nail varnish.

If staining of the X-chromatin after quinacrine staining is required, place the preparation in 96% ethanol for 30 min. The nail varnish will dissolve and the coverslip falls off. The quinacrine is eluted, and the preparation is air-dried and then ready for any desired staining.

4. Feulgen Staining

Hydrolysis in HCl requires good adhesion of the cells to the slides. Air-drying after fixation will be sufficient for cell smears. Membrane preparations and sections must be secured with celloidin:

1. Dehydration of preparation in ascending alcohol series. Final stage: ethyl alcohol twice every 3 min.

2. Ether-alcohol (1:1), 5 min.

3. 1% celloidin (in ether), 5 min.

4. Remove slide, allow celloidin to drip off.

5. 70% Alcohol for at least 20 min (to harden the celloidin).

6. Rinse in distilled water.

The actual staining takes place in the following steps:

1. Place in 5 N HCl at 22° C for exactly 20 min; or in 1 N HCl at 60° C for 12 min.

2. Rinse twice in distilled water, each time for 3 min.
3. Place in Schiff's reagent for 1.5 hours in darkness at room temperature (about 20° C).
4. Rinse in SO_2 solution (changed 3 times) each time for 5 min.
5. Wash in running tap water for 30 min.
6. Dehydration in ascending alcohol series (70, 96, 100%) and mount (e.g. xylene and a neutral synthetic resin).

Preparation of Schiff's reagent:

To obtain 1 liter of reagent, dissolve 4 g pararosaniline base (Merck or Chroma) in a mixture of 800 ml distilled water and 240 ml 1 N HCl at about 60° C. This is then kept simmering gently for 3 min. Filter after cooling to 60° C. Cool further to about 30° C, then add 15 g $Na_2S_2O_5$ and stir well. Seal bottle properly and keep 12–24 hours in the dark. The solution is then slightly yellowish in colour. Then add 3 g of activated charcoal and filter again. The filtrate must be absolutely clear. If stored refrigerated in the dark, the reagent will keep for several months. It can be used until a reddish shimmer appears.

Preparation of SO_2 solution:

Add 10 ml of a 10% $Na_2S_2O_5$ solution and 10 ml of 1 N HCl to 200 ml distilled water.

5. *Thionine Staining (Klinger and Ludwig, 1957)*

Prepare slides as for the Feulgen stain (see above).
1. Wash in distilled water for several minutes.
2. 5 N HCl, 10 min at 22° C (room temperature).
3. Tap water, change several times.
4. Distilled water, several min.
5. Thionine solution, 30 min.
6. Briefly (seconds) rinse with distilled water.
7. Briefly differentiate in 70% alcohol.
8. 96% Alcohol, 1 min, absolute alcohol 2 changes (each 1–2 min), xylene, twice, mount.

Solutions:

Stock solutions
1. Saturated thionine: 20 g thionine in 500 ml of 50% alcohol. Filter before mixing with buffer solution.
2. Buffer solution (Michaelis)
   Na acetate   9.714 g
   Na barbiturate 14.714 g
   adjusted with double distilled water to 500 ml.
   Can be stored for several months in refrigerator.

Staining solution (pH 5.7)

Buffer solution (2)	28 ml
1/10 N HCl	32 ml
Saturated thionine (1)	40 ml

The staining solution can be kept for several months.

6. *Carbol Fuchsin Stain (Eskelund, 1956; Barr, 1965)*

Place air-dried fixed preparation directly into the stain solution, preparations removed from alcohol are passed through distilled water.

1. Place in carbol fuchsin for 10 min.
2. 95% Ethyl alcohol 1–3 min (light differentiation).
3. Absolute alcohol, change twice, each for 1 min.
4. Xylene, and mount.

Preparation of the carbol fuchsin solution.

Stock solution: Pararosaniline 3 g in 100 ml of 70% ethyl alcohol (can be stored for several months).
Working solution: 5% phenol 90 ml

Stock solution 10 ml
Formaldehyde (37%, "conc. formalin") 10 ml
Glacial acetic acid 10 ml.

Allow to stand for 24 hours, then filter (can be stored for up to 1 month).

7. *Cresyl Violet Stain (Moore and Barr, 1955)*

Place air-dried fixed preparation directly into the stain solution (preparations removed from alcohol are passed through distilled water).

1. Cresyl violet solution (0.5% in distilled water for 8 min).
2. 95% Ethyl alcohol for differentiation. Microscopic monitoring is necessary.
3. Absolute alcohol, two changes, each 1 min.
4. Xylene. Then mount.

8. *Fixation of Larger Pieces of Tissues for Paraffin or Celloidin Embedding*

Prepare blocks of tissue with a maximum of 10 mm in their smallest dimension. In larger embryos, open body and skull cavities.

Fixation according to Davidson (concentrated formol i.e. 40% Formalin 20 parts, 95% ethanol 40 parts, glacial acetic acid 10 parts, distilled water 30 parts) for a maximum of 48 hours. Then transfer to 70% ethyl alcohol, three changes, 24 hours each. Finally, pass through ascending series of alcohol and embed.

References

Abbo, G., Zellweger, H.: Prenatal determination of fetal sex and chromosomal complement. Lancet 1970 I, 216–217.

Arrighi, F. E., Hsu, T. C.: Localization of heterochromatin in human chromosomes. Cytogenetics 10, 81–86 (1971).

Atkin, N. B.: Die chromosomale Basis von Sex-Chromatinabweichungen in menschlichen Tumoren. Wien. klin. Wschr. 49, 859–862 (1964).

Barlow, P., Vosa, C. G.: The Y-chromosome in human spermatozoa. Nature (Lond.) 226, 961–962 (1970).

Barr, M. L.: Sex chromatin techniques. In: Yunis, J. J. (ed.), Human chromosome methodology. London-New York: Academic Press 1965.

Barr, M. L.: The significance of the sex chromatin. Int. Rev. Cytol. 19, 35–95 (1966).

Barr, M. L., Bertram, E. G.: A morphological distinction between neurons of the male and female, and the behaviour of the nucleolar satellite during accelerated nucleoprotein synthesis. Nature (Lond.) 163, 676–677 (1949).

Barr, M. L., Carr, D. H.: Correlations between sex chromatin and sex chromosomes. Acta cytol. (Philad.) 6, 34–35 (1962).

Bobrow, M., Madan, K., Pearson, P. L.: Staining of some specific regions of human chromosomes, particularly the secondary constriction of No. 9. Nature New Biol. (Lond.) 238, 122–124 (1972).

Bobrow, M., Pearson, P. L., Pike, M. C., El-Alfi, O. S.: Length variation in the quinacrine banding segment of human Y-chromosomes of different sizes. Cytogenetics 10, 190–198 (1971).

Carpentier, P. J.: Sex chromatin in smears from the reproductive and urinary tracts. In: Moore, K. L. (ed.), The sex chromatin. Philadelphia-London: W. B. Saunders Co. 1966.

Caspersson, T., Zech, L., Johansson, C.: Differential binding of alkylating fluorochromes in human chromosomes. Exp. Cell Res. 60, 315–319 (1970).

Caspersson, T., Zech, L., Johansson, J., Lindsten, J., Hultén, M.: Fluorescent staining of heteropycnotic chromosome regions in human interphase nuclei. Exp. Cell Res. 61, 472–474 (1970).

Cuadrillero, C. B.: Stains for sex chromatin: silver impregnation in tissues and blood films. Stain Technol. 34, 290–292 (1959).

DeMars, R.: Sex chromatin mass in living, cultivated human cells. Science 138, 980–981 (1962).

Engel, E., Culbertson, J., Mahley, R. W.: Nuclear sexing from hair root cells. In: Dorfman, A. (ed.), Antenatal diagnosis. Chicago: Chicago Univ. Press 1972.

Eskelund, V.: Determination of genetic sex by examination of epithelial cells in urine. Acta endocr. (Kbh.) 23, 246–250 (1956).

Fraccaro, M., Lindsten, J.: Observation on the so-called "sex-chromatin" in human somatic cells cultivated in vitro. Exp. Cell Res. 17, 536–539 (1959).

Fuchs, F., Riis, P.: Antenatal sex determination. Nature (Lond.) 177, 330 (1956).

Gagné, R., Laberge, C.: Specific cytological recognition of the heterochromatic segment of number 9 chromosome in man. Exp. Cell Res. 73, 239–242 (1972).

George, K. P.: Cytochemical differentiation along human chromosomes. Nature (Lond.) 226, 80–81 (1970).

Gropp, H., Pera, F., Lohmann, H., Wolf, U.: Untersuchungen über die Anzahl der X-Chromosomen beim Mammacarcinom. Z. Krebsforsch. **69**, 326–334 (1967).

Gropp, H., Wolf, U., Pera, F.: Sex-Chromatin und Chromosomenstatus beim Mammacarcinom. Dtsch. med. Wschr. **90**, 637–642 (1965).

Grumbach, M. M., Morishima, A., Taylor, J. H.: Human sex chromosome abnormalities in relation to DNA replication and heterochromatinization. Proc. nat. Acad. Sci. (Wash.) **49**, 581–589 (1963).

Guard, H. R.: A new technique for differential staining of the sex chromatin and the determination of its incidence in exfoliated vaginal epithelial cells. Amer. J. clin. Path. **34**, 145–151 (1959).

Hamerton, J. L.: Sex chromosomes and their abnormalities in man and mammals. In: Lima-de-Faria, A. (ed.), Handbook of cytology. Amsterdam-London: North-Holland Publishing Co. 1969.

Hamerton, J. L.: Human cytogenetics, vol. I: General cytogenetics. New York-London: Academic Press 1971.

Hellriegel, K. P., Reitz, H., Deh, C., Zinser, H. K.: Y-Chromatin-Untersuchungen zur Frage des foetomaternalen Leukozytentransfers. Humangenetik (in press).

Hienz, H. A.: Die zellkernmorphologische Geschlechtserkennung in Theorie und Praxis. Heidelberg: Dr. Alfred Müthig 1959.

Hsu, L. Y. E., Klinger, H. P., Weiss, J.: Influence of nuclear selection criteria on sex chromatin frequency in oral mucosa cells of newborn females. Cytogenetics **6**, 371–382 (1967).

James, J.: Observations on the so-called sex chromatin. Z. Zellforsch. **51**, 597–616 (1960).

Kallenberger, A., Wenner, R.: Geschlechtschromatin bei Mammacarcinom. Schweiz. med. Wschr. **96**, 80–84 (1966).

Klinger, H. P.: Morphological characteristics of the sex chromatin. In: Moore, K. L. (ed.), The sex chromatin. Philadelphia-London: W. B. Saunders Co. 1966.

Klinger, H. P., Davis, J., Goldhuber, P., Ditta, T.: Factors influencing mammalian X chromosome condensation and sex chromatin formation. I. The effect of in vitro cell density on sex chromatin frequency. Cytogenetics **7**, 39–57 (1968).

Klinger, H. P., Hammond, D. O.: Rapid chromosome and sex-chromatin staining with pinacyanol. Stain Technol. **46**, 43 (1971).

Klinger, H. P., Lindsten, J., Fraccaro, M., Barrai, L., Dolinar, Z. J.: DNA content and area of sex chromatin in subjects with structural and numerical aberrations of the X chromosome. Cytogenetics **4**, 96–116 (1965).

Klinger, H. P., Ludwig, K. S.: A universal stain for the sex chromatin body. Stain Technol. **32**, 235–244 (1957).

Klinger, H. P., Moser, G. C.: Improved chromatin-fluorescence technique. Lancet **1972 II**, 1366.

Klinger, H. P., Schwarzacher, H. G., Weiss, J.: DNA content and size of sex chromatin positive female nucleic during the cell cycle. Cytogenetics **6**, 1–19 (1967).

Lennox, B.: A ribonuclease-gallocyanin stain for sexing skin biopsies. Stain Technol. **31**, 167–172 (1956).

Majewski, F., Bier, L., Pfeiffer, R. A.: Fluoreszenzmikroskopischer Nachweis des menschlichen Y-Chromosoms in Interphasekernen durch Acridinderivate (Atebrin, Acranil). Klin. Wschr. **49**, 814–818 (1971).

Marberger, E., Boccabella, R., Nelson, W. O.: Oral smear as a method of chromosomal sex detection. Proc. Soc. exp. Biol. (N.Y.) **89**, 488–489 (1955).

McKee-Katz, M., Wright, S. W.: The use of hair root sheath for X-chromatin determination. J. Pediat. **76**, 292–295 (1970).

Moore, K. L., Barr, M. L.: Smears from the oral mucosa in the detection of chromosomal sex. Lancet **1955 II**, 57.

Müller, H. J., Bühler, E. M., Voegelin, M. G., Stalder, G. R.: Eine neue Methode der Geschlechtsbestimmung in Leukozyten aus eingetrockneten Blutflecken. Schweiz. med. Wschr. **101**, 1171–1174 (1971).

Mukherjee, A. B., Blattner, P. Y., Nitowsky, H. M.: Quinacrine mustard fluorescence of sex chromatin in human amniotic fluid cell cultures. Nature (Lond.) **235**, 226–229 (1972 a).

Mukherjee, A. B., Moser, G., Nitowsky, H. M.: Fluorescence of X and Y chromatin in human interphase cells. Cytogenetics **11**, 216–227 (1972 b).

Pearson, P. L., Bobrow, M.: Fluorescent staining of the Y chromosome in meiotic stages of the human male. J. Reprod. Fertil. **22**, 177–179 (1970).

Pearson, P. L., Bobrow, M., Vosa, C. G.: Technique for identifying Y-chromosomes in human interphase nuclei. Nature (Lond.) **226**, 78–80 (1970).

Polani, P. E., Mutton, D. E.: Y-Fluorescence of interphase nuclei, especially circulating lymphocytes. Brit. med. J. **1971 I**, 138–142.

Riis, P., Fuchs, F.: Sex chromatin and antenatal sex diagnosis. In: Moore, K. L. (ed.), The sex chromatin. Philadelphia-London: W. B. Saunders Co. 1966.

Sachs, L., Serr, D. M., Danon, M.: Prenatal diagnosis of sex using cells from the amniotic fluid. Science **123**, 548 (1956).

Sanderson, A. R.: Rapid nuclear sexing. Lancet **1960 I**, 1252.

Schmid, W.: Sex chromatin in hair roots. Cytogenetics **6**, 342–349 (1967).

Schnedl, W.: Untersuchungen über das Sex-Chromatin in menschlichen Fibroblastenkulturen. Acta anat. (Basel) **57**, 52–65 (1964).

Schnedl, W.: Fluoreszenzuntersuchungen über die Längenvariabilität des Y-Chromosoms beim Menschen. Humangenetik **12**, 188–194 (1971).

Schwarzacher, H. G.: Sex chromatin in living cells *in vitro*. Cytogenetics **2**, 117–128 (1963).

Schwarzacher, H. G.: Sexchromatin in polyploiden Zellen. Humangenetik **2**, 28–35 (1966).

Schwinger, E., Pera, F.: On the splitting of the Y-fluorescent body in man. Humangenetik **14**, 107–111 (1972).

Schwinger, E., Rakebrand, E., Müller, H. J., Bühler, E., Tettenborn, U.: Y-body in hair roots. Humangenetik **12**, 79–80 (1971).

Smith, D. W., Marden, P. M., McDonald, M. J., Speckhard, M.: Lower incidence of sex chromatin in buccal smears of newborn females. Pediatrics **30**, 707–711 (1962).

Sumner, A. T., Robinson, J. A., Evans, H. J.: Distinguishing between X, Y and YY-bearing human spermatozoa by fluorescence and DNA content. Nature New Biol. (Lond.) **229**, 231–233 (1971).

Taylor, A. E.: Sex chromatin in the newborn. Lancet **1963 I**, 912–914.

Therkelsen, A. J., Petersen, G. B.: Frequency of sex-chromatin-positive cells in the logarithmic and post-logarithmic growth phases of human cells in tissue culture. Exp. Cell Res. **28**, 588–623 (1962).

Zech, L.: Investigation of metaphase chromosomes with DNA-binding fluorochromes. Exp. Cell Res. **58**, 463 (1969).

CHAPTER X

The Diagnosis of X-Chromatin
by the Leukocyte Test

Marlis Tolksdorf

With 6 Figures

1. Introduction

The morphology of *X-chromatin* in peripheral *blood* differs from that of
other tissues. The so-called leukocyte test for the diagnosis of X-chromatin
is based on the identification of specific nuclear formations which were
first described and recognized as important by Davidson and Smith in
1954.

These authors described an appendage shaped like a drumstick that
can be differentiated from other nuclear protrusions unrelated to sex
in nuclei of polymorphonuclear leukocytes of normal females.

These findings were confirmed, and form the basis of a diagnostic
test with extensive applications (Romatowski *et al.*, 1955; Tolksdorf
et al., 1955; Riis, 1955; Wiedemann *et al.*, 1955; Lüers, 1956; Sun *et al.*,
1956; van Harnack *et al.*, 1956; Lupatkin *et al.*, 1956; Wiedemann *et al.*,
1956a; Peiper *et al.*, 1956; Tenczar *et al.*, 1956, and others).

The *Y-chromatin* in leukocytes can be demonstrated by staining with
the fluorochromes quinacrine mustard and quinacrine dihydrochloride.
The details of this staining technique are described in Chapter V; the
evaluation of the Y-chromatin in leukocytes is discussed in Chapter IX.
In this chapter only the X-chromatin in leukocytes (the drumstick) is
considered.

2. Methods

The initial procedure for the leukocyte test is simple. It is merely necessary to make good blood smears – not too thin or too thick – and to

stain them according to May-Grünwald-Giemsa (Pappenheim). Other stains (Feulgen, toluidine blue, hematoxylin-eosin or Giemsa) have no practical advantage.

One can either employ the usual smear technique (spreading a small drop of blood on a grease-free slide with another slide or a coverslip with smooth edges at an angle) or the cover-glass method (Wintrobe, 1946; Davidson, 1961). The latter yields excellent preparations of evenly distributed well-spread blood cells: a drop of blood is placed between two cover slips which are carefully slid apart, dried dust-free, stained, and mounted on slides. Such preparations are ready for microscopic examination and can easily be transported.

A number of procedures exist for preparing concentrates of white blood cells to facilitate the time-consuming evaluation of the usual smears (Klima *et al.*, 1949; Kosenow, 1956a; Procopio-Valle, 1958; Hienz, 1965). However, this may lead to a poorer quality of the preparation. Therefore, leukocyte concentrates should be handled with great care and such smears examined with particular caution. In addition, regular blood smears should be prepared and kept for comparison.

3. Evaluation

The significant part of the nuclear structure in the leukocyte-X-chromatin-test is the "drumstick", which is a sharply defined round to oval nuclear appendage with a thin stalk and a head of about 1.5 μ in

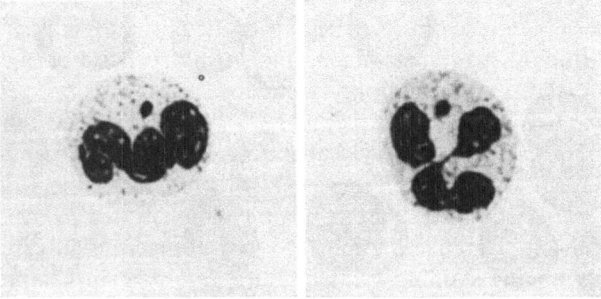

Fig. 1. Granulocytes with a typical drumstick on a thin stalk, May-Grünwald-Giemsa stain

diameter. It is densely stained and can be found in neutrophil, eosinophil and basophil granulocytes (Fig. 1).

Drumsticks are present in the peripheral blood of individuals carrying more than one X chromosome in at least part of their granulocytes.

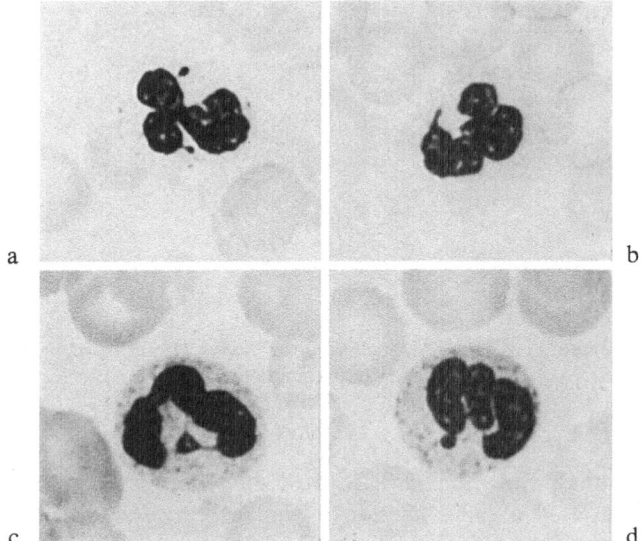

Fig. 2. a Granulocyte nuclei with small clubs. b Granulocyte nuclei with a spike- or hook-like tag. c Small nuclear lobe with double stalk (minor lobe). d Broadly-based nodule without stalk (sessile nodule)

This is referred to as X-chromatin positive.

Normal males with XY sex chromosomes have no drumsticks. This is referred to as X-chromatin negative.

Mature granulocytes may exhibit a number of different and unspecific nuclear appendages.

They include:

1. small clubs that have a thin stalk and resemble drumsticks in their shape, except that they are smaller. They may occur frequently on granulocyte nuclei together with a drumstick (Fig. 2a);

2. Spike, hook or thread-like tags occurring alone or in groups in a granulocyte nucleus (Fig. 2b).

3. Minor lobes which are frequently connected to the adjacent segment of the nucleus by two thread-like bridges (Fig. 2c).

4. Club-shaped appendages like a tennis racket with a lightly stained center.

5. Broadly-based sessile nodules or nodules with a slim stalk and the shape of a droplet, corresponding to a drumstick in size and chromatin density (Fig. 2d). Sessile nodules are found in both sexes, but are more common in females (Tolksdorf, 1963).

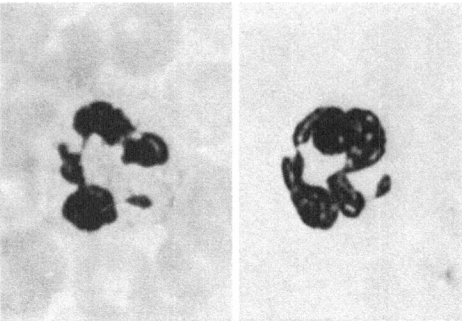

Fig. 3. Artifactual, misleading nuclear appendages resulting from technically inadequate smears

The studies with quinacrine showed that the Y-chromatin is sometimes situated in a small nuclear appendage corresponding to a small club or a spike (see Chapter IX, Fig. 2b). Such small appendages, which are found in both sexes, may generally represent smaller heterochromatic regions.

Artifacts due to technical faults must be mentioned. Poorly cleaned, greasy slides, poor smear techniques or fixation, enrichment of leukocytes; in short, the process of preparation, can result in a number of indistinct and unspecific nuclear bodies. Fig. 3 shows cells that are e.g. squashed or lie too close together, and which should not be considered because they are misleading.

Drumsticks may be concealed by toxic granulation, e.g. Alder's granulation anomaly, which leads to diagnostic difficulties. In this case, it is advisable to use a pure Giemsa stain.

The frequency of drumsticks varies in normal females between 1.5 and 5% (Hienz, 1964). Comparisons of different age groups show a higher frequency of premature and newborn infants (von Harnack et al., 1956; Peiper et al., 1956; Romatowski et al., 1956; Wiedemann et al., 1956b; Mosler, 1957). The degree of nuclear segmentation influences the frequency of drumsticks (Davidson et al., 1954; Kosenow et al., 1956; Müller, 1959). Blood smears with a high proportion of polymorphonuclear neutrophils (shift to the right) will reveal more drumsticks than those (shift to the left) containing more juvenile neutrophils with less segmentation of the nucleus. The correlation in the frequency of drumsticks with the degree of maturity of polymorphonuclear leukocytes also explains why drumsticks are decreased in the Klinefelter and Down syndrome as well as in other chromosomal disturbances showing a left shift on differential blood smears, possibly due to a genetically deter-

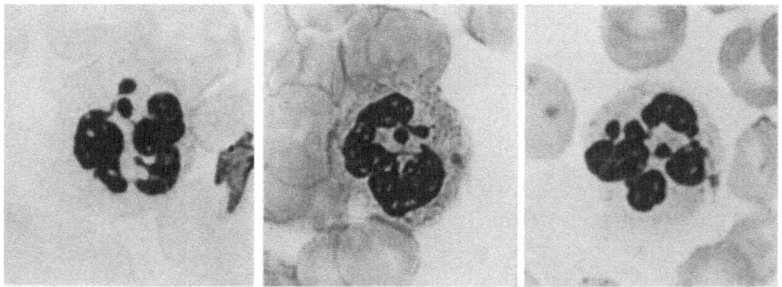

Fig. 4. Granulocyte nuclei with "double drumsticks"

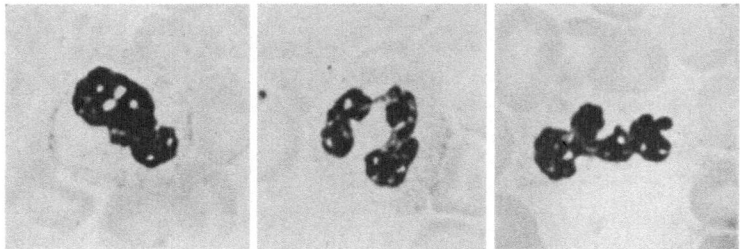

Fig. 5. Chromatin structure of granulocyte nuclei in an XXXXY/XXXY sex
chromosome mosaic

mined impairment of the leukocyte lobing (Mittwoch, 1964). The influence of exogenous factors on the frequency of drumsticks is at present uncertain.

Sex chromosomal anomalies cause changes in the number and appearance of the nuclear appendages.

The correlation between numerical and structural anomalies of the sex chromosomes and the chromatin pattern in interphase nuclei in terms of the Barr body are well established (see Chapter IX). Although this is less obvious and at times difficult to detect in peripheral blood, it is beyond doubt that Barr bodies and drumsticks are identical formations in different cell types. Davidson himself describes the drumsticks as "a characteristic variation of the Barr body type of sex chromatin" (Davidson, 1965).

Drumsticks can be found in various sex chromosomal aberrations, e.g. in the XXY Klinefelter syndrome, and if three or four X chromosomes are present, two (Fig. 4) or even three drumsticks can be found, although this is rather rare.

The frequency of drumsticks in XXY patients falls below the range of normal females. In Klinefelter syndrome with three or more X chromo-

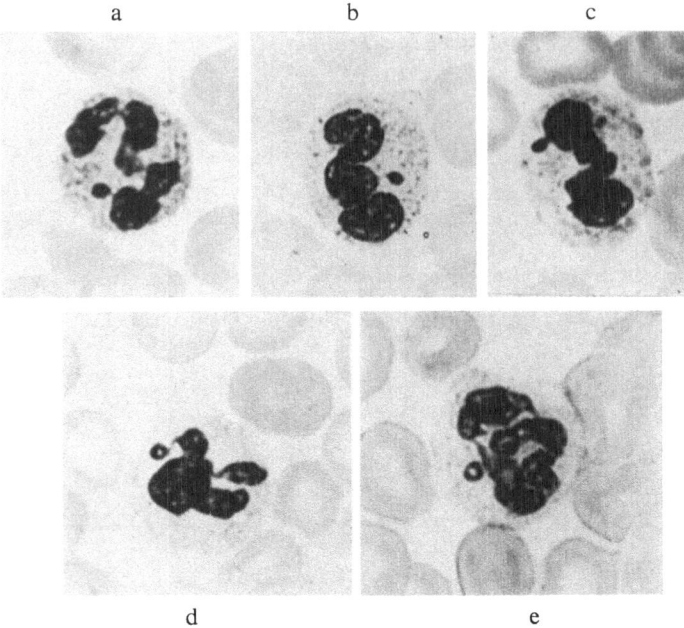

Fig. 6. a Large drumstick in an X isochromosome for the long arm. b Normal drumstick. c Small drumstick-like appendage in a deletion on the short arm of an X chromosome. d Drumstick with light central area in a patient with an X isochromosome. e X isochromosome-type appendage with a stalk scarceyl visible and typical light area

somes it is still less, so that intractable diagnostic difficulties may result. Mosaics within this group cannot always be diagnosed with certainty by the leukocyte test. However, the leukocyte nucleus displays in its well delineated chromatin "free" light areas appearing as if punched out from the dense intranuclear structures enclosing it (Tolksdorf, 1970), as shown in Fig. 5.

Reports on nuclear appendages in X polysomics in the absence of a Y chromosome are not consistent. Mittwoch (1963) found a higher frequency of sessile nodules in XXX females. There seem to be fewer drumsticks than expected. Two-stemmed drumsticks on one granulocyte nucleus are rare in contrast to either two sessile nodules or one drumstick and one sessile nodule.

With structural anomalies of the X chromosome, the situation is even more complex, because their appearance in the interphase nucleus is difficult to interpret (Maclean, 1962; Bamford et al., 1963, 1964; Fraccaro et al., 1970).

Drumsticks may be larger in patients with an isochromosome for the long arm and smaller in patients with deletion of the X chromosome (Fig. 6a–c). In the case of X isochromosomes, the drumsticks are not only obviously larger than normal drumsticks, but they are more frequent and may be characterized by centrally or slightly laterally localized light areas (Fig. 6d and e); these may be so large and sharply delineated that one cannot distinguish the appendage in question from a so-called tennis racket.

Extra large drumsticks may occasionally be found in blood smears from patients with increased numbers of X chromosomes, e.g. in XXX or XXXY karyotypes, regardless of the presence or absence of a Y chromosome (Mittwoch, 1967; Tolksdorf, 1970).

In practice, the diagnostic evaluation of a blood smear still rests on the guiding principles set forth in the original work of Davidson and Smith (1954). They showed that six typical drumsticks must be demonstrated in 500 granulocytes examined to justify the diagnosis "female pattern". The number six was chosen for statistical reasons, because it corresponds to 5 intervals of frequency (Davidson, 1961).

This is still accepted today as the basic requirement for the diagnosis "X-chromatin positive". However, we also know that the presence of even a single typical drumstick is diagnostic for the presence of two X chromosomes in at least some cells of hematopoietic origin. It does not occur as a false positive finding in otherwise X-chromatin negative individuals with only one X chromosome, such as normal males or abnormal females with a 45,X karyotype. Insistence on six drumsticks as a criterion for reliable diagnosis excludes a mosaic pattern with reasonable certainty in most instances.

Generally 500 granulocytes should be scored for typical drumsticks. Their shape, size and density as well as uncharacteristic nuclear appendages and the overall structure of the nucleus are analyzed and recorded.

The absence of drumsticks does not necessarily constitute an X-chromatin negative finding. The presence of a single drumstick would cast doubt on this diagnosis, so it may be necessary to score 1,000, 2,000, and even 3,000 or more cells to be certain of the diagnosis "X-chromatin negative" in some cases. This particularly applies to preparations exhibiting conspicuous chromatin structures, small clubs, sessile nodules or even suspicious drumstick-like appendages that for technical reasons cannot be accurately classified.

If the guidelines outlined above are taken into consideration, the leukocyte drumstick test is still a reliable method for the diagnosis of X-chromatin in peripheral blood.

References

Bamford, S. B., Cassin, C. M., Dilba, D. L., Mitschell, G. W.: Neutrophil appendages as indicators of sex chromosome aberrations. Acta cytol. (Philad.) 8, 323–331 (1964).

Bamford, S. B., Cassin, C. M., Mitschell, B. S.: Sex chromatin determination in selected cases of developmental sex abnormalities with an assessment of results. Acta cytol. (Philad.) 7, 151–158 (1963).

Briggs, D. K., Kupperman, H. S.: Sex differentiation by leukocyte morphology. J. clin. Endocr. 16, 1163–1179 (1956).

Davidson, W. M.: Sexing the blood leukocytes in abnormalities of the sex chromosomes. Minerva pediat. 17, 585–587 (1965).

Davidson, W. M., Smith, D. R.: A morphological sex difference in the polymorphonuclear neutrophil leukocytes. Brit. med. J. 1954 II, 6–7.

Davidson, W. M., Smith, D. R.: Das Kerngeschlecht der Leukocyten. In: Die Intersexualität (Hrsg. C. Overzier), S. 78–85. Stuttgart: Georg Thieme 1961.

Fraccaro, M., Lindsten, J., Mittwoch, U., Zonta, L.: Size of drumsticks in patients with abnormalities of the X-chromosome. Lancet 1964 II, 43–44.

Harnack, G. A. v., Strietzel, H. N.: Die Altersabhängigkeit der geschlechtsbedingten Leukocytenmerkmale. Klin. Wschr. 34, 401–402 (1956).

Hienz, H.: Die Beziehungen zwischen zellkernmorphologischem und chromosomenmorphologischem Befund. Z. menschl. Vererb.- u. Konstit.-Lehre 37, 378–394 (1963).

Hienz, H.: Pers. Comm. 1965.

Klima, R., Beyreder, J., Lampar, J.: Zur Methodik der morphologischen Blutuntersuchung im Leukocytenkonzentrat. Wien. med. Wschr. 99, 358–360 (1949).

Kosenow, W.: Untersuchungen zur hämatologischen Geschlechtsbestimmung: Kernanhangsdifferenzierung im Leukocytenkonzentrat. Ärztl. Wschr. 14/15, 320–325 (1956a).

Kosenow, W.: Geschlechtsdiagnose mit Hilfe von Kernmerkmalen der Leukocyten. Triangel 2, 321–327 (1956b).

Kosenow, W., Scupin, R.: Die Bestimmung des Geschlechts mit Hilfe einer Kernanhangsformel der Leukocyten. Acta haemat. (Basel) 15, 349–363 (1956).

Lüers, T.: Ein morphologisches Geschlechtsmerkmal in Leukocytenkernen. Berl. med. Z. 7, 120–121 (1956).

Lupatkin, M., Prader, A.: Welches ist die einfachste Methode zur Bestimmung des chromosomalen Geschlechts? Schweiz. med. Wschr. 86, 928–930 (1956).

Maclean, N.: The drumsticks of polymorphonuclear leucocytes in sex-chromosome abnormalities. Lancet 1962 I, 1154–1158.

Mittwoch, U.: The incidence of drumsticks in patients with three X chromosomes. Cytogenetics (Basel) 2, 24–33 (1963).

Mittwoch, U.: Frequency of drumsticks in normal woman and in patients with chromosomal abnormalities. Nature (Lond.) 201, 317–319 (1964).

Mittwoch, U.: Sex chromosomes, p. 175–216. New York-London: Academic Press 1967.

Mosler, W.: Zur Frage der Geschlechtsbestimmung aufgrund morphologischer Leukocytenveränderungen. Zbl. Gynäk. H. 18, 696–701 (1957).

Müller, D.: Zur Entwicklung der geschlechtsspezifischen sog. Drumsticks an segmentkernigen Leukocyten. Ärztl. Wschr. 14, 260–262 (1958).

Peiper, U., Oehme, J.: Die Abhängigkeit geschlechtsgebundener Leukocytenmerkmale bei Feten und Frühgeborenen vor der Reife. Klin. Wschr. 34,
1067–1068 (1956).

Procopio-Valle, J., Chagas, W. A., Freitas, A. de: Determination of sex
chromatin in the polymorphonuclear neutrophils of enriched smears.
J. clin. Endocr. 18, 1432–1433 (1958).

Riis, P.: On the morphological sex difference in neutrophilic and eosinophilic
granulocytes. Dansk. med. Bull. 2, 190–192 (1955).

Romatowski, H., Tolksdorf, M., Bungart, K., Wiedemann, H.-R.: Zur Frage
der Altersabhängigkeit blutmorphologischer geschlechtscharakteristischer
Kernmerkmale bei Gesunden und bei Probanden mit Abnormitäten auf
dem Gebiet der Sexualentwicklung. Mschr. Kinderheilk. 105, 141–142
(1957).

Romatowski, H., Tolksdorf, M., Wiedemann, H.-R.: Geschlechtsbestimmung
aus dem Blutausstrich. Klin. Wschr. 33, 911 (1955).

Stoeckenius, M.: Der Chromatinbefund bei einem Fall von Turner-Syndrom
mit X-Isochromosom. Z. menschl. Vererb.- u. Konstit.-Lehre 37, 440–446
(1964).

Sun, L. C. Y., Rakoff, A. E.: Evaluation of the peripheral blood smear in the
detection of chromosomal sex in the human. J. clin. Endocr. 16, 55–61
(1956).

Tenczar, F. J., Streitmatter, D. E.: Sex difference in neutrophils. Amer. J.
clin. Path. 26, 384–387 (1956).

Tolksdorf, M.: Zur Problematik der Kerngeschlechtsdiagnostik. Ber. d.
8. Tagg. d. Dtsch. Ges. f. Anthropologie 1963, S. 31–37 (1965).

Tolksdorf, M.: Störungen der Geschlechtschromosomen beim Menschen.
Habilitationsschrift, Universität Kiel, Germany (1970).

Tolksdorf, M., Romatowski, H., Saile, M., Wiedemann, H.-R.: Über Geschlechtsbestimmung aus dem Blutbilde und deren Anwendung beim
Hermaphroditismus. Ärztl. Wschr. 10, 1029–1034 (1955).

Wiedemann, H.-R., Romatowski, H., Tolksdorf, M.: Unsere bisherigen Ergebnisse mit der Geschlechtsbestimmung aus dem Blutausstrich bei krank-
haften Zuständen. Hermaphroditismus und „Ovarialgenesie". Medizinische
Welt 50, 1734–1736 (1955).

Wiedemann, H.-R., Romatowski, H., Tolksdorf, M.: Geschlechtsbestimmung
aus dem Blutbilde. Grundlagen − Anwendung − Bedeutung. Münch.
med. Wschr. 98, 1090–1093, 1108–1112 (1956a).

Wiedemann, H.-R., Romatowski, H., Tolksdorf, M., Prediger, F.: Zur Frage
pränataler Geschlechtsbestimmung sowie zur blutmorphologischen Geschlechtsdiagnose bei Frühgeborenen und Feten. Medizinische Welt 16,
631–632 (1956b).

Wintrobe, M. M.: Clinical hematology, 2th ed., p. 273–275. Philadelphia:
Lea & Febiger 1946.

CHAPTER XI

Chromosomes in Meiosis[1]

SUSUMU OHNO

With 3 Figures

1. Introduction

Within the past few years, a considerable number of papers dealing with human meiotic figures has been published (see Ref.). Microphotographs published with these papers are of good quality and reveal that technical difficulty in obtaining analyzable meiotic figures from human gonads no longer exists. At the same time, these studies make it clear that meiotic process in man is similar to that of other placental mammals. Further analysis of human meiotic figures can be rewarding only if study is appropriately orientated.

In addition to describing various procedures for obtaining analyzable meiotic figures, this chapter includes a brief summary of current knowledge about the meiotic process in man.

Analysis of meiotic metaphase figures of phenotypically normal subjects suspected of having continuously produced genetically unbalanced gametes should be revealing. They may carry reciprocal translocations, insertions, or inversions, but the nature of the human diploid complement is such that some of these changes are likely to escape notice if only mitotic metaphase figures are analyzed, even if the analysis were aided by various banding techniques.

2. Review of the Meiotic Process in Man (Fig. 1)

The ultimate goal of meiosis is to segregate two haploid sets of chromosomes from a diploid germ cell. Although two cell divisions are involved in the meiotic process, replication of chromosomal materials (DNA) takes place only once, at the very beginning of meiosis. When either

[1] This work was supported in part by grant CA-05138-13 from the U.S. Public Health Service.

spermatogonium or oogonium completes its last mitosis, the meiotic process begins in each of the two daughter cells. The cell begins to grow in size, while its nucleus maintains the appearance of mitotic interphase. DNA replication takes place during this growth period; at its end, the cell attains a size noticeably larger than its predecessors and enters first meiotic prophase.

At the leptotene stage, each chromosome still exists as a separate entity. However, individual contours of threadlike chromosomes appear better defined than those of mitotic prophase. In the male, there is another characteristic which distinguishes leptotene from mitotic prophase. In leptotene, both the X and Y stand out by virtue of positive heteropyknosis along their entire length, while in mitotic prophase, only the Y chromosome manifests this condition.

When synapsis of homologs begins, the cell is said to be in zygotene. A necessary prelude to synapsis appears to be orientation of both ends of every chromosome toward the direction of the centriole. This particular configuration assumed by the threadlike chromosomes gives the "bouquet" appearance to zygotene nuclei. Indeed, "bouquet" nuclei can readily be observed on histological sections of sexually mature testes and fetal ovaries in man. In squash preparations, however, the "bouquet" configuration cannot be seen, due to disruption of spatial relationships. As all the chromosomal ends are placed in close proximity to each other, the opportunity is afforded for each chromosome end to find its homologous counterpart and unite. Lengthwise synapsis of two homologs appears to begin at both ends and proceeds toward the middle in a zipperlike fashion. Both X chromosomes of the female are in a fine threadlike state, and they seem to synapse in the same manner as do the autosomes. The positively heteropyknotic X and Y of the male, however, appear to fuse with each other much in the same manner as two oil drops fusing to form one larger drop, becoming one amorphous mass of heterochromatin.

When longitudinal synapsis of homologs is completed, the cell is said to be in pachytene. With the exception of the XY bivalent of the male, each bivalent now appears as a rather thick thread with a number of transverse bands called chromomeres. A chromomere adjacent to the centromere locus of each bivalent stands out by virtue of heavier staining (centromeric heterochromatin), and so does a terminal chromomere at each end (telomeric heterochromatin). Pachytene bivalents show the tendency to associate with each other at centromeric as well as telomeric heterochromatin, a tendency more pronounced in female pachytene. Two or three bivalents which are members of D and G groups usually share a common nucleolus. The XY bivalent of this stage forms a so-called sex vesicle resembling a minute interphase nucleus.

Longitudinal contact between the homologs is broken at diplotene, except at several points where chiasmata are formed. The larger bivalents of the A and B group may demonstrate ten to twelve chiasmata, while members of the E, F, and G groups may demonstrate only one or two. Each chiasma of diplotene is said to represent the point where crossing-over between the homologs has taken place. Thus, chiasma counts on diplotene bivalents should be very revealing. In practice, however, diplotene bivalents are so slender and prone to mechanical distortion that it is nearly impossible to distinguish true chiasmata from artificial twists introduced during the squashing procedure. Genetic crossing-over is believed to take place at random along the entire length of the two homologs. On the other hand, there is a preferential location for chiasma formation. At diplotene, one chiasma is almost invariably seen at the region adjacent to the centromeric heterochromatin of each bivalent. In the male, the sex vesicle begins to disintegrate during diplotene; however, individual contours of the positively heteropyknotic XY bivalent are not yet discernible.

Diplotene marks the end of first meiotic prophase. Between the end of last mitosis and diplotene, the cell has replicated its chromosomal DNA, and synapsis of the homologs is followed by the exchange of genetic material between the two; only the mechanical separation of one haploid set from the other remains. First meiotic prophase takes a rather long time, estimated to be several days.

In the testis, meiosis begins at puberty, and the entire meiotic process of individual cells, once started, proceeds uninterrupted. In the female, on the contrary, meiosis progresses up to the diplotene stage in each oocyte during fetal life, and the meiotic process is suspended at diplotene. The remainder of the meiotic process is carried out by the individual oocyte within a mature ovarian follicle and completed after ovulation and fertilization within the Fallopian tube.

Diakinesis of meiosis corresponds to prometaphase of mitosis; the advanced state of contraction is already achieved by individual chromosomes, yet they remain within the nuclear membrane. In addition, repulsion of homologous centromeres takes place in each bivalent. As homologous centromeres move away from each other, interstitial chiasmata on both sides of the centromere begin to slip toward the terminal ends. The X and the Y of the male are now clearly in end-to-end association.

The nuclear membrane and nucleoli soon simultaneously disappear, and the spindle apparatus is formed. All the bivalents which are in the maximum state of contraction are aligned on the equatorial plate. The cell is now in first meiotic metaphase. As hypotonic pretreatment disrupts both the nuclear membrane and the spindle apparatus, and squashing as well as the air-drying procedure disturbs the topographical arrange-

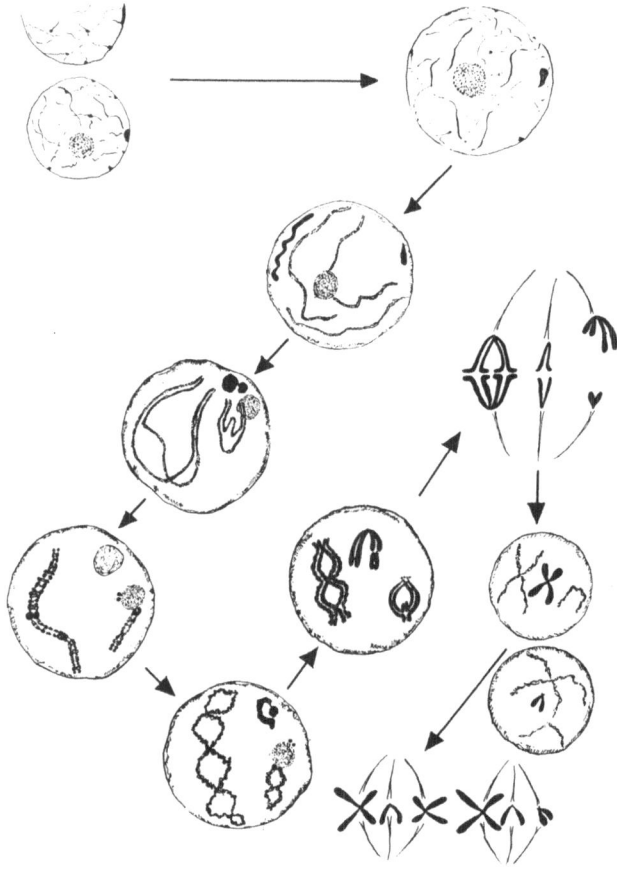

Fig. 1. Schematic representation of the meiotic process in the human male.
The X and Y are shown, and the autosomes are represented by a pair of large
mediocentrics and a pair of small acrocentrics carrying the nucleolus organizer.
The nuclear membrane is indicated by a circle. The chromosomes in an extended state are shaded, while those in the condensed state are solid black.
The nucleolus is indicated by a round shaded body. Two daughter nuclei
produced by the last spermatogonial mitosis (upper left corner) enter the
growth period and complete DNA replication. At the end of this period,
the nucleus attains considerable size (upper right corner). Then following the
arrow, leptotene, zygotene, pachytene, diplotene, and diakinesis. At the end
of diakinesis, the nuclear membrane disappears and the spindle apparatus
is formed. The cell is now in first meiotic metaphase. Interkinesis and second
meiotic metaphase follow. There are two fundamental differences between
male and female meiosis. Firstly, in the female there is a long suspended period
between diplotene and diakinesis. Secondly, two X chromosomes of the female
do not manifest positive heteropyknosis and do not associate end-to-end,
behaving instead in exactly the same manner as two homologous autosomes

ment of bivalents, the distinction between diakinesis and first meiotic metaphase cannot be made with absolute certainty in these preparations. A priori considerations suggest that the bivalents of diakinesis should appear slightly more slender than those of first meiotic metaphase, and fewer interstitial chiasmata should remain at the latter stage.

It is believed that each of the two haploid cells formed at the end of first meiosis does reconstitute its nuclear membrane for a short period, when a pair of nuclei may be observed, in which the X in one nucleus and the Y in the other is prominent. Although individual contours of all the chromosomes in the interkinesis nucleus are not completely lost, all 22 autosomes are in an extended state, while either the X or the Y appears as though it had been taken away from a heavily colchicinized mitotic metaphase and placed in the center of an interkinesis nucleus.

Subsequent to interkinesis, the process should proceed through prophase to metaphase. However, prophase II and metaphase II cannot be easily distinguished. From the very beginning of second meiosis, no attraction between two sister chromatids of each chromosome can be detected. This lack of attraction may be due to the fact that, as a result of crossing-over, various parts of sister chromatids of each chromosome are derived not from that chromosome but from its homolog. Thus, while two sister chromatids of each mitotic prophase chromosome are in relational coiling, a metacentric chromosome at prophase II already assumes an X shape. Individual chromosomes of prophase II resemble those of colchicinized mitotic anaphase, while those of metaphase II correspond to colchicinized mitotic metaphase chromosomes. The only distinguishing feature is the degree of condensation.

At second meiotic anaphase, two sister chromatids of each member of the haploid set segregate from each other, and the entire process of meiosis is completed.

3. Preparations for Meiotic Prophase Figures

As mentioned, the entire process of first meiotic prophase is carried out within the nuclear membrane, and individual chromosomes of leptotene, zygotene, pachytene, and diplotene stages are in a fine threadlike state. While hypotonic pretreatment causes swelling of whole nuclei, it also hydrates the already threadlike chromosomes and eliminates morphological landmarks of individual chromosomes. The disadvantages caused by hypotonic pretreatment outweighs the advantages it offers. The use of hypotonic pretreatment is not recommended for preparations of meiotic prophase figures. Admittedly, a piece of gonadal tissue which has undergone hypotonic pretreatment often yields first meiotic prophase figures of good quality; these prophase figures are situated in the center of a tissue fragment protected from the hypotonic solution.

3.1. Male

(See Ferguson-Smith, 1964; Eberle, 1963; Sasaki and Makino, 1965; Hungerford, 1971; Williams *et al.*, 1971; Luciani *et al.*, 1971; Dutrillaux, 1971; Matte and Sasaki, 1971.)

While meiotic metaphase figures deteriorate very rapidly after removal from the body, first meiotic prophase figures are relatively stable.

Thus, meiotic prophase figures of good quality can be obtained not only by biopsy but also by necropsy performed several hours after death. Healthy males as young as 12 and as old as 75 yield a sufficient number of meiotic figures.

To obtain meiotic prophase figures from the male, cut the testicular tissue into pieces no larger than $2 \times 2 \times 2$ mm and suspend them in isotonic solution, such as Medium 858 or Hanks' solution at $4°$ C. After 10 min, transfer the pieces to fresh cold solution; repeat this twice. The pieces are now ready to be fixed.

Fresh fixative is prepared by mixing equal volumes of glacial acetic acid and distilled water. At least 50 times the volume of tissue to be fixed is necessary for adequate fixation.

Small cubes ($2 \times 2 \times 2$ mm) of pretreated gonadal tissue are immersed in fresh fixative for at least 15 min, and at most 45 min. One cube of fixed tissue is placed on a glass slide along with about 0.1 cm³ of fixative. To release most of the free cells, the cube is gently tapped with a blunt metal instrument, such as the end of a diamond glass pencil. The stringy connective tissue remaining in the free cell suspension is removed with watchmaker's forceps. If clumps of tissue are left, no amount of pressure will effectively flatten individual cells. After removal of the connective tissue, less than 0.1 cm³ of free cell suspension remains on the slide, which is gently covered with a No. 1 cover slip (24×40 mm), great care being taken to avoid formation of air bubbles.

For protection during the squashing process, and to provide a surface which the finger can grip, the covered slide is placed in the center of four layers of No. 1 filter paper folded in the middle. So enveloped, the slide is placed on a completely flat surface and both thumbs aligned on top of the paper-covered slide, and even pressure directed straight downward for at least one min, preferably two. At this crucial stage, a preparation can be completely ruined by the slightest movement of the thumbs, or by directing the pressure obliquely. After squashing, when the slide is removed from the filter paper, the presence of Newton rings between the cover slip and the slide indicate a successful squash.

For one min. the slide is immersed in a mixture of dry ice and methanol in a large beaker; this permits the cover slip to be pried off the frozen fixative with a razor blade and discarded. (Several years of experience have taught us that neither coating the cover slip with silicon nor the slide with albumin is necessary to keep well-squashed cells on the slide. We simply rub a clean slide vigorously with a coarse paper towel for several min before its use.)

After the slide dries in the air, it is immersed in methanol for 15 min to extract the fatty substances usually found in gonadal tissue. The slide is again dried, washed in tap water, and hydrolyzed in 1 N HCl at 60° C for 15 min to remove most of the RNA from the chromosomes and flattened cytoplasm.

The slide is now ready for staining. In our laboratory, we obtain equally good results by staining for three hours with Feulgen reagent, 5 min with Giemsa solution, or one min with 0.25 percent basic fuchsin solution. After drying in air, the slide is mounted with a No. 1 coverslip and synthetic balsam.

If wet preparations are preferred, fixed pieces of tissue may be transferred to 2% aceto-orcein before squashing. All the edges of a cover slip of this preparation should be sealed with a mixture of paraffin and beeswax.

If freshly prepared mixture of 3:1 ethyl alcohol and glacial acetic acid mixture is substituted as a fixative, air-dry preparations from free cell suspension may be tried (Ferguson-Smith, 1964).

Hungerford (1971) used the following technique for his subsequent analysis of human individual pachytene chromomere banding patterns.

Seminiferous tubules freshly obtained by biopsy are transferred to BSS (Earle's balanced salt solution) and minced with scissors at 37° C. The resulting suspension is pipetted into 15 ml graduated conical centrifuge tubes. Large supernatant suspension is pipetted into a second centrifuge tube and centrifuged at 150 g for 4 min to obtain a pellet which is resuspended in excess of 0.125 M KCl, heparin added (20,000 U/l) and incubated at 37° C for 1 hr. After a second centrifugation and removal of hypotonic KCl, the cells are fixed in a 3:1 mixture of methanol-glacial acetic acid for 10–15 min. The slides are prepared according to the standard air-drying method, and stained for 2 hr in a 1 percent solution of neutral orcein (G. T. Gurr) in 60 percent acetic acid.

The pretreatment of seminiferous tubules with hypotonic KCl or NaCl has also been utilized by Williams et al. (1971) and Luciani et al. (1971) among others. Their preparations seem to have yielded meiotic prophase as well as metaphase figures of good quality.

For autoradiographic and other cell kinetic studies of human male germ cells, testicular tissue or organ culture would be essential. A certain degree of success appears to have been obtained. Matte and Sasaki (1971) first cultured a mass of minced seminiferous tubules representing half of a testis obtained by orchidectomy in 80 ml of culture medium (80 percent NCTC-109 and 20 percent calf serum) for 15 hr at 30° C in the presence of tritiated thymidine. The pellet obtained subsequently by centrifugation at 1,000 rpm for 10 min was equally divided into 11 culture tubes, each containing 7 to 8 ml fresh medium and cultures for 35 additional days. Since tritiated thymidine given at the beginning of culture was detected in early spermatid nuclei 35 days later, it appeared that at least some of the male germ cells completed 1st as well as 2nd meiosis *in vitro*. Dutrillaux (1971) also succeeded in culturing human male germ cells in a slightly simpler medium containing calf serum and $MgCl_2$.

3.2. Female

(See Ohno *et al.*, 1962; Baker, 1963; Manotaya and Potter, 1963; Stahl and Luciani, 1971.)

Numerous first meiotic prophase figures of oocytes are contained in ovaries of fetuses from the third month of development to near parturition. However, in more developed fetuses, each oocyte in prophase is already enveloped by a single layer of follicular cells which interfere with squashing of individual oocytes. Meiotic prophase figures of good quality are more easily obtained from younger fetuses in the fourth and fifth month of gestation.

Longitudinal slices about 2 mm thick should be cut from a freshly removed fetal ovary using a razor blade. Pieces no larger than $2 \times 2 \times 2$ mm are cut from the cortical area (directly adjacent to the surface) of these slices. They should then be treated exactly in the same manner as testicular pieces. The pretreatment with hypotonic KCl followed by the air-drying method may also give good results (Stahl and Luciani, 1971).

4. Preparations for Meiotic Metaphase Figures

4.1. Male

(See Ford and Hamerton, 1956; Eberle, 1963; Böök and Kjessler, 1964; Evans *et al.*, 1964; Sasaki and Makino, 1965; Schleiermacher, 1966; Meredith, 1969; Pearson and Bobrow, 1970; McDermott, 1971; Chen and Falek, 1971; Sperling and Kaden, 1971; Schnedl, 1972.)

4.1.1. Squash Method

To obtain well-spread diakinesis and first meiotic metaphase figures, pieces of testicular tissue no larger than $2 \times 2 \times 2$ mm should undergo hypotonic pretreatment prior to fixation. However, it has been our experience that slides prepared for meiotic prophase figures often yield second meiotic metaphase figures of better quality than those prepared from pieces which have undergone hypotonic pretreatment. Thus, hypotonic pretreatment, while essential for first meiotic metaphase figures does not appear to improve the quality of second meiotic metaphase figures.

Either distilled water or 1.0–1.2 per cent citrate solution can be used for hypotonic pretreatment. The only hazard of using distilled water is that, unless periodically adjusted, it tends to become excessively acidic due to absorption of atmospheric carbon dioxide. In our laboratory, distilled water adjusted to pH 7.0 is preferred to citrate solution.

As human seminiferous tubules are enveloped by a rather thick sheath of connective tissue, prolonged hypotonic treatment of 30 min duration is preferable to the customary 12–15 min. The volume of distilled water should be at least 100 times that of the tissue.

The subsequent procedure of fixation, squashing, and staining is the same as that already described for meiotic prophase figures.

4.1.2. Air-drying Method

There can be many modifications of the air-drying technique first described by Evans et al. (1964). The following procedure has been used with satisfactory results in various laboratories (Chandley, 1972).

Testicular biopsies are collected in 1 percent hypotonic sodium citrate. After about 30 min, seminiferous tubules are minced with scissors into fine cubes in a slightly tilted glass Petri dish. Large pieces are allowed to settle to the bottom and discarded. The supernatant suspension is drawn and pipetted into a centrifuge tube and spun down for 10 min at 1,500 rpm. If the suspension appears too concentrated, it is desirable to dilute it by adding additional 1 percent sodium citrate before centrifugation. After centrifugation, discard most of the supernatant, but leave enough fluid for resuspension of the pellet by gently tapping the side of the tube. Twice the volume of the fixative (3:1 methanol-glacial acetic acid) is added to a small volume of this suspension by pouring slowly down the side of the tube. Then the mixture is repeatedly drawn

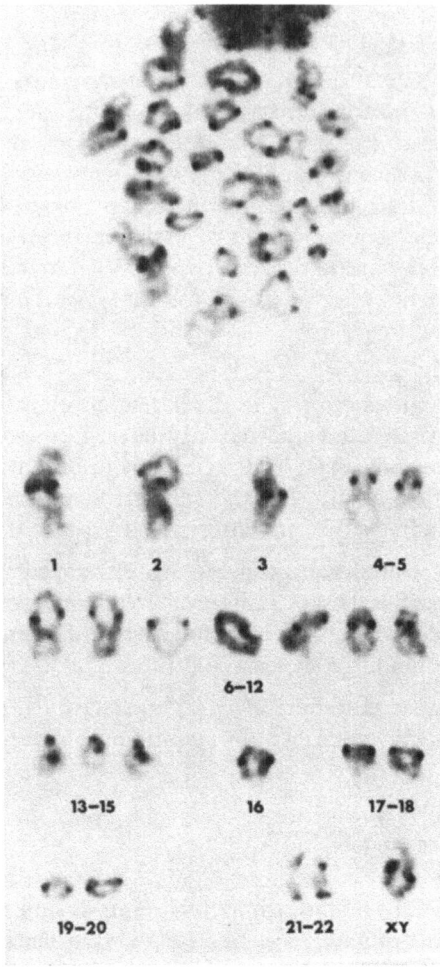

Fig. 2. A microphotograph of a human male first meiotic metaphase figure
stained for the centromeric heterochromatin. The visualization of the centromere positions permits the alignment of 22 autosomal bivalents and the
XY pair. (Courtesy of Dr. Ann C. Chandley, Edinburgh)

in and expelled out of a pipette until clumps of cells which may have
formed by the fixation are dispersed. After this procedure, the total
volume of the cell suspension is increased to about 5 ml by adding more
fixative, and then the suspension is subjected to centrifugation at 1,500
rpm for 5–8 min. The pellet is resuspended in 5 ml of fresh fixative
and allowed to stand at room temperature for 30 min. From this, a

pellet is obtained by centrifugation at 1,500 rpm for 10 min. The pellet is resuspended in 1 ml of fresh fixative. The milky suspension is now ready to be air-dried on the slide glass. The air-dried preparations can be stained by Giemsa, aceto-orcein or carbol fuchsin.

4.1.3. Staining for Banding Patterns of Meiotic Chromosomes (Fig. 2)

If the positions of centromeres could be positively identified on each bivalent, the analysis of 1st meiotic metaphase figures would become considerably more precise. The centromeric heterochromatin can now be visualized by various C banding techniques discussed in Chapter VI (Fig. 2). The application of quinacrine mustard banding techniques to meiotic bivalents is also useful. The following technique developed by Chandley (1972) appears to be the simplest means of visualizing the centromeric heterochromatin on meiotic bivalents. Place the squash or air-dried preparations in 0.2 N HCl at room temperature for 1 hr. After rinsing in deionized water, incubate the microscopic slides in 5 percent barium hydroxide for 30 sec at 50° C. After the second rinse with de-ionized water, incubate the slides in $2 \times SSC$ for 1 hr at 60° C. The rinsed slides can now be stained by Giemsa. Staining for 1 hr in the dilute Giemsa solution (G. T. Gurr's Giemsa "R 66", 1 ml diluted by 50 ml of pH 6.8 buffer made with Gurr's buffer tablets) is recommended.

4.1.4. Recovery of Meiotic Figures from the Ejaculated Seminal Fluid

Testicular biopsy cannot readily be done as it is an unpleasant procedure. Thus, the availability of materials limits the progress in studying human male meiotic process. Sperling and Kaden (1971) have shown that an-alyzable meiotic figures can regularly be obtained from ejaculated seminal fluids of normal males as well as of patients suffering from oligospermia.

4.2. Female

(See Jagiello, 1965; Edwards, 1965; Tarkowski, 1966; Yuncken, 1968; Jagiello and Polani, 1969.)

Only when ovarian follicles of sexually-mature women attain full growth and are ready for ovulation do the oocytes resume meiosis. Since only one ovarian follicle usually grows to full size and ovulates during each estrous cycle, the chance of obtaining a human female diakinesis or 1st and 2nd meiotic metaphase figures appears extremely small. Fortunately, Edwards (1965) found that if oocytes are dissected out from follicles, even those from relatively immature follicles would resume meiosis *in vitro* and go through 1st and 2nd meiosis, extruding polar bodies. Because of this discovery, it is now possible to carry out the meiotic analysis of human oocytes on almost any ovarian materials obtained by biopsy.

The ovaries, or parts of ovaries obtained by biopsy, are collected in TC 199 stabilized with 5 percent CO_2 in air (pH 7.2) and taken to the laboratory at ambient temperature. The large follicles are dissected out intact, and then punctured into culture medium to liberate oocytes. TC 199 (Microbiological Associates Inc.) supplemented with 15 percent fetal calf serum, 100 units penicillin, and 100 µg streptomycin per ml is used as the culture medium and a gas phase of 5 percent CO_2 in air at pH 7.1 is maintained (alkaline medium destroys the ability of oocytes to mature in culture).

Many of the oocytes recovered after 25–36 hr of *in vitro* incubation are either in diakinesis or 1st meiotic metaphase, whereas 2nd meiotic figures can be obtained by harvesting the oocytes after between 48 and 70 hr of culture.

For the preparation of oocyte meiotic figures, the air-drying method developed by Tarkowski (1966) appears to work best. After an appropriate duration of culture, individual oocytes in the culture medium are pipetted out under the microscope and placed in hypotonic 1 percent sodium citrate and left to stand at room temperature for 5–15 min. A microdrop of this hypotonic solution containing several oocytes is placed on a grease-free microscopic slide, a few drops (0.06 ml) of the fixative (absolute ethanol:glacial acetic acid, 3:1) are immediately expelled from another pipette, of which the tip is brought just over the microdrop which contains the oocytes. Air-dried preparations thus obtained are stained in lactic-acetic-orcein, a 2 percent aqueous solution of toluidine blue or Giemsa.

5. Interpreting the Findings

5.1. Polyploid Germ Cells?

In testicular squash preparations of any healthy mammal, including man, apparently tetraploid and even octaploid spermatogonial metaphase figures are frequent; a similar percentage of first meiotic metaphase figures also appear to be polyploid. These observations do not necessarily indicate that polyploid germ cells are normal in the mammalian testis, as shown by examination of histological sections of these materials. Clusters of two, four or even eight spermatogonial metaphase figures as well as first meiotic metaphase figures are frequently observed, suggesting that the last two maturation mitoses of spermatogonia and subsequent meiosis of their daughter cells often proceed in absolute synchrony. A squashing or air-drying procedure might make two or four metaphase figures in close proximity to each other appear as a single metaphase.

Extreme caution should be exercised in interpreting the significance of these apparently polyploid germ cells in testicular squashes. Only when an apparently tetraploid germ cell in first meiosis contains a number of quadrivalents is its tetraploidy confirmed.

5.2. Chiasma Frequency

As stated earlier, each chiasma of early diplotene might well represent an actual point of chromatid exchange between two homologous chromosomes. However, two component homologs of each bivalent at this stage are in an extremely elongated state and prone to mechanical distortion. True chiasmata and artificial intertwining of two threads introduced by squashing or air-drying cannot be distinguished with absolute certainty. Estimation of mean chiasma frequency at this stage encounters apparently intractable difficulties.

At diakinesis and first meiotic metaphase, each chiasma can be recognized unequivocally. However, chiasma terminalization has already occurred. Estimated chiasma frequency at these late stages may merely reflect the efficiency or inefficiency of the terminalizing mechanism of a given species, rather than revealing the frequency of genetic crossing-over.

Genetic crossovers occur more frequently in females than in males. However, this may not readily be reflected as a sex difference in chiasma frequency. Interstitial chiasmata are genetically more significant than terminal chiasmata.

5.3. Precocious Disjunction

In squash and air-dried preparations, as much as 15 percent of human spermatocytes in diakinesis and first meiotic metaphase contain the X and the Y as two separate univalents. This apparently precocious disjunction is also observed on components of the few smallest autosomal bivalents. If precocious separation of two components occurs with high frequency before the formation of the spindle apparatus, grave consequences are expected, as independent movements of two homologous centromeres would result in the movement of both homologs to the same division pole in 50 percent of the cases.

On the other hand, it is more probable that these figures which include apparent univalents are actually already in the earliest part of first meiotic anaphase. In sectioned materials of various organisms, it has been observed that the separation of the X from the Y and their movement toward the opposite division poles occurs before the separation of autosomal bivalents. Among autosomal bivalents, the separation of two components similarly appears to occur first in the smallest bivalents.

The true frequency of first meiotic non-disjunction or precocious disjunction, or both cannot be estimated by observing first meiotic figures. Careful analysis of pairs of second meiotic metaphase figures should yield an answer.

5.4. Trivalents

(See Sasaki, 1965; Hamerton et al., 1961; Kjessler, 1964.)

The presence of three homologs in the same nucleus might be supposed to result automatically in the formation of a trivalent. Thus, the testis of a trisomy 21 individual should contain 21 autosomal bivalents and one trivalent in each first meiotic figure. This assumption, however, is not borne out by observation. The formation of one bivalent, one univalent, and one trivalent is to be expected in an individual with a simple autosomal trisomy.

In the case of a phenotypically normal individual who is heterozygous for a Robertsonian translocation, on the other hand, the formation of a trivalent is unavoidable, if the two arms of a newly-formed metacentric are not homologous to each other.

The formation of a trivalent by one metacentric and two acrocentrics, however, do not necessarily produce genetically unbalanced gametes. Intraspecific chromosomal polymorphism due to Robertsonian translocations occur widely among insects as well as among reptiles and certain mammals. That these species with built-in chromosomal polymorphism do not show reproductive failure indicates that three centromeres of such a trivalent are oriented in such a way on the first meiotic spindle that unfailing segregation of two acrocentrics to one division pole and one metacentric to the other is assured. In this respect, the meiotic mechanism of a human female heterozygous for such a translocation should be viewed as exceptionally untidy, since it produces a genetically unbalanced gamete resulting in a trisomic child (translocation mongolism).

5.5. Quadrivalents (Lindsten et al., 1965) (Fig. 3)

Provided that both chromosomal segments exchanged between two non-homologs are of substantial size, consistent appearance of a quadrivalent is to be expected in meiotic figures of an individual carrying a reciprocal translocation.

If two large autosomes are involved in translocation, the quadrivalent should be easily recognized in each first meiotic metaphase figure, as it substantially exceeds the size of the largest autosomal bivalent.

If one participant in translocation is a sex element, either the X or the Y, the XY autosome quadrivalent should again be recognizable as such, even if the other partner is a small autosome, because of the characteristic end-to-end association between the X and the Y during male meiosis.

On the other hand, if both autosomes involved in translocation are small in size, the distinction between the quadrivalent and a bivalent of similar size is difficult in the human. In a species where almost complete terminalization of chiasmata occurs, a ring quadrivalent can easily be singled out from ring bivalents of corresponding size at first meiotic metaphase as four terminalized chiasmata are noted on the former, while the latter demonstrates only two (Fig. 3).

In the human many autosomal bivalents unfortunately appear to maintain interstitial chiasmata until the very end of first meiotic metaphase. Thus both a small quadrivalent and a large bivalent are apt to

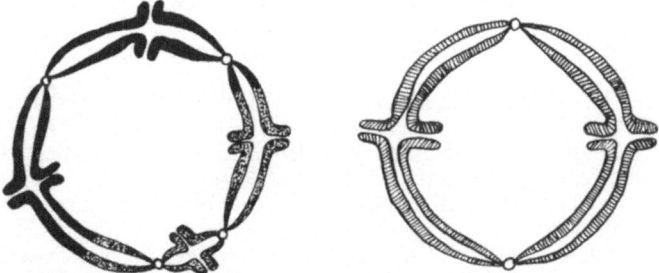

Fig. 3. If each of the two chromosomes involved in reciprocal translocation (black and shaded) are rather small in size, the quadrivalent formed cannot be distinguished by size alone from the large bivalent in a translocation heterozygote

present similar configurations, such as figures of eight or loops of three. Visualization of centromeric heterochromatin of meiotic chromosomes would, therefore, be extremely helpful in identifying the quadrivalent (Fig. 2).

6. Summary

Morphological analysis of colchicinized mitotic metaphase chromosomes has demonstrated numerous types of chromosomal abnormalities in man since 1959. The recent discovery of various banding techniques greatly increased the precision of somatic chromosome analysis. Advances have also been made in the analysis of meiotic metaphase figures. Both male germ cells and female oocytes can now be analyzed, and the application of new banding techniques would no doubt facilitate the identification of quadrivalents and other abnormal components in meiotic figures.

References

Baker, T. G.: A quantitative and cytological study of germ cells in human ovaries. Proc. roy. Soc. B **158**, 417–433 (1963).

Böök, J. A., Kjessler, B.: Meiosis in the human male. Cytogenetics 3, 143–147 (1964).

Chandley, A. C.: Personal communication to Prof. Dr. H. G. Schwarzacher 1972.

Chen, A. T. L., Falek, A.: Cytological evidence for the association of the short arms of the X and Y chromosomes in the human male. Nature (Lond.) **232**, 555–556 (1971).

Dutrillaux, B.: La culture de cellules germinales males: Methodes et applications. Ann. Génét. **14**, 157–159 (1971).

Eberle, P.: Meiotische Chromosomen des Mannes. Klin. Wschr. **17**, 848–856 (1963).

Edwards, R. G.: Maturation *in vitro* of human ovarian oöcytes. Lancet **1965 II**, 926–929.

Evans, E. P., Breckon, G., Ford, C. E.: An air-drying method for meiotic preparations from mammalian testes. Cytogenetics 3, 289–294 (1964).

Ferguson-Smith, M. A.: The sites of nucleolus formation in human pachytene chromosomes. Cytogenetics 3, 124–134 (1964).

Ford, C. E., Hamerton, J. L.: The chromosomes of man. Nature (Lond.) **178**, 1020–1023 (1956).

Hamerton, J. L., Cowie, V. A., Giannelli, F., Briggs, S. M., Polani, P. E.: Differential transmission of Down's syndrome (mongolism) through male and female translocation carriers. Lancet **1961 II**, 956–958.

Hungerford, D. A.: Chromosome structure and function in man. I. Pachytene mapping in the male, improved methods and general discussion of initial results. Cytogenetics **10**, 23–32 (1971).

Jagiello, G.: A method for meiotic preparations of mammalian ova. Cytogenetics **4**, 245–250 (1965).

Jagiello, G., Polani, P. E.: Mouse germ cells and LSD-25. Cytogenetics **8**, 136–147 (1969).

Kjessler, B.: Meiosis in man with a D/D translocation and clinical sterility. Lancet **1964 I**, 1421–1423.

Lindsten, J., Fraccaro, M., Klinger, H. P., Zetterqvist, P.: Meiotic and mitotic studies of a familial reciprocal translocation between two autosomes of group 6–12. Cytogenetics **4**, 45–64 (1965).

Luciani, J. M., Devictor-Vuillet, M., Stahl, A.: Hypotonic KCl: An improved method of processing human testicular tissue for meiotic chromosomes. Clin. Genet. **2**, 32–36 (1971).

Manotaya, T., Potter, E. L.: Oöcytes in prophase of meiosis from squash preparations of human fetal ovaries. Fertil. and Steril. **14**, 378–392 (1963).

Matte, R., Sasaki, M.: Autoradiographic evidence of human male germ cell differentiation *in vitro*. Cytologia (Tokyo) **36**, 298–303 (1971).

McDermott, A.: Human male meiosis: chromosome behavior at pre-meiotic and meiotic stages of spermatogenesis. Canad. J. Genet. Cytol. **13**, 536–549 (1971).

Meredith, R.: A simple method for preparing meiotic chromosomes from mammalian testis. Chromosoma (Berl.) **26**, 254–258 (1969).

Ohno, S., Klinger, H. P., Atkin, N. B.: Human oögenesis. Cytogenetics **1**, 42–51 (1962).

Pearson, L., Bobrow, M.: Definitive evidence for the short arm of the Y chromosome associating with the X chromosome during meiosis in the human male. Nature (Lond.) **226**, 959–961 (1970).

Schleiermacher, E.: Über den Einfluß von Trenimon und Endoxan auf die Meiose der männlichen Maus. I. Methodik der Präparation und Analyse meiotischer Teilungen. Humangenetik **3**, 127–133 (1966).

Schnedl, W.: End-to-End association of X and Y chromosomes in mouse meiosis. Nature New Biol. (Lond.) **236**, 29–30 (1972).

Sperling, K., Kaden, R.: Meiotic studies of the ejaculated seminal fluid of humans with normal sperm count and oligospermia. Nature (Lond.) **232**, 481 (1971).

Stahl, A., Luciani, J. M.: Individualisation d'un stade preleptotene de condensation chromosomique au début de la meiose chez l'ovocyte fetal humain. C. R. Acad. Sci. (Paris) **272**, 2041–2044 (1971).

Tarkowski, A. K.: An air-drying method for chromosome preparations from mouse eggs. Cytogenetics **5**, 394–400 (1966).

Williams, D. L., Hagen, A. A., Runyan, J. W., Lafferty, D. A.: A method for the differentiation of male meiotic chromosome stages. J. Hered. **62**, 17–22 (1971).

Yuncken, C.: Meiosis in the human female. Cytogenetics **7**, 234–238 (1968)

CHAPTER XII

Marginal Notes on the Handling of Tissue Culture Cells for Biochemical Analysis

WINFRID KRONE

1. Introduction

The cytogeneticist who uses tissue culture cells primarily for the purpose of chromosome analysis and cytological examination must adopt a quite different attitude when turning to biochemicial research. More often than he will perform the biochemical tests himself, he will be asked to provide cells for biochemical analysis. In both cases, the cytogeneticist becomes confronted with requirements and demands with which he is unfamiliar or which might even interfere with his own methods designed to increase the yield and to improve the quality of metaphases.

On the other hand, the biochemist who becomes involved with cultured cells is well advised to acquire familiarity with tissue culture procedures. This will not only facilitate communication with the cytogeneticist and the personnel trained in routine tissue culture procedures, but also enable him to implement his intentions and to accept the full responsibility for failure or success.

In this chapter, an attempt is made to describe some of the basic rules that should be followed if successful biochemical analysis of cell cultures – of human homonuclear fibroblasts in particular – is to be carried out. The reader should, however, appreciate that almost all of the methods currently used can potentially be improved, and that the optimum procedure for the determination of any single biochemical parameter in fibroblasts needs to be worked out by meticulous methodological research.

There has been growing interest in the application of tissue culture for the investigation and diagnosis of inherited diseases, and several review articles on this subject were published recently, notably Mellman (1971). In some of them methodological aspects are discussed. The reader is further referred to the following publications: Gartler and

Pious (1966), Krooth *et al.* (1968), Davidson (1970), Krooth and Sell (1970), Mellman and Kohn (1970), Boyle and Seegmiller (1971), Priest (1971), Raivio and Seegmiller (1972), Cristofalo and Kabakjian (1973).

Articles with special emphasis on culture of amnion cells and prenatal diagnosis: Milunsky *et al.* (1970, 1972), Littlefield (1971). Two classic publications by DeMars (1964) and by Hayflick and Moorhead (1961) are still very worthwhile reading.

2. Selected General Notes

2.1. Some Peculiarities of Homonuclear Fibroblasts

To the biochemist, cultured fibroblasts present a number of particular difficulties.

2.1.1. Limitation of the Amount of Material

A maximum of 50–60 µg of cell protein is obtained from 1 cm² of cell layer at saturation density, if multilayering is not induced deliberately by continuous feeding in perfusion cultures (Kruse and Miedema, 1965). For studies performed during the exponential growth phase, much lower amounts have to suffice. Mass cultivation can be carried out in roller flasks which are commercially available with surface areas between several hundred and more than a thousand square centimeters. Comparative research with several genetically different cell strains not only requires extensive space, but is also expensive because of the amounts of medium consumed. The same pertains to the perfusion system mentioned above. Some of these difficulties can be alleviated with an elegant device which was contrived by Wöhler *et al.* (1972). However, the growth potential of individual cell strains derived from small skin biopsies is rapidly exhausted during mass cultivation because of the limited in vitro life span of fibroblasts. This fundamental limitation seems to be exacerbated by the suboptimal media which are commonly used. Certain modifications of the culture medium result in a considerable extension of the in vitro life span of fibroblasts (Macieira-Coelho, 1966; Litwin, 1972).

2.1.2. Fibroblasts Do Not Grow in Suspension

Taking samples at various time intervals from one growing cell suspension is a routine procedure in biochemical research. Fibroblasts do not usually allow this method of sampling, because they grow attached to a surface. Multiple sampling must be performed by setting up many sub-cultures in parallel. At this step, the introduction of variation with respect to the initial cell density and slight differences in the culture flasks used (especially in the case of conventional prescription bottles or milk dilution bottles having uneven surfaces) cannot be avoided.

There are ways of overcoming this difficulty, particularly when only small samples are required. Cells can be seeded on several coverslips lying in a Petri dish or on plastic membranes (Hösli, 1970; Munder et al., 1971) from which pieces can be cut off.

2.1.3. Growth of Human Homonuclear Fibroblasts is Impossible or Unsatisfactory in Chemically Defined Media

The composition of the culture medium cannot be standardized because of the necessity of adding serum, as a rule fetal calf serum which varies with respect to its composition and its growth promoting potential. Attempts to find chemically defined substitutes for the fetal calf serum have not been successful in achieving equivalent growth rates and cloning efficiencies. Exhaustively dialysed fetal calf serum, also commercially availalbe, assures constancy of the kind and concentrations of the low molecular weight components of the culture medium. It becomes a necessity in quantitative work with radioactive precursors and e.g. in studies requiring the elimination of glucose or its substitution by other hexoses.

The major difficulties for biochemical studies with fibroblasts arise from the numerous parameters that are known or must be suspected to influence the metabolism of these cells.

Some of the factors which require control to produce reproducible results are discussed in the ensuing section. Many of these factors are mutually interdependent. Control of all conceivable parameters would not, however, completely eliminate interindividual variability between control strains.

2.2. Composition of the Medium

The following components of the culture medium have been shown to influence biochemical properties of human homonuclear fibroblast cultures:

concentration of glucose (Griffiths, 1970a, Schulz et al., 1970);

concentration and variety of amino acids and vitamins (Litwin, 1972;

Griffiths, 1970a; Schafer *et al.*, 1967; Ryan *et al.*, 1972);

source and concentration of the serum supplement (Rhode and Ellem, 1968);

other supplementary factors, like hydrocortisone, serotonin, insulin, unsaturated fatty acids (Macieira-Coelho, 1966; Boucek and Alvarez, 1971; Griffiths, 1970b);

hydrogen ion concentration (pH) (references, see below).

Some of these factors will be briefly discussed.

The glucose content of some commercial tissue culture media is listed below (glucose concentration in mg per 1,000 ml):

Eagle BME and MEM	1,000
Dulbecco modified Eagle	1,000
Medium 199	1,000
NCTC 135	1,000
CMRL 1066	1,000
Ham F10	1,100
Puck N16	1,100
Ham F12	1,802
RPMI 1634 and 1640	2,000
RPMI 1603 and 1630	2,500
RPMI 1629	3,000
McCoy 5a	3,000
Waymouth MB752/1	5,000

As a consequence of these large differences in the glucose content, different rates of glycolysis will be obtained with various growth media, especially during the later stages of a growth period. The Crabtree effect (i.e.: the inhibition of respiration by high concentrations of glucose; see: Paul, 1965) will also influence the rate of oxidative phosphorylation in the mitochondria. Therefore, when parameters of the oxidative and glycolytic pathways are studied, it may be desirable to measure the utilization of glucose to ascertain the comparability of results.

Large differences between culture media also exist in the numbers and concentrations of amino acids. The true optimal ratios of the concentrations of several amino acids have not been thoroughly investigated in terms of competition for and stimulation of shared transport mechanisms. Ling *et al.* (1968) are probably correct in considering the existing media as bad approximations of a physiological optimum. Supplementation of the medium with serine (0.4–1 mM) at low cell densities increases survival and shortens the lag phase. Supplementation with serine has been found to be as effective as the addition of commercial preparations of the non-essential amino acids.

One of the most important parameters influencing cell growth and cell function is the pH of the culture medium. Its strong influence on the rate of glycolysis, discovered by Zwartouw and Westwood (1958) and confirmed by Paul and his coworkers (1965), is one of the mechanisms responsible for its pronounced effects on cultured cells. It also affects the catabolism of mucopolysaccharides (Lie *et al.*, 1972), contact inhibition of growth, the protein content per cell (Ceccarini and Eagle, 1971 a, b) and the efficiency of cell hybridization with Sendai virus (Croce *et al.*, 1972).

For many years control of the pH of the culture medium was almost exclusively exercised with the volatile buffer system HCO_3^-/CO_2 in CO_2-incubators. The introduction of a series of zwitterionic buffers by Good (1965) not only facilitated pH control in cell culture media but also stimulated the investigation of pH effects. Some useful buffer combinations were published by Eagle (1971), all of them being used in conjunction with 24 mM sodium bicarbonate.

Some of the systematic names used in the publication of Eagle are incorrect and, therefore, do not designate the substances referred to by their correct abbreviations. For the correct names, see Good (1965). Valuable information on the physical properties of these buffers and their suitability for biochemical test systems (e.g. interference with protein determination) is given by Good and Izawa (1972).

The combinations resulting in the highest buffer-capacity (corresponding to the flattest titration curve: Δ pH vs. unit amount of acid or alkali added) given by Eagle are:

Above pH 7.5: 20 mM HEPES
10 mM TES
10 mM Tricine.

Below pH 7.5, equimolar BES is substituted for TES.

It is not necessary to use Good buffers for routine propagation of cell strains if a reliable CO_2 incubator is employed. These buffers are, however, of great value for pH control during experiments. Their possible effect on the parameter studied should be tested in each case. Development of zwitterionic buffer substances with even lower toxicity than that of the Good buffers is required.

2.3. Stage within the "Lag-log-stationary" Cycle

It has become a commonplace (although still not universally accepted) that enzyme activities and the intracellular concentrations of metabolites fluctuate during the growth cycle of fibroblasts. Therefore, comparative determinations of biochemical parameters in different strains require the analysis of its behaviour through all stages of the growth cycle consisting of an initial lag phase, a logarithmic growth phase, and a so-called stationary phase. This cycle is, therefore, designated as the "lag-log-stationary cycle" (Littlefield, 1971). These phases in turn largely depend on the experimental conditions and on the particular strains used. The duration of the lag phase is inversely related to the initial cell density. In poorly pH-controlled environments, the initial increase of the pH of the medium lengthens the lag phase and decreases cell survival. The growth rate and the duration of the exponential growth phase are strongly influenced by the culture medium used (especially by its pH; see above), the feeding schedule, the age of the culture in terms of the number of in vitro generations, and the age of the donor. Some unexplained variability between strains remains even under the most carefully controlled experimental conditions. This variability could in part reflect unknown genetic differences between strains with respect to optimum growth requirements. Also subject to interindividual variability is the cell density at which exponential growth subsides (Raff and Houck, 1969). The enormous variations of final cell densities achieved by diploid fibroblasts which are reported in the literature (between $2.5 \times 10^4/cm^2$ and $2 \times 10^5/cm^2$ under normal feeding conditions!) reflects its strong dependence on experimental conditions; the inaccuracy of cell counting and the unevenness of the inner surface of some kinds of culture flasks might also contribute to this broad range. Frequent feedings or continuous perfusion of stationary cultures with medium brings about a continuation of growth with a population doubling time two to three times longer than during log-phase (Kruse and Miedema, 1965; Kruse et al., 1969). Multiple cell layers are obtained under these conditions corresponding to up to five monolayer equivalents. The tremendous metabolic alterations occurring in the postconfluent phase in homonuclear fibroblasts are emphasized by the studies of Griffiths (1971, 1972) who showed that, besides the well known slowdown of RNA and DNA synthesis, an increased amino acid requirement for the attainment of protein synthesis arises in the postconfluent phase. Thus, pre- and postconfluent growth requirements are basically different, and the medium routinely used for the propagation of the cells during log-phase rather rapidly becomes limiting for growth and survival during the stationary phase.

The fluctuations of enzyme activities during the lag-log-stationary cycle are increasingly recognized as a source of variation in experimental results. Examples were discussed by DeMars (1964) and Mellman (1971). The causes of these fluctuations are presently being investigated in many laboratories (e.g. Mellman et al., 1972); shifts in the equilibrium between enzyme synthesis and degradation as well as reversible mechanisms of activation and inactivation must be considered. Since the pattern of change during the lag-log-stationary cycle varies with the enzyme studied, stages showing the smallest fluctuations must be found anew for the purpose of comparative determinations for each enzyme. Plateau of activity are frequently maintained during early stationary phase. On the basis of what has been said earlier about the biology of the post-confluent phase, the feeding schedule must be controlled carefully since, with each feeding, a new semisynchronous wave of postconfluent growth is elicited. If the stage of relatively constant specific activity happens to be the exponential growth phase, inadvertent partial synchronization by feeding or by subculturing of uniform G_1 populations (late stationary phase) should be avoided.

2.4. Contamination of Cultures with Mycoplasma

As opposed to infection of cell cultures by bacteria and fungi, which in general are easily detectable, the presence of mycoplasma contamination can remain undetected for long periods of time. This is probably one of the reasons why these insiduous parasites, which adversely affect research as well as diagnostic use of cell cultures, are still ignored by many investigators. There is an abundant literature on the effects of mycoplasma on cell cultures and many authors have expressed serious warnings concerning the interpretation of results obtained with cell cultures that have not been monitored for the presence of mycoplasma. There is general agreement that "... no one working with tissue cultures can afford to ignore them" (Macpherson, 1966).

Mycoplasma strains have been shown to interfere with biochemical, virological, immunological, and cytogenetic properties of cultured cells. Some of the biochemical effects may be related to the consumption by mycoplasma of essential nutrients (like arginine, glutamine, glutamic acid) from the medium. An impressive example of interference by myco-plasma with RNA synthesis and of contamination of host cell cyto-plasmic RNA with a RNA component of the parasite was reported by Levine et al. (1967, 1968). By production of lactate from glucose, many strains of mycoplasma contribute considerably to acidification of the medium.

Overt cytopathic effects are produced by some mycoplasma species, with a complete cessation of growth and extensive vacuolization of the cells. Less noxious forms may cause increased cell granularity, a tendency of the cells to detach from the substratum, slowed growth and failure to form a complete monolayer. As already mentioned, other strains may pass completely undetected.

Various methods for the detection of mycoplasma infections have been published. They can be classified according to the techniques employed as microbiological, immunological and biochemical methods. A very efficient means of recognizing mycoplasma infections is electron microscopy.

The immunological methods use staining of the cells with fluorescent antibodies against mycoplasma-associated antigens. Strain-specific as well as polyvalent antibodies have been used.

Inoculation in special mycoplasma broths and colony formation on appropriate agar plates is the most widely used method. The variability of the growth requirements and of the optimum growth conditions of different mycoplasma strains render it advisable to have the tests performed by an experienced microbiologist. The ability to recognize mycoplasma colonies on agar plates can hardly be acquired from textbook descriptions.

The fact that many species of mycoplasma deplete the culture medium of arginine by an enzyme reaction catalyzed by arginine deiminase (L-arginine iminohydrolase, E.C.3.5.3.6), which does not occur in the host cell, led Barile and Schimke (1963) to design the first biochemical test for mycoplasma infection. The discovery by Levine (1972) that the abnormal labeling behaviour of mycoplasma-infected cultures with radioactive nucleosides like uridine and thymidine is partly due to the almost universal occurrence in mycoplasma strains of high nucleoside phosphorylase activity has become the basis of an elegant rapid procedure to detect the presence of these organisms.

Eradicating mycoplasma from contaminated cultures is a difficult problem. Since many mycoplasma species live and multiply both extra- and intracellularly, the application of rather high concentrations of antibiotics like aureomycin, kanamycin and tylosin is required. Toxicity to the cells and stability in the culture medium at 37° C influence the choice of the proper antibiotic (see Table 1, p. 557, in Brown and Officer, 1968). One or two passages in the presence of the antibiotic should be performed. Freshly trypsinized infected cells contain a lower titer of mycoplasma particles than older cultures. With regard to the prevention of mycoplasma infection, it must be noticed that primary cultures are very rarely infected. The most probable source of mycoplasma contamination is the upper respiratory tract of the laboratory personnel. As preventive measures, the most stringent aseptic conditions and the omission of antibiotics from the culture medium are recommended by several authors.

In some cases, the presence of a low level of mycoplasma infection might allow determination of biochemical parameters of the host cells. It should, however, be demonstrated in these cases, that the parameter under concern is not affected by the presence of the parasite.

3. Discussion of Some Procedures

3.1. Establishment of Cultures and Method of Subculturing

The comparative nature of experiments in biochemical genetics necessitates the simultaneous establishment of several primary cultures. This requirement can only be satisfied on the basis of careful planning in relation to the capacity of the tissue culture personnel and facilities available. It is desirable to obtain skin punch biopsies from the patient, his sibs and his parents. Furthermore, a number of cultures from unrelated probands, age- and sex-matched with the patient, should be set up. Experience gained with several control cultures before the cultures of the affected family are studied under identical conditions facilitates the experimental approach to the latter. Since interindividual variability is regularly encountered during the establishment of primary cultures, the growth behaviour of the explants should be scored with respect to the size and persistence of the epitheloid outgrowth often observed around the explants and the time and vigor of the emergence of the first fibroblasts. Biochemical studies with fibroblasts should be performed during "phase II" of the culture (Hayflick and Moorhead, 1961) except those studies which are actually concerned with the growth phases of homonuclear cells. Phase II, during which the doubling time is fairly constant, may already begin in the primary culture in some cases, while in others it commences only after the first few passages. This introduces differences between strains in the number of generations which pass before experiments are performed. The tissue culture protocol, containing (most properly on suitable form sheets) information about the number and time of the passages and the number of subcultures made, must always be consulted, when the biochemical experiments are planned and their results are scored. The number of subcultures made upon passage is of course not necessarily a measure of the split ratio applied, which must be calculated from the ratio of the seeded surfaces of the culture vessels used.

Although there is an abundant literature on cell adhesion after dispersal of cell cultures by chelating agents and/or proteolytic enzymes, surprisingly little effort has been devoted to comparative studies of the effects of e.g. versene and trypsin on fibroblasts under various conditions of treatment. Thus, the recommendations given below are based primarily on theoretical rather than on empirical facts.

A good approach to subculturing for biochemical purposes can be acquired if one realizes that upon removal from the surface to which they are attached the cells are exposed to highly adverse conditions. The problem is to achieve a relatively rapid detachment and dispersal of the cells under the gentlest possible conditions. Speed is required, because a prolonged trypsinization procedure will introduce heterogeneity into the cell population with respect to the extent of the exposure to the protease.

Because of the possible intracellular effects of chelating agents like versene (ethylene diamine tetraacetic acid; EDTA), proteolytic methods should be preferred. Simple phosphate-buffered isotonic solutions of sodium chloride as a solvent for the trypsin should be avoided. If 0.25% Difco trypsin is used, absence of calcium ions is not required for the enzyme to exert its effect. Thus, Hanks' salt solution is a good solvent. Dulbecco's balanced salt solution does not contain glucose, which should be added (1 g/l) if equivalence with Hanks' is desired. Trypsinization in serum-free culture medium has been reported, but the benefits of this theoretically optimal method remain to be demonstrated. Litwin (1973) recommends 0.5% lactalbumin hydrolysate in Hanks' solution.

There are two basically different trypsinization procedures:

a) After removing the medium and two washes with Hanks' solution, the cell layer is exposed for 30–60 sec to about 1 ml of trypsin solution per 40 cm² at 37° C. The bulk of the trypsin solution is poured off and cell detachment is allowed to proceed at 37° C with the small amount adhering to the cell layer. After five to seven min, the cells can be loosened completely by striking the culture flask a few times against the palm, the direction of the movement being tangential with respect to the cell layer. After this step, the cells can be suspended in culture medium by pipetting in the usual way.

b) The more conventional method of trypsinization involves the exposure of the cell layer to a larger volume of trypsin solution (5–10 ml per 40 cm²) and incubation at 37° C until the cells are detached. This procedure usually does not require washing of the cell layer after removal of the medium, which may, however, become necessary for the rapid trypsinization of older "mono-layers", especially if collagen synthesis had been stimulated by ascorbic acid. The cells are suspended gently in the trypsin solution by pipetting and then centrifuged. At this stage it is desirable to stop the action of the trypsin by addition of a small volume of culture medium (2 ml per 10 ml of suspension) or by acidification with a brief exposure to a stream of CO_2.

Vigorous resuspending of the cells in culture medium by pipetting can not be recommended. To avoid foaming, the tip of the pipette should remain below the meniscus all the time and air should not be blown into the suspension. Continuation of the pipetting for one to 2 min after the visible cell clumps have disappeared usually results in a single cell suspension.

Foley and Aftonomos (1970) found 0.1% pronase dissolved in a balanced salt solution superior to trypsin with respect to its speed of action and the production of a single cell suspension. In contrast to trypsin, however, traces of pronase interfere with cell attachment even in the presence of serum. This necessitates a higher number of wash centrifugations. Since the action of trypsin can be accelerated by complete removal of the culture medium prior to trypsinization and by other means, its substitution by pronase does not seem advantageous.

One of the prerequisites for the achievement of reproducible results is equal cell density in the subcultures to be compared with each other in the course of the experiment. Counting of the cells and their distribution over a large number of subcultures – often with two or more batches in parallel – requires that the cells be kept in suspension. This can be achieved by resuspending them in serumfree culture medium and agitating the suspensions on magnetic stirring plates with teflon-coated stirring bars. The fetal calf serum can be pipetted into the empty culture flasks which are to receive aliquots of the cell suspension afterwards, or, it can be added to the bulk suspension immediately before its splitting into the subcultures. The initial cell proteins measured after the attachment of the cells in subcultures prepared in this way do not differ by more than 10% from each other. Initial cell densities below 2,500 per cm² result in very long lag phases (time before exponential growth commences).

3.2. Harvest of Fibroblast Cultures

The method of termination of the culture for subsequent biochemical analysis depends to some extent on the kind of test to be performed. Most assay systems, however, require the prior removal of the medium. There are two possible ways of washing the cells: one of them (a) takes advantage of the fact that the cells are firmly attached to the substratum; the other one (b) is based on trypsinization and centrifugation of the cell suspension.

a) After the medium has been poured off, the culture flask is allowed to stand upside down on a sheet of filter paper for three minutes to remove the last drops of medium. Two washings of the entire inner surfaces of the culture flask with 0.25 ml of rinsing fluid per cm² of grown surface and a three minutes exposure of the cell layer to the fluid each time suffice to reduce the protein content of the second wash to less than 5 µg/ml. Whether or not the entire procedure needs to be carried out in the cold again depends on the type of study to be made. The optimum conditions must be found anew for each test system, since some enzymes are stabilized and others (especially some of the allosterically regulated enzymes) are inactivated in the cold. It is selfevident that isotonic solutions must be used for washing of the cells. Again, the glucose-containing complex balanced salt solutions are superior to simple phosphate-buffered saline. Washing with non-isotonic buffers is entirely inappropriate. Washing of cell layers grown in Petri dishes should not be carried out according to the procedure just described. Because of the large exposed surface of the Petri dish, residual medium as well as rinsing-fluid rapidly evaporate, during the few minutes needed for complete drainage, leaving behind a crust which osmotically damages the cells. A faster washing procedure should therefore be adopted for Petri dishes. The same applies to cell layers grown on cover slips or on membranes. A number of qualitative enzymatic tests can be carried out with the washed cell layers on coverslips without preparation of homogenates. Some examples are:

Glucose-6-phosphate dehydrogenase (DeMars, 1968)
Acid phosphatase   (DeMars, 1964)
β-Galactosidase   (Sloan *et al.*, 1969).

In principle, this type of histochemical test should be applicable to the detection of many lysosomal enzyme deficiencies.

To prepare homogenates, the washed cells are scraped from the surface with a chemically inert rubber policeman in a small volume of the buffer subsequently used for homogenization. Volumes between 2 and 3 ml per 40 cm² cell layer yield protein concentrations (0.1–1 mg/ml) which allow the measurement of enzyme activities with small aliquots of the homogenate. Dense monolayers usually produce large flakes which tend to adhere to the inside of the pipette upon transfer of the suspension. This can be avoided by pipetting the slurry up and down rapidly in the lower tenth of a 2 ml pipette before transfer. The total volume of the suspension will exceed the volume of the buffer added by up to 0.2 ml, depending on the cell density of the culture. Precise determinations of the total enzyme activity per culture requires the measurement of the total volume which can be done with micropipettes during transfer of the suspension into the homogenization tube.

b) The disadvantage of the method described under *a* is that cell counts are not possible. Mechanical dislodging of the cells destroys a substantial proportion especially if a complete harvest of the culture is intended. Cell counts can, on the other hand, be easily made with trypsinized cell suspensions. Similar precautionary measures must be taken upon harvesting by trypsinization as those mentioned under the section on subculturing: Use isotonic glucose-containing solutions for the trypsin as well as for the wash centrifugations. Stop the action of the trypsin by addition of culture medium or trypsin-inhibitor during the dispersal of the cells.

Avoid foaming when the cells are suspended or resuspended before each centrifugation.

Three wash centrifugations with 10 ml each per 10^6 cells should suffice for complete removal of the trypsin.

3.3. Homogenization

The choice of a method of homogenization of fibroblasts depends on the biochemical parameter which is to be studied. The most suitable procedure for a particular purpose must be determined empirically. Methods that have been successfully applied are:

a) Repeated freezing and thawing;
b) Ultrasonic disintegration ("sonication");
c) Coaxial homogenization with pestle and tube homogenizers;
d) Nitrogen cavitation;
e) Lyophilization (freeze drying);
f) Lysis with detergents;
g) Combination of methods.

Because of the small sample size, Waring blendors and other high speed mixers have rarely been employed.

a) This is a gentle method of extraction rather than a true method of homogenization. Undispersed cell fragments must be removed by centrifugation of the lysate if it is to be used in a spectrophotometric test system. The number of freeze-thawing cycles needed to obtain a reproducible percentage of the total cell protein and an essentially complete recovery of the enzyme activity in the supernatant must be ascertained experimentally. Mitochondria and lysosomes are disrupted by freezing and thawing, though complete release of soluble mitochondrial enzymes may require a higher number of freeze-thawing cycles than the release of lysosomal enzymes. Additional brief sonication saves the time spent on lengthy freeze-thawing procedures but also leads to the disruption of nuclei. The procedure is carried out by rapid freezing of the suspension in acetone-dry ice (if this is not available, precooled baths of organic solvents in a deep freeze can be used) and slow thawing with occasional gentle agitation in a water bath at below 10° C.

b) Sonication has become the most widely used method of disrupting tissue culture cells for subsequent biochemical analysis. With the Raytheon Sonic Oscillator, operating at 10 Kc/s and 200 W up to 5 min of sonication are required for complete disruption of the cells. In tissue culture work, this instrument has essentially been replaced by the Branson Sonicator with its exchangeable probes which are submerged in the sample. This instrument operates at 20 kc/s and between 0 and 150 W.

Mitochondria, lysosomes and cell nuclei are broken by this method. Some enzymes are reported to become inactivated by sonication. The formation of traces of peroxides during ultrasonic treatment calls for the presence of mercaptoethanol, dithiothreitol, or other SH-protective agents. A considerable amount of heat is produced in the sample exposed to ultrasonic radiation. The sample must therefore be cooled with an ice-water bath and several short bursts of sonication with 5–15 sec breaks between them are superior to continuous treatment. Total sonication times of 0.5–1.5 min with samples between 1 and 4 ml are sufficient for complete disruption of the cells. Slightly longer exposure times are required if isotonic sucrose is used as the homogenization medium. Preservation of mitochondria during sonication of fibroblasts in the presence of 0.25 M sucrose has been described (Hakami and Pious, 1967). Samples of up to 5×10^5 cells per 2 ml of buffer yield a slightly opaque sonicate from which aliquots can be used in spectrophotometric assays without prior centrifugation. The homogeneity of sonicates is unstable: the aliquots for the measurement of enzyme activity and for protein determination should be taken within an hour after sonication. The success of sonication should be checked by phase contrast microscopy.

c) The classic homogenization method by hand- or motordriven coaxial pestle and tube devices is particularly useful for the conservation of subcellular fractions. The types of homogenizers available for this purpose have been summarized by Allfrey (1959). Particularly suitable are homogenizers with teflon pestles and the handoperated Dounce homogenizers with defined clearances (Kontes Glass Co., Vineland, N.J. and Braun, Melsungen, West Germany). For sample sizes below 1 ml, microhomogenizers have been constructed, for instance by E. Schütt, Göttingen, West Germany. Coaxial homogenization is often performed in the presence of 0.25 M sucrose. Swelling of the cells for a few minutes in distilled water at 0° C before the addition of concentrated solutions of buffer and sucrose to the desired final concentrations facilitates homogenization.

The methods described for the homogenization of HeLa cells for subsequent fractionation of subcellular particles are not applicable to fibroblasts whithout modification. The isolation of clean intact nuclei, especially, requires a much milder procedure since fibroblast nuclei do not resist the high concentrations of Triton X100 (up to 0.2 %) used for the preparation of HeLa nuclei. Pure nuclei can also be isolated after Dounce-homogenization of fibroblasts by the method of Chaveau (see Wang, 1967).

Clearance, time, and speed of operation of coaxial homogenizers are important parameters which influence the morphological integrity or the functional activity, or both, of subcellular fractions like microsomes

(Ganoza *et al.*, 1965) mitochondria, lysosomes (Conchie *et al.*, 1961) and nuclei (Busch, 1967).

The fractionation of subcellluar components is performed by differential centrifugation techniques, some of which utilize sucrose gradients or discontinuous multiple layers of sucrose (Allfrey, 1959; Hogeboom, 1955; DeDuve, 1964). Discussion of modern, more sophisticated methods of cell fractionation is outside the scope of this article; the reader is referred to literature on biochemical methodology.

The isolation of lysosomes represents a special problem because of their heterogeneity and the latency of their enzymes (Beaufay, 1972). The size range of these particles, and hence, their sedimentation behaviour overlaps that of heavy microsomes and of light mitochondria, respectively. It is therefore meaningless to speak of the "lysosomal fraction" on the basis of its sedimentation behaviour alone: The success of purification must rather be indicated by the specific activities of lysosomal enzymes on the one hand, and of a mitochondrial marker enzyme on the other. Release of enzymes from lysosomes is usually achieved with the nonionic detergent Triton X100 (1 % in appropriate buffer).

d) In the nitrogen cavitation method, a cell suspension in an isosmotic solution is exposed to high pressures of nitrogen in a stainless steel bomb (Artisan Metal Products, Waltham, Mass., U.S.A.) designed for this purpose. After equilibration of the liquid phase with the gas phase which, under gentle stirring, takes 20–30 min, the suspension is expelled through an outlet at a controlled rate, so that excessive shearing forces are avoided. This procedure is related to the method introduced by Fraser (1951) for the homogenization of bacteria, in which, however, shearing forces produced by rapid release of the sample exert the homogenizing effect.

Nitrogen cavitation under isosmotic conditions is a very gentle method which proves useful for the preparation of intact subcellular fractions, including nuclei and mitochondria (Hunter and Commerford, 1961). The method is most effective in the isolation of fragmented plasma membranes from mammalian cells (Wallach and Kamat, 1964; Ferber *et al.*, 1972) and it may therefore contribute in the future to the purification of cell surface antigens. Nadler *et al.* (1969) applied this method successfully to lymphocytes and to fibroblasts (Nadler and Egan, 1972). A description of the procedure as used in the preparation of plasmamembrane fragments was given by Wallach and Kamat (1966).

e) Freeze drying after isolation of animal tissues has become a technique to preserve the intracellular distribution of enzymes and metabolites which can be subsequently analysed by cell fractionation procedures in non-aqueous solution (Siebert, 1964). This method can easily be applied to tissue culture cells. If aqueous solutions are used for extraction of the lyophilized cells, the method becomes essentially a gentle homogenization procedure related in its effects to repeated freezing-thawing cycles.

Lyophilized cell layers may be a very useful storage form of enzymes. Small clones of fibroblasts and even single cells grown on plastic films are lyophilized in situ by Hösli (1970) for subsequent direct application of the sample to cellulose acetate strips, on which a micro-electrophoretic enzyme analysis is performed. This method has proven to be applicable to prenatal diagnosis with small amounts of amniotic fluid cells (Hösli, personal communication).

A different type of dry powder, the acetone powder, provides a stable storage form of lysosomal enzymes (Hall et al., 1973).

f) Lysis and extraction of cells and tissues with diluted solutions of surface active substances (surfactants, detergents) is one of the oldest methods of homogenization. The natural detergent sodium deoxycholate (DOC) is used to solubilise the microsomal membranes. Fibroblasts were lysed with DOC by Hakami and Pious (1967) for the measurement of cytochrome oxidase activity. Many other enzymes are inactivated by DOC. The choice of a suitable detergent also depends on its effects on the method used for protein determination. Thus, while 1% sodium-dodecyl sulfate (SDS) completely inactivates glucose-6-phosphate de-hydrogenase, it does not interfere with the Lowry method for protein determination. The nonionic detergent Brij 58, on the other hand, precludes this analytical method while leaving glucose-6-phosphate de-hydrogenase intact at a concentration of 0.7%.

Sodium dodecyl sulfate is routinely used for the dissociation of nucleic acids from proteins. The complete removal of proteins from nucleic acids requires repeated extractions with phenol and treatment with proteolytic enzymes. The following procedure for the isolation of DNA from fibroblasts is used in the author's laboratory: The cells are suspended in H_2O and a stock solution of pronase and salts is added to adjust to the following concentrations:

$$0.25 \text{ mg pronase E/ml}$$
$$0.005 \text{ M CaCl}_2$$
$$0.1 \text{ M NaCl}$$
$$1.0 \text{ M LiCl}.$$

After a one hour incubation at 37° C with occasional gentle agitation, SDS is added to give a concentration of 0.5%. The mixture is again incubated for one hour at 37° C and DNA is prepared from the aqueous phase after at least three successive phenolizations in the presence of m-kresol. RNA is finally removed by treatment with pankreatic RNAase and with RNAase T_1.

4. Presentation of Results

Publications on biochemical studies with cell cultures should contain the following information:

a) Number and kind of control strains established, including age, sex, and race of the donors from whom they are derived, and the site of the skin biopsy. If control biopsies are taken from persons affected by a disease unrelated to the disorder studied, the precise diagnosis should be specified.

b) Number of passages before the strains were used in the experiments.

c) Composition of the culture media, including the kind and concentration of antibiotics, the source of the serum and the method of pH control.

d) Feeding schedule during routine propagation and in the course of the experiments.

e) Whether or not the cultures were monitored for the presence of mycoplasma and if so, by which method.

f) At which stage of the "lag-log-stationary" cycle the enzyme activities determined showed minimal variations. For diagrammatic representation of the alterations of enzyme activities during the growth cycle, specific or total activity (units/culture) can be plotted vs. time, total protein (mg/culture), or total number of cells per culture, respectively.

The activities of enzymes should always be given in international units, as recommended by the Commision on Enzymes (1965): "One unit (U) of any enzyme is that amount which will catalyse the transformation of 1 micromole of the substrate per minute under standard conditions". The specific activity of an enzyme is expressed as units of enzyme per mg of protein. For the low activities often encountered in fibroblasts, milli-units (mU) or micro-units (μU) may be used.

A meaningful parameter to which enzyme activities may be related is the number of cells. Statements on cell counts, whether performed with an electronic counter or with a hemocytometer, should always include information about the total number of cells counted and the accuracy achieved. With the hemocytometer, a reasonable accuracy is usually obtained with a total of 400 cells counted.

Some parameters, like the efficiency of cloning, must be related to the number of viable cells. Viability counts with trypsinized fibroblasts by means of dye-exclusion tests have to be performed in glucose-containing complex balanced salt solutions because very high non-viable scores are obtained with isotonic sodium chloride.

With regard to the parameters discussed in the previous sections which influence biochemical properties of fibroblast cultures, failure to bring them under control should be admitted rather than asserting that "they did not seem to be of importance" for the particular property studied, especially, if broad ranges of variation must be reported thereafter.

References

Allfrey, V.: The isolation of subcellular components. In: The cell, Brachet, J., Mirsky, A. E. (eds.), vol. I, p. 193–290, 1959.

Barile, M. F., Schimke, R. T.: A rapid chemical method for detecting PPLO contamination of tissue cell cultures. Proc. Soc. exp. Biol. (N.Y.) **114**, 676–679 (1963).

Beaufay, H.: Methods for the isolation of lysosomes. In: Lysosomes, a laboratory handbook, Dingle, J. T. (ed.), p. 1–32. Amsterdam: North Holland Publ. Co. 1972.

Boucek, R. J., Alvarez, T. R.: Increase in survival of subcultured fibroblasts mediated by serotonin. Nature (Lond.) New Biol. **229**, 61–62 (1971).

Boyle, J. A., Seegmiller, J. E.: Preparation and processing of small amounts of human material. In: Colowick, S. P., Kaplan, N. O. (eds.), Methods in enzymology, vol. XXII, p. 149–168. New York-London: Academic Press 1971.

Brown, A., Officer, J. E.: Contamination of cell cultures by mycoplasma (PPLO). In: Methods in virology, Maramorosch, K., Koprowski, H. (eds.), vol. 4, p. 531–564. New York-London: Academic Press 1968.

Busch, H.: Isolation and purification of nuclei. In: Methods in enzymology, vol. 12A, p. 421–448, Grossman, L., Moldave, K. (eds.). New York: Academic Press 1967.

Ceccarini, C., Eagle, H.: Induction and reversal of contact inhibition of growth by pH modification. Nature New Biol. (Lond.) **233**, 271–273 (1971a).

Ceccarini, C., Eagle, H.: pH as a determinant of cellular growth and contact inhibition. Proc. nat. Acad. Sci. (Wash.) **68**, 229–233 (1971b).

Commission on Enzymes: In: Comprehensive biochemistry, Florkin, M., Stotz, E. H. (eds.), vol. 13, 2nd ed., p. 6–10. Amsterdam: Elsevier Publ. Co. 1965.

Conchie, J., Hay, A. J., Levry, G. A.: Mammalian glycosidases. 3. The intracellular localization of ß-glucuronidase in different mammalian tissues. Biochem. J. **79**, 324–330 (1961).

Cristofalo, V. J., Kabakjian, J.: Processing cells for enzyme analysis. In: Tissue culture methods and applications, Kruse P. F., Patterson, M. K., Jr. (eds.), p. 204–207. New York-London: Academic Press 1973.

Croce, C. M., Koprowski, H., Eagle, H.: Effect of environmental pH on the efficiency of cellular hybridization. Proc. nat. Acad. Sci. (Wash.) **69**, 1953–1956 (1972).

Davidson, R. G.: Application of cell culture techniques to human genetics. In: Modern trends in human genetics, Emery, A. E. H. (ed.), vol. I, p. 143–180. London: Butterworths 1970.

DeDuve, C.: Principles of tissue fractionation. J. theor. Biol. **6**, 33–59 (1964).

DeMars, R.: Some studies of enzymes in cultivated human cells. Nat. Cancer Inst. Monogr. **13**, 181–193 (1964).

DeMars, R.: A temperature sensitive glucose-6-phosphate dehydrogenase in mutant cultured human cells. Proc. nat. Acad. Sci. (Wash.) **61**, 562–569 (1968)

Eagle, H.: Buffer combinations for mammalian cell culture. Science **174**, 500–503 (1971).

Ferber, E., Resch, K., Wallach, D. F. H., Imm, W.: Isolation and characterization of lymphocyte plasma membranes. Biochim. biophys. Acta (Amst.) **266**, 494–504 (1951).

Foley, J. F., Aftonomos, B.: The use of pronase in tissue culture: a comparison with trypsin. J. Cell Physiol. **75**, 159–161 (1970).

Ganoza, M. C., Williams, C. A., Lipman, F.: Synthesis of serum proteins by a cell-free system from rat liver. Proc. nat. Acad. Sci. (Wash.) **53**, 619–622 (1965).

Gartler, S., Pious, D. A.: Genetics of mammalian cell cultures. Humangenetik **2**, 83–114 (1966).

Good, N. E.: Hydrogen ion buffers for biochemical research. Biochemistry **5**, 467–477 (1966).

Good, N. E., Izawa, S.: Hydrogen ion buffers. In: Methods in enzymology, vol. XXIV B, San Pietro, A. (ed.), p. 53–68. New York-London: Academic Press 1972.

Griffiths, J. B.: The quantitative utilization of amino acids and glucose and contact inhibition of growth in cultures of the human diploid cell, WI-38. J. Cell Sci. **6**, 739–749 (1970a).

Griffiths, J. B.: The effect of insulin on the growth and metabolism of the human diploid cell, WI-38. J. Cell Sci. **7**, 575–585 (1970b).

Griffiths, J. B.: The effect of cell population density on nutrient uptake and cell metabolism: a comparative study of human diploid and heteroploid cell lines. J. Cell Sci. **10**, 515–524 (1972).

Hageboom, G. H.: Fractionation of cell components of animal tissue. In: Methods in enzymology, Colowick, S. P., Kaplan, N. O. (eds.), vol. I, p. 16–19, 1955.

Hakami, N., Pious, D. A.: Regulation of cytochrome oxidase in human cells in culture. Nature (Lond.) **216**, 1087–1090 (1967).

Hall, C. W., Cantz, M., Neufeld, E. F.: A ß-glucuronidase deficiency mucopolysaccharidosis: studies in cultured fibroblasts. Arch. Biochem. Biophys. **155**, 32–38 (1973).

Hayflick, L., Moorhead, P. S.: The serial cultivation of human diploid cell strains. Exp. Cell Res. **25**, 585–621 (1961).

Hösli, P.: Personal communication, 1970.

Hunter, M. J., Commerford, S. L.: Pressure homogenization of mammalian tissues. Biochim. biophys. Acta (Amst.) **47**, 580–586 (1961).

Krooth, R. S., Darlington, G. A., Velazquez, A. A.: The genetics of cultured mammalian cells. Ann. Rev. Genet. **2**, 141–164 (1968).

Krooth, R. S., Sell, E. K.: The action of Mendelian genes in human diploid cell strains. J. cell. Physiol. **76**, 311–330 (1970).

Kruse, P. F., Miedema, E.: Production and characterization of multiplelayered populations of animal cells. J. Cell Biol. **27**, 273–279 (1965).

Kruse, P. F., Whittle, W., Miedema, E.: Mitotic and nonmitotic multiple layered perfusion cultures. J. Cell Biol. **42**, 113–121 (1969).

Levine, E. M.: Mycoplasma contamination of animal cell cultures: a simple rapid detection method. Exp. Cell Res. **74**, 99–109 (1972).

Levine, E. M., Burleigh, I. G., Boone, C. W., Eagle, H.: An altered pattern of RNA synthesis in serially propagated human diploid cells. Proc. nat. Acad. Sci. (Wash.) **57**, 431–438 (1967).

Levine, E. M., Thomas, L., McGregor, D., Hayflick, L., Eagle, H.: Altered nucleic acid metabolism in human cell cultures infected with mycoplasma. Proc. nat. Acad. Sci. (Wash.) **60**, 583–589 (1968).

Lie, S. O., McKusick, V. A., Neufeld, E. F.: Simulation of genetic mucopolysaccharidoses in normal human fibroblasts by alteration of pH of the medium. Proc. nat. Acad. Sci. (Wash.) **69**, 2361–2363 (1972).

Ling, C. T., Gey, G. O., Richters, V.: Chemically characterized concentrated corodies for continuous cell culture. Exp. Cell Res. **52**, 469–489 (1968).

Littlefield, J. W.: Problems in the use of cultured amniotic fluid cells for biochemical diagnoses. Birth Defects: Original Article Series, Vol. VII, No. 5, 15–17 (1971).

Litwin, J.: Human diploid cell response to variations in relative amino acid concentrations in Eagle medium. Exp. Cell Res. **72**, 566–568 (1972).

Litwin, J.: Trypsinization of diploid human fibroblasts. In: Tissue culture methods and applications, Kruse, P. F., Patterson, M. K., Jr. (eds.), p. 188–192. New York-London: Academic Press 1973.

Macieira-Coelho, A.: Action of cortisone on human fibroblasts in vitro. Experientia (Basel) **22**, 390–391 (1966).

Macpherson, I.: Mycoplasmas in tissue culture. J. Cell Sci. **1**, 145–168 (1966).

Mellman, W. J.: A biochemical genetic view of human cell culture. Adv. Human. Genet. **2**, 259–306 (1971).

Mellman, W. J., Kohn, G.: Human cell cultures, their use in the investigation and diagnosis of disease. Med. Clin. N. Amer. **54**, 701–712 (1970).

Mellman, W. J., Schimke, R. T., Hayflick, L.: Catalase turnover in human diploid cell cultures. Exp. Cell Res. **73**, 399–409 (1972).

Milunsky, A., Littlefield, J. W.: The prenatal diagnosis of inborn errors of metabolism. Ann. Rev. Med. **23**, 57–76 (1972).

Milunsky, A., Littlefield, J. W., Kaufer, J. N., Kolodny, E. H., Shih, V. E., Atkins, L.: Prenatal genetic diagnosis. New Engl. J. Med. **283**, 1370–1381 (1970).

Munder, P. G., Modolell, M., Hoelzl-Wallach, D.: Cell propagation on films of polymeric fluorocarbon as a means to regulate pericellular pH and pO_2 in cultured monolayers. FEBS Letters **15**, 191–196 (1971).

Nadler, H. L., Dowben, R. M., Hsia, D. Y.-Y.: Enzyme changes and polyribosome profiles in phytohaemagglutinine-stimulated lymphocytes. Blood **34**, 52–62 (1969).

Nadler, H. L., Egan, Th. J.: Deficiency of lysosomal acid phosphatase; a new familial metabolic disorder. New Engl. J. Med. **282**, 302–307 (1972).

Paul, J.: Carbohydrate and energy metabolism. In: Cells and tissues in culture, Willmer, E. N. (ed.), vol. 1, p. 239–276. London:Academic Press 1965.

Priest, J. H.: Human cell culture in diagnosis of disease. Springfield (Ill.): Charles C. Thomas Publ. 1971.

Raff, E. C., Houck, J. C.: Migration and proliferation of diploid human fibroblasts following "wounding" of confluent monolayers. J. cell. Physiol. **74**, 235–244 (1969).

Raivio, K. O., Seegmiller, J. E.: Genetic diseases of metabolism. Ann. Rev. Biochem. **41**, 543–576 (1972).

Rhode, S. L., Ellem, K. A. O.: Control of nucleic acid synthesis in human diploid cells undergoing contact inhibition. Exp. Cell Res. **53**, 184–204 (1968).

Ryan, C. A., Lee, S. Y., Nadler, H. L.: Effect of culture conditions on enzyme activities in cultivated human fibroblasts. Exp. Cell Res. **71**, 388–392 (1972).

Schafer, I. A., Silverman, L., Sullivan, J. C., Robertson, W. v. B.: Ascorbic acid deficiency in cultured human fibroblasts. J. Cell Biol. **34**, 83–95 (1967).

Siebert, G.: Gewinnung von Zellkernen und anderen Zellfraktionen in nicht-wäßrigen Medien. In: Biochemisches Taschenbuch, Rauen, H. M. (Hrsg.), Bd. II, S. 541–546. Berlin-Göttingen-Heidelberg-New York: Springer 1964.

Sloan, H. R., Uhlendorf, B. W., Jacobson, C. B.: ß-Galactosidase in tissue culture derived from human skin and bone marrow: enzyme defect in GM1 gangliosidosis. Pediat. Res. **3**, 532–537 (1969).

Wallach, D. F. H., Kamat, V. B.: Plasma and cytoplasmic membrane fragments from Ehrlich ascites carcinoma. Proc. nat. Acad. Sci. (Wash.) **52**, 721–728 (1964).

Wallach, D. F. H., Kamat, V. B.: Preparation of plasma-membrane fragments from mouse ascites tumor cells. In: Methods in enzymology, vol. 8, Neufeld, E. F., Ginsburg, V. (eds.), p. 167–168. New York-London: Academic Press 1966.

Wang, T. Y.: The isolation and purification of mammalian cell nuclei. In: Methods in enzymology, vol. XIIA, Grossman, L., Moldave, K. (eds.), p. 420–421, 1967.

Wöhler, W., Rüdiger, H. W., Passarge, E.: Large scale culturing of normal diploid cells on glass beads using a novel type of culture vessel. Exp. Cell Res. **74**, 571–573 (1972).

Zwartouw, H. T., Westwood, J. C. N.: Factors affecting growth and glycolysis in tissue culture. Brit. J. exp. Path. **39**, 529–539 (1958).

Subject Index

Springer-Verlag
Berlin
Heidelberg
New York

München Johannesburg
London New Delhi Paris
Rio de Janeiro Sydney
Tokyo Utrecht Wien

Comparative
Mammalian
Cytogenetics

An International Conference Held at Dartmouth Medical School
in Hanover,
New Hampshire,
July 29 - August 2, 1968
Editor: K. Benirschke
1969. Cloth DM 99,20;
US $40.70
ISBN 3-540-04442-6

R. Rieger, A. Michaelis,
M.M. Green:

A Glossary of
Genetics and
Cytogenetics

Classical and Molecular.
Third completely revised
edition. 1968
Cloth DM 66,—;
US $27.10

ISBN 3-540-04316-0
(Distribution rights for
U.K., Commonwealth,
and the Traditional
British Market (excluding
Canada): Allen &
Unwin Ltd., London).

W. Fuhrmann, F. Vogel:

Genetic Counseling

A Guide for the Practicing Physician. Translated
by S. Kurth. 1969
(Heidelberg Science
Library, Vol. 10)
DM 12,—; US $5.00
ISBN 3-540-90011-X
(Distribution rights
for U.K., Commonwealth, and the
Traditional British
Market (excluding
Canada): English
Universities Press Ltd.,
London).

Chemical Mutagenesis
in Mammals and Man

Editors: F. Vogel,
G. Röhrborn. 1970
Cloth DM 124,—;
US $50.90
ISBN 3-540-05063-9

T.C. Hsu, K. Benirschke:

An Atlas of
Mammalian
Chromosomes

Available: 7 volumes
loose-leaf boxed

S. Ohno:

Sex Chromosomes
and Sex linked Genes

1967. (Monographs on
Endocrinology, Vol. 1).
Cloth DM 38,—;
US $15.60
ISBN 3-540-03934-1

S. Ohno:

Evolution by Gene
Duplication

1970. Cloth DM 36,—;
US $14.80
ISBN 3-540-05225-9
(Distribution rights for
U.K., Commonwealth,
and the Traditional
British Market (excluding
Canada): Allan & Unwin
Ltd., London).

H.T. Lynch:

Hereditary Factors
in Carcinoma

1967. (Recent Results in
Cancer Research, Vol. 12)
Cloth DM 26,—;
US $10.70
ISBN 3-540-03960-0